LONDON MATHEMATICAL SOCIETY STUDENT TEXTS

Managing editor: Dr C.M. Series, Mathematics Institute
University of Warwick, Coventry CV4 7AL, United Kingdom

London Mathematical Society Student Texts. 17

Aspects of Quantum Field Theory in Curved Space-Time

STEPHEN A. FULLING
Professor, Department of Mathematics
Texas A&M University

CAMBRIDGE
UNIVERSITY PRESS

Published by the Press Syndicate of the University of Cambridge
The Pitt Building, Trumpington Street, Cambridge CB2 1RP
40 West 20th Street, New York, NY 10011–4211, USA
10 Stamford Road, Oakleigh, Melbourne 3166, Australia

© Cambridge University Press 1989

First published 1989
Reprinted 1991, 1996

Library of Congress cataloguing in publication data available

British Library cataloguing in publication data available

ISBN 0 521 34400 X hardback
ISBN 0 521 37768 4 paperback

Transferred to digital printing 2003

Contents

Preface

During the fall semester of 1984 I visited the State University of New York at Stony Brook and taught an advanced graduate course on wave equations and quantum fields in curved space-times. This book is based on the notes for that course. I am very grateful to the Mathematics Department of SUNY, particularly Professor Michael Taylor, for arranging my temporary faculty appointment there.

The audience for the course consisted of graduate students and faculty members, mostly in mathematics but some in physics. They were assumed to have some knowledge of differential geometry and general relativity, and therefore not much time was spent on expounding those subjects. (The major exception is a chapter on connections on vector bundles and the Synge–DeWitt formalism needed for curved-space renormalization.) More time was spent on establishing a background in quantum theory and certain aspects of analysis, notably eigenfunction expansions.

Addressing such a mixed audience forces two difficult decisions — the relatively superficial one of what language to adopt, and the deeper one of what background knowledge to assume. I have an easy way out of the first problem: because of my own mixed background, the terminology which comes most naturally to me is a roughly equal mixture of the standard vocabularies of mathematicians and physicists. I have tried to say things in several different ways, to keep as many readers comfortable as possible. When it comes to content, there is no way to avoid putting in some elementary material which many readers will find superfluous, and also some technical remarks which many readers will find beyond their depth. I can only ask each reader to skip over inappropriate patches without feeling either guilt or resentment.

Although the book is self-contained, it makes no pretense of providing a balanced and complete treatment of quantum field theory in curved space. A semester is short, and so is the time available for writing additional material later — and so, I know, is people's time for reading. I have chosen to emphasize certain aspects of the subject on which I feel particularly qualified to speak. I leave to other authors detailed astrophysical applications, fermion fields, efforts to quantize the gravitational field itself, and interacting field theories generally. I do regret not being able to include a chapter on two-dimensional models, a topic on which I published quite a lot in years past. I had intended to have a chapter on thermal effects associated with black holes and other horizons, but I have already published two lengthy, largely pedagogical articles on that topic [Fulling 1977; Fulling & Ruijsenaars 1987], and there seems little point in repeating them.

One topic not covered in the lectures has been added to the book. In 1975, at King's College, I wrote a review article on the various instabilities of the vacuum induced by time-independent but very strong external fields, electric or gravitational, with application to the "superradiance" and particle-creation phenomena associated with rotating or charged black holes. That unpublished paper has been widely distributed and widely cited, and many people have urged me to publish it somehow. It is attached here as an appendix.

The body of the book begins with some background in general quantum theory and in eigenfunction expansions for second-order elliptic differential operators.* This provides the basis for a thorough treatment of a scalar field satisfying a time-independent, linear equation of motion, including the construction of the various kinds of two-point functions, and the renormalization (by vacuum subtraction) of the energy-momentum tensor in a flat-space region with nontrivial boundary conditions (the Casimir scenario).

The second half of the book explains why the structure so carefully built up in the first half is inadequate, and reports some of the progress in the fascinating and still unfinished task of replacing or supplementing it. After an introduction to the formalism of scalar field theory in an external gravitational field, the conceptual problems of field theory in a time-dependent gravitational background are explored. Bogolubov transformations are expounded, and they also are found somewhat wanting. We find ourselves in the almost vicious circle of needing to understand renormalization in order to identify the physical states of the system, and vice versa. Finally, knowledge of the

---

* It may seem to some readers that I place inordinate emphasis on the latter. After all, it may be argued, physicists seem to do just fine assuming that all normal-mode expansions are discrete sums, while the mathematician quantizing a linear field needs only enough of the abstract functional calculus to define the square root and the exponential of an operator. I believe, however, that the mathematician will never be able to read the physical literature well without acquiring an appreciation for the language of normal modes, and that a competent physicist must know some genuine spectral theory. "$\int d\mu(\omega)$" is just as easy to write as "$\sum_{j=1}^{\infty}$", and, while that level of rigor does not instantly solve all technical problems, it may protect one from publishing a formula which is in error by a factor $\delta(0)^2$.

short-distance behavior of the two-point function is seen to lead to a geometrically covariant prescription for renormalization of the energy-momentum tensor; I give an introduction to the technology of doing such calculations, and present the arguments for believing the results to be essentially unique and physically relevant.

This book was produced almost entirely on IBM and compatible personal computers. The text is written in PC-T_EX; the figures (except those in the appendix) are drawn with the P_ICT_EX macros of M. J. Wichura, with preliminary assistance from the programs DesignCAD and MathCAD. I thank my Texas A&M colleagues Norman Naugle, Michael Pilant, Jon Pitts, and Al Sato for indispensable help during the learning process, and Jan Want and the Texas A&M University Department of Mathematics for producing the final copy.

Finally, special thanks to Leonard Parker, Paul Davies, Bob Wald, Steve Christensen, and all the other collaborators and professional colleagues who have helped build our current understanding of these aspects of quantum field theory in curved space-time.

Chapter 1

A Quick Course in Quantum Mechanics

This book is concerned with the quantum theory of fields satisfying linear equations of motion. As a prerequisite, one needs to understand the quantum theory of particles, which is related to field theory as finite-dimensional linear algebra is related to functional analysis. In particular, we need to treat the simplest linear quantum system, the harmonic oscillator.

CLASSICAL MECHANICS: THE CANONICAL FORMALISM

The dynamics of a typical mechanical system is described by a system of second-order ordinary differential equations, the *Newtonian equations of motion*:

$$\frac{d^2\mathbf{x}}{dt^2} = \mathbf{F}\left(\mathbf{x}, \frac{d\mathbf{x}}{dt}, t\right). \tag{1.1}$$

Often the force $\mathbf{F}$ is the negative of the gradient of a *potential* function, $V(\mathbf{x})$. In the simplest situation, $\mathbf{x}(\cdot)$ is a function from $\mathbf{R}$ (the domain of the time variable) to $\mathbf{R}^d$ (the *configuration space* of the system), where d equals $3n$ for a system of n particles in the real world of dimension 3. (More generally, $\mathbf{R}^d$ may be replaced by some other manifold, and the time may need to be restricted to a subinterval of $\mathbf{R}$, but such complications are not relevant to our present concerns.)

Let $\dot{\mathbf{x}} \equiv d\mathbf{x}/dt$. Suppose that (1.1) arises as the Euler–Lagrange equation of a problem in the calculus of variations: to extremize the *action functional*

$$S \equiv \int_{t_1}^{t_2} L\big(\mathbf{x}(t), \dot{\mathbf{x}}(t), t\big)\, dt.$$

L is called the *Lagrangian* of the system. The functions $\mathbf{x}$ which extremize S satisfy

$$\frac{d}{dt}\left(\frac{\partial L}{\partial \dot{x}^j}\right) = \frac{\partial L}{\partial x^j} \qquad (j = 1, \ldots, d). \tag{1.2}$$

These are the *Lagrangian equations of motion.*

For example, from the Lagrangian (with $d = 1$)

$$L = \tfrac{1}{2}m\dot{x}^2 - \tfrac{1}{2}m\omega^2 x^2$$

we obtain $\frac{d}{dt}(m\dot{x}) = -m\omega^2 x$, which is equivalent to the Newtonian equation

$$\ddot{x} + \omega^2 x = 0.$$

This system is the one-dimensional *harmonic oscillator* of mass m and frequency ω; if $\omega = 0$ it becomes a free particle (in one dimension), and if ω^2 is replaced by a negative number, one has a "runaway" particle (the solutions being exponential).

The *canonical momentum conjugate* to x^j is

$$p_j \equiv \frac{\partial L}{\partial \dot{x}^j}.$$

The *Hamiltonian* is defined by

$$H \equiv -L + \dot{x}^j p_j\,,$$

where a sum over j is understood. For the harmonic oscillator, for example, we have $p = m\dot{x}$ and $H = \frac{1}{2m}p^2 + \frac{1}{2}m\omega^2 x^2$. In the more general class of problems usually studied in quantum mechanics,

$$H = \frac{1}{2m}\mathbf{p}^2 + V(\mathbf{x}).$$

H can be interpreted as the energy of the physical system, but its main significance for the general formulation of mechanics lies elsewhere: Let us regard H as a function of $\mathbf{x}$ and $\mathbf{p}$ (rather than of $\mathbf{x}$ and $\dot{\mathbf{x}}$). (This is a *Legendre transformation*; a mathematically precise reference on Legendre transformations (in a different context) is Maslov & Fedoriuk 1981, Sec. 1.3.) Then (1.2) is easily seen to be equivalent to the *Hamiltonian equations of motion,*

$$\frac{dx^j}{dt} = +\frac{\partial H}{\partial p_j}\,, \qquad \frac{dp_j}{dt} = -\frac{\partial H}{\partial x^j}\,. \tag{1.3}$$

Phase space is $\mathbf{R}^{2d}$ with the coordinates $(\mathbf{x}, \mathbf{p})$. (More generally, the phase space is the cotangent bundle of the manifold which serves as configuration space; $\mathbf{p}(t)$ is a cotangent vector and $\dot{\mathbf{x}}(t)$ a tangent vector at the point $\mathbf{x}(t)$. This accounts for the placement of the index on x and $\dot{x}$ as a superscript.) Thus, the second-order differential system (1.1) or (1.2) has been replaced by a first-order system, (1.3), in a space of twice the dimension. A *pure state* of a classical system is a point in phase space (as opposed to an impure or mixed state, which is a probability distribution on phase space). An *observable* is just a function of $\mathbf{x}$ and $\mathbf{p}$. We measure the observable $A(\mathbf{x}, \mathbf{p})$ simply by "looking" at the system and observing A to have its value at the point of phase space where the system happens to be! By measuring $2d$ independent observables, such as $\mathbf{x}$ and $\mathbf{p}$ themselves, exactly and simultaneously, we completely determine the state.

QUANTUM STATES AND OBSERVABLES

In microscopic physics, observations are inherently probabilistic. A *pure state* of a quantum system is a one-dimensional subspace of a complex Hilbert space, $\mathcal{H}$. The subspace is traditionally represented by one of the vectors in it, normalized (i.e., $\|\Psi\| = 1$). Ψ is determined by the state only up to phase, but one speaks informally of "the state Ψ" when there is no chance of confusion. An *observable* is represented by a self-adjoint operator in $\mathcal{H}$. Often the operator is unbounded, and hence its domain is a dense subspace of $\mathcal{H}$, not the whole space. When the observable A is measured, its *expectation value* is

$$\langle A \rangle = \langle \Psi, A\Psi \rangle \equiv \langle \Psi | A | \Psi \rangle. \qquad (1.4)$$

The fundamental postulate of the physical interpretation of quantum theory is that the average of the results of many experiments to measure A on identical systems under identical conditions is given by the formula (1.4).

Notational remarks: The middle member of (1.4) is an inner product in $\mathcal{H}$, reexpressed in the last member in *Dirac notation*. I use the conventions standard in physics, wherein the complex conjugate of a number z is denoted z^* (not $\bar{z}$), and the adjoint of a matrix or operator A is $A^\dagger$ (not A^*). Furthermore, the inner product is linear in the *right* variable, not the left:

$$\langle \Psi, z\Phi \rangle = z \langle \Psi, \Phi \rangle, \quad \langle z\Psi, \Phi \rangle = z^* \langle \Psi, \Phi \rangle.$$

In the Dirac notation, where $\langle \Psi, \Phi \rangle \equiv \langle \Psi | \Phi \rangle$, the angular brackets are thought of as permanently attached to the vectors; thus "$|\Phi\rangle$" is a typical element of $\mathcal{H}$, and "$\langle \Psi |$" is the linear functional defined by taking the inner product of the argument vector with Ψ. (By the Riesz theorem, all linear functionals on $\mathcal{H}$ (elements of the dual space of $\mathcal{H}$) are of this form.)

As a first example, consider a purely canonical system, where $\mathcal{H} = \mathcal{L}^2(\mathbf{R}^d)$. That is, the vectors are complex-valued, square-integrable functions $\Psi(\mathbf{x})$ (the famous "wave functions"). The operator representing the position observable x^j is simply multiplication by the variable x^j. The observable p_j is represented by the operator $-i\,\partial/\partial x^j$, in appropriate units (that is, units in which Planck's constant, $\hbar$, equals 1 — which I shall always use).

This system has an alternative description by functions $\hat{\Psi}(\mathbf{p})$. Now the momentum observable is represented by multiplication by p_j, and the position by $+i\,\partial/\partial p_j$. $\hat{\Psi}$ is just the Fourier transform of Ψ; these two distinct square-integrable functions are two different representations of the same vector in the abstract Hilbert space of physical states.

A more complicated example is presented by a (nonrelativistic) particle of spin $\frac{1}{2}$. In this case the Hilbert space is $\mathcal{L}^2(\mathbf{R}^3) \oplus \mathcal{L}^2(\mathbf{R}^3)$. That is, the vectors are *pairs* of square-integrable functions, $\Psi_a(\mathbf{x})$ ($a = 1$ or 2). In addition to $\mathbf{x}$ and $\mathbf{p}$, there are some fundamental observables of the theory which are represented by matrices acting on the index a (leaving $\mathbf{x}$ alone):

$$S_1 = \frac{1}{2}\begin{pmatrix} 0 & 1 \\ 1 & 0 \end{pmatrix}, \qquad S_2 = \frac{1}{2}\begin{pmatrix} 0 & -i \\ i & 0 \end{pmatrix}, \qquad S_3 = \frac{1}{2}\begin{pmatrix} 1 & 0 \\ 0 & -1 \end{pmatrix}.$$

These observables are the components of the particle's spin, or intrinsic angular momentum, in the three directions of physical space. They have no counterpart in the classical phase-space formalism.

The *commutator* of two operators is

$$[A, B] \equiv AB - BA.$$

In the examples, the commutators of the basic observables are

$$[x^j, p_k] = i\delta^j_k$$

and

$$[S_1, S_2] = iS_3$$

with its cyclic permutations; also, the various components of $\mathbf{x}$ or of $\mathbf{p}$ commute among themselves, and the positions and momenta commute with the spins. If $[A, B] \neq 0$, then A and B can't be measured simultaneously. (Here "measured" means not only that a certain numerical value is experimentally obtained, but also that the physical system is left in such a state that the identical value would with certainty be obtained if the measurement were repeated immediately.) That is, states characterized by "sharp" values of A necessarily have spread-out probability distributions for measurements of B. A pure state can, at best, be characterized by the values of "a complete set of commuting observables", not of *all* observables.

If these commuting observables all have totally discrete spectra, then there is a orthonormal basis for $\mathcal{H}$ whose elements are labelled by their possible eigenvalues. (This is what it means for the set to be "complete".) That is, if the set comprises, say, three observables, $\{A_1, A_2, A_3\}$, then there are vectors $|\alpha_1, \alpha_2, \alpha_3\rangle$ (for each α_j in the spectrum $\sigma(A_j)$) such that

$$A_1|\alpha_1, \alpha_2, \alpha_3\rangle = \alpha_1|\alpha_1, \alpha_2, \alpha_3\rangle, \qquad \text{etc.,}$$

and for any normalized $\Psi \in \mathcal{H}$,

$$\Psi = \sum_{\alpha_j \in \sigma(A_j)} \psi(\alpha_1, \alpha_2, \alpha_3)|\alpha_1, \alpha_2, \alpha_3\rangle,$$

$$\sum_{\alpha_j \in \sigma(A_j)} |\psi(\alpha_1, \alpha_2, \alpha_3)|^2 = 1.$$

Finally, $|\psi(\alpha_1, \alpha_2, \alpha_3)|^2$ is the probability of finding A_j to have the value α_j when the three observables are simultaneously measured.

If continuous spectra occur, we can no longer speak of basis vectors, but the representation of Ψ by a coefficient function ψ is still valid. For example, the spin-$\frac{1}{2}$ particle has a complete commuting set $\{x^1, x^2, x^3, S_3\}$, and the wave function $\Psi_a(\mathbf{x}) \equiv \psi(\mathbf{x}, a) \in \mathcal{L}^2(\mathbf{R}^3) \oplus \mathcal{L}^2(\mathbf{R}^3)$ is the representation of a state relative to that choice of basic observables. (An alternative choice might be $\{x^1, p_2, p_3, S_1\}$.) Also,

$$\sum_{a=1}^{2} \int_V |\psi(\mathbf{x}, a)|^2 \, d^3x$$

is the probability that the particle will be found to be in the set $V \subset \mathbf{R}^3$ with either of the allowed values $(\pm\frac{1}{2})$ of S_3.

Note that an equation of motion is inadequate to specify a quantum system. We must also know what the observable operators are. In many cases the commutation relations essentially uniquely determine the operators. More precisely, the abstract algebra defined by the commmutation relations has a unique irreducible representation up to unitary equivalence. For example, the *Stone–Von Neumann theorem* for the canonical commutation relation, $[x, p] = i$, and its *finite-dimensional* generalizations state that (under technical assumptions which are not entirely innocent, since they eliminate the elementary example of "a particle in a box") the only quantum system supporting those commutation relations is the one built on the Hilbert space $\mathcal{L}^2(\mathbf{R}^d)$ as described above. (See, for instance, Putnam 1967, Chap. 4.) In other elementary cases the contrary is true, however. For instance, the two-dimensional (spin $\frac{1}{2}$) representation is only one of the infinite sequence of representations of the SU(2) Lie algebra, $[S_1, S_2] = iS_3$ etc. Different spin representations represent *different physical systems* (different kinds of particle), rather than different states of a single system. Although quantum field theory, as we'll see, can be regarded as an infinite-dimensional generalization of the canonical commutation relations, the Stone–Von Neumann theorem does not apply there, and inequivalent representations do exist. (See Segal 1967, and Wightman 1967, Secs. 6–7.) (That is, there are distinct sets of operators satisfying the commutation relations; these sets are *essentially* different, not transformable into one another by recoordinatizations of the Hilbert space like the Fourier transform of $\mathcal{L}^2(\mathbf{R}^d)$.) Therefore, the passage from a formal algebra of observables to a full quantum theory in a Hilbert space is a very nontrivial step in field theory. Apart from the thorny technical problems, the choice of a physically appropriate representation is a major conceptual issue, especially in the context of curved space-time. In field theory the inequivalent representations can sometimes represent different physical configurations of the same system, although by no means are all representations physically meaningful.

QUANTUM DYNAMICS: THE HEISENBERG PICTURE

Quantum field theory is usually developed in the *Heisenberg picture*, a formulation in which the operators satisfy equations of motion like those of the classical observable quantities they represent.

Let us solve the equation of motion of the harmonic oscillator,

$$\ddot{x} + \omega^2 x = 0,$$

with *operator initial data*. Because of the linearity of the equation, this can be done by simply writing down the standard general solution and interpreting the arbitrary constants in it as operators. We have

$$x(t) = x(0) \cos \omega t + \frac{1}{\omega} \dot{x}(0) \sin \omega t, \qquad (1.5)$$

where $\dot{x}(0) = p(0)/m$. (When $\omega = 0$, this reduces to

$$x(t) = x(0) + \frac{p(0)}{m} t, \qquad (1.6)$$

which classically is the trajectory of a free particle.) The derivative of (1.5), times m, provides a formula for $p(t)$. At $t = 0$ we take the canonical operators to have their usual (*Schrödinger*) representation, discussed above:

$$x(0) = x \quad \text{(the multiplication operator)}, \qquad p(0) = -i \frac{\partial}{\partial x}. \qquad (1.7)$$

Then (1.5) defines $x(t)$ and $p(t)$ for each $t \in \mathbf{R}$ as operators in $\mathcal{L}^2(\mathbf{R})$. It is easy to see that they satisfy the canonical relation, $[x, p]\big|_t = i$.

By the Stone–Von Neumann theorem, these operators are unitarily equivalent to those in (1.6); I shall exhibit later the unitary operator connecting them. However, they are quite distinct as concrete operators in the space of wave functions, $\mathcal{L}^2$. The physical interpretation of this situation is that $\langle \Psi | x(t) | \Psi \rangle$ and $\langle \Psi | p(t) | \Psi \rangle$ are the expectation values of the position and momentum if those quantities were to be measured at time t with the particle in the state Ψ. The state itself is a time-independent concept (at least so long as the system evolves under its internal dynamics, without interaction with external agencies). The state Ψ is an abstract object in a Hilbert space, $\mathcal{H}$. It is *represented* by a function $\Psi(x) \in \mathcal{L}^2$; this representation has been arbitrarily chosen to be the one which gives directly the probability density for position measurements at $t = 0$. The probability densities for momentum measurements and for position measurements at other times are given similarly in other representations; in this one, at least the expectation values for such measurements can be calculated from (1.5), using the concrete form of the inner product, $\langle \Psi, \Phi \rangle = \int_{-\infty}^{\infty} \Psi(x)^* \Phi(x) \, dx$.

It is interesting to note that (when ω is defined as I have done) the classical equation of motion is independent of the mass, m, but different values of m denote physically different *quantum* systems: The relation between $\dot{x}$ and p depends on m, and hence so will the expected results of measurements of x at later times, for a given initial wave function. On the other hand, m may be eliminated from the formalism by redefining x to absorb a factor $\sqrt{m}$. To concentrate attention on the ω dependence, I henceforth set $m = 1$.

Let's treat the harmonic oscillator in the Hamiltonian approach. Recall that we have $H = \frac{1}{2}p^2 + \frac{1}{2}\omega^2 x^2$ and hence from (1.3) the equations of motion

$$\frac{dx}{dt} = p, \qquad \frac{dp}{dt} = -\omega^2 x.$$

We shall solve this by the method of *creation and annihilation operators*: Let

$$x = \frac{1}{\sqrt{\omega}}(a + a^\dagger), \qquad p = -i\sqrt{\omega}(a - a^\dagger), \qquad (1.8)$$

so that

$$a \equiv \frac{1}{2}\left(\omega^{\frac{1}{2}}x + i\omega^{-\frac{1}{2}}p\right), \qquad a^\dagger = \frac{1}{2}\left(\omega^{\frac{1}{2}}x - i\omega^{-\frac{1}{2}}p\right).$$

(Classically all these quantities are complex numbers, and $a^\dagger$ is simply a^*. In quantum mechanics, x and p arc self-adjoint operators, so $a^\dagger$ is indeed the adjoint of the operator a — at least if we ignore technicalities about the domains of unbounded operators.) It is easy to see that the equations of motion become

$$\frac{d}{dt}a = -i\omega a, \qquad \frac{d}{dt}a^\dagger = +i\omega a^\dagger,$$

with solution

$$a(t) = e^{-i\omega t}a(0), \qquad a(t)^\dagger = e^{+i\omega t}a(0)^\dagger.$$

We then arrive at

$$x(t) = \frac{1}{\sqrt{\omega}}\left[a(0)e^{-i\omega t} + a(0)^\dagger e^{i\omega t}\right],$$
$$p(t) = \dot{x}(t) = -i\sqrt{\omega}\left[a(0)e^{-i\omega t} - a(0)^\dagger e^{i\omega t}\right], \qquad (1.9)$$

which are equivalent to (1.5).

This construction is the prototype of the quantization of a field theory satisfying a linear, time-independent equation of motion. Note, however, that it does not apply to the case $\omega = 0$, where the definition of a breaks down. Nor does it apply when ω^2 is replaced by a negative number. As we'll soon see, it is through the annihilation-creation operators that the notion of "particles" enters the formalism. We must therefore be prepared to encounter situations where the particle concept is not applicable. If our most fundamental model of nature is a field theory, then we are saying that fields are more fundamental than particles.

Now consider a more general quantum-mechanical system, corresponding to equations of motion which are time-independent (autonomous) but not necessarily linear. Ordinarily it is assumed that there is a self-adjoint *Hamiltonian operator*, H, such that the equation of motion for each observable operator is equivalent to

$$\frac{dA}{dt} = i[H, A]. \tag{1.10}$$

This is the *Heisenberg equation of motion*. This much can be true even for a time-dependent dynamics, in which case H is a function of time as well as of the basic observables of the theory. If H is independent of time, however, the solution of the Heisenberg equation can be written

$$A(t) = U(t)^{-1} A(0) U(t), \qquad U(t) \equiv e^{-itH}.$$

$U(t)$ is a unitary operator, so $U(t)^{-1} = U(t)^\dagger = e^{+itH}$.

In purely canonical cases, H is obtained from the classical Hamiltonian function on phase space, $H(\mathbf{x}, \mathbf{p})$, by substituting $-i\, \partial/\partial x^j$ for p_j and interpreting x^j as a multiplication operator. This prescription makes elementary sense in the Schrödinger representation as long as H is polynomial in its dependence on $\mathbf{p}$, and one can then verify that

$$i[H, x^j] = \frac{\partial H}{\partial p_j}, \qquad i[H, p_j] = -\frac{\partial H}{\partial x^j},$$

as needed for consistency of (1.10) and (1.3).

Advanced remarks: Nonpolynomial terms in $H(\mathbf{x}, \mathbf{p})$ may be interpreted via the Fourier transformation (if the term is independent of $\mathbf{x}$) or the calculus of pseudodifferential operators [Petersen 1983, Treves 1980, Taylor 1981]. However, there is an *ordering ambiguity*: Since the operators x and p don't commute, it is unclear whether the classical function $x^2 p^2$ should be interpreted as $\frac{1}{2}(x^2 p^2 + p^2 x^2)$ or $x p^2 x$ or $\frac{1}{2}(xpxp + pxpx)$ or something else. (Since the operator should be formally self-adjoint, one can at least restrict attention to expressions with palindromic symmetry, as I have here.) Consequently, there are in principle *many* quantum systems corresponding to a given classical system; their Hamiltonians differ by terms involving the canonical commutator, $[x, p] = \hbar i$, and the presence of $\hbar$ (in general units) indicates why such terms should disappear in the classical limit. H. Weyl's attempt at a resolution of this ambiguity [Weyl 1927; Moyal 1949] was a precursor of what is now called the *Weyl calculus for pseudodifferential operators* [Grossmann *et al.* 1968; Hörmander 1979]. Such issues can be ignored in elementary quantum mechanics, because the Hamiltonians of most physical systems are quadratic in $\mathbf{p}$ with quadratic term independent of $\mathbf{x}$. (An $\mathbf{x}$-dependent linear term is not considered to present a problem: $f(\mathbf{x})\mathbf{p}$ is interpreted in the first way that comes to mind, $\frac{1}{2}(f\mathbf{p} + \mathbf{p}f)$.)

THE SCHRÖDINGER PICTURE

The nonrelativistic quantum mechanics of particles is usually conducted in a different formalism, where the operator representing an observable A is the same at all times, while the *states* evolve. (This *Schrödinger picture* is not the same thing as the "Schrödinger [or position] representation", which is a particular realization of the states as wave functions.) The equation of motion of the states is the *(time-dependent) Schrödinger equation*,

$$i \frac{\partial \Psi}{\partial t} = H\Psi.$$

In a purely canonical theory with H polynomial in $\mathbf{p}$, this is a linear partial differential equation. For example, for the harmonic oscillator (with $m = 1$) it is

$$i \frac{\partial \Psi(x, t)}{\partial t} = -\frac{1}{2} \frac{\partial^2 \Psi}{\partial x^2} + \frac{1}{2}\omega^2 x^2 \Psi.$$

In most cases of interest the linear partial differential operator H is of the *elliptic* type; by hypothesis, it is self-adjoint.

When H is independent of t, the solution of the initial-value problem for the Schrödinger equation is

$$\Psi(t) = e^{-itH}\,\Psi(0).$$

What this means in concrete terms is that the equation can be solved by separation of variables: If one looks for solutions of the form $\Psi(t,x) = \psi(x)T(t)$, one finds that $T(t) = e^{-iEt}$ and

$$H\psi = E\psi$$

for some number E. (This is called the *time-independent Schrödinger equation*.) Therefore, the problem reduces, in principle, to finding the eigenvectors of the self-adjoint operator H, or, more generally, the spectral representation of H:

Suppose, first, that H possesses a complete set of eigenvectors $\{\psi_j\}$,

$$H\psi_j = E_j\psi_j, \qquad \|\psi_j\| = 1.$$

Then the initial data for the Schrödinger equation can be expanded as

$$\Psi(0) = \sum_{j=1}^{\infty} c_j\psi_j, \qquad c_j = \langle \psi_j | \Psi(0) \rangle.$$

(In the Schrödinger representation the ψ_j will be functions of $\mathbf{x}$, and the coefficients have the formula $c_j = \int \psi_j(\mathbf{x})^*\Psi(0,\mathbf{x})\,d^d x$.) Therefore,

$$\Psi(t) = \sum_{j=1}^{\infty} c_j\psi_j e^{-iE_j t} \equiv e^{-itH}\Psi(0). \tag{1.11}$$

If H has continuous spectrum, then the solutions of the partial differential equation $H\psi = E\psi$ generally are not square-integrable and hence are not vectors in $\mathcal{H} = \mathcal{L}^2(\mathbf{R}^d)$. Nevertheless, at least in simple cases one can complete the classical procedure of solution by separation of variables to obtain the general solution as an integral over such solutions, regarded as "generalized eigenvectors". (This construction provides a concrete realization of the abstract *spectral theorem*, which generalizes the eigenvector expansion to deal with the most general self-adjoint operators.) We delay a detailed discussion to Chapter 2, and consider a few fundamental examples here.

Example 1: **The free particle in three-dimensional space.**
The time-independent Schrödinger equation is the Helmholtz equation,

$$H\psi \equiv -\frac{1}{2m}\nabla^2\psi = E\psi.$$

Its basic solutions are

$$\psi_{\mathbf{k}}(\mathbf{x}) \equiv e^{i\mathbf{k}\cdot\mathbf{x}}, \qquad E = \frac{k^2}{2m}$$

($k \equiv |\mathbf{k}|$). It follows that the general solution of the time-dependent
equation is

$$\Psi(t,\mathbf{x}) = (2\pi)^{-\frac{3}{2}} \int_{\mathbf{R}^3} d^3k \, e^{i\mathbf{k}\cdot\mathbf{x}} \, e^{-k^2t/2m} \, \hat{\psi}(\mathbf{k}),$$

where

$$\hat{\psi}(\mathbf{k}) \equiv (2\pi)^{-\frac{3}{2}} \int_{\mathbf{R}^3} d^3x \, e^{-i\mathbf{k}\cdot\mathbf{x}} \, \Psi(0,\mathbf{x}).$$

Thus the eigenfunction expansion in this case is the Fourier transform
(and we refer to the general theory of the Fourier transform to justify
the normalization factors $(2\pi)^{-\frac{3}{2}}$).

Since $(\mathbf{p}\psi)^{\hat{}}(\mathbf{k}) = \mathbf{k}\,\hat{\psi}(\mathbf{k})$, what we have really done here is to
pass to a representation of the Hilbert space in which the momentum
observable, $\mathbf{p}$, is diagonal. The three components of $\mathbf{p}$ make up the
complete set of commuting observables. But since E is a function of
$\mathbf{p}$ in this problem, we can also regard the eigenfunction expansion as
an integral over E, supplemented by an integration over the angles in
$\mathbf{k}$-space:

$$(2\pi)^{-\frac{3}{2}} \int d^3k = (2\pi)^{-\frac{3}{2}} \int_0^\infty k^2 \, dk \int_0^{2\pi} d\phi_{\mathbf{k}} \int_0^\pi \sin\theta_{\mathbf{k}} \, d\theta_{\mathbf{k}}$$

$$\equiv \int d\mu(E) \int d\Omega_{\mathbf{k}},$$

where, since $k = \sqrt{2mE}$, we calculate that $d\mu(E) = (2\pi)^{-\frac{3}{2}}m^{\frac{3}{2}} \times$
$\sqrt{2E}\,dE$. Returning momentarily to the general problem, we antici-
pate in analogy with this example an eigenfunction expansion with the
schematic structure

$$\Psi(t,\mathbf{x}) = \int_{\sigma(H)} d\mu(E) \sum_\alpha c_\alpha(E) \, e^{-iEt} \, \psi_{E,\alpha}(\mathbf{x}).$$

Here α is the eigenvalue of another observable that goes together with H to make a complete commuting set, or an n-tuple of eigenvalues of a set of such observables. The summation range of α may depend on E, and, as the example shows, α may itself be a continuous variable. (The notation has been chosen to make this equation a generalization of the discrete eigenfunction expansion (1.11), but in terms of a notation used earlier in this chapter, $\psi_{E,\alpha}$ is $|E, \alpha_2, \dots\rangle$ and $c_\alpha(E)$ is $\psi(E, \alpha_2, \dots)$.) In general, "$\int_{\sigma(H)} d\mu(E)$" entails a summation over discrete eigenvalues of H and an integration over the continuous spectrum of H; thus μ is a Stieltjes measure with support in $\sigma(H)$, the spectrum of H. As the example of a free particle shows, even when the spectrum is completely (and, for the experts, absolutely) continuous, it may be convenient to normalize the measure as something different from the Lebesgue measure.

Example 2: **The harmonic oscillator.** With $m = 1$, the equation is

$$H\psi \equiv -\frac{1}{2}\frac{\partial^2\psi}{\partial x^2} + \frac{1}{2}\omega^2 x^2\psi = E\psi.$$

I summarize two well known methods of solving this problem (Messiah 1961, Chap. 12 and App. B3).

The first method is to solve the differential equation directly in terms of known special functions. The square-integrable eigenfunctions are

$$\psi_n(x) \equiv N_n\, H_n(\sqrt{\omega}x)\, e^{-\omega x^2/2},$$

with eigenvalues

$$E_n = \left(n + \tfrac{1}{2}\right)\omega, \qquad n = 0,\ 1,\ 2,\ \dots.$$

H_n is the *Hermite polynomial* of degree n, and N_n is a certain normalization constant. This operator has a discrete spectrum; (1.11) applies. The nth eigenfunction has $n - 1$ nodes (zeros), and it is oscillatory in the region where E_n exceeds the potential, exponentially decaying outside that region. The figure shows the potential ($V \equiv \frac{1}{2}\omega^2 x^2$) and two typical eigenfunctions, one of them raised on the graph for clarity.

The second method is an algebraic trick peculiar to this potential. In studying the oscillator in the Heisenberg picture, we introduced the non-self-adjoint operator

$$a = \frac{1}{2}\left(\omega^{\frac{1}{2}}x + i\omega^{-\frac{1}{2}}p\right).$$

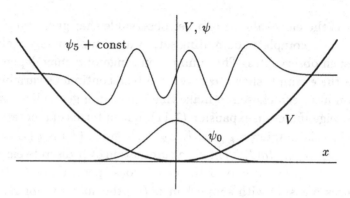

A bit of calculation shows

$$[a, a^\dagger] = 1, \qquad H = \omega a^\dagger a + \tfrac{1}{2}\omega.$$

Define

$$N \equiv a^\dagger a.$$

Then we find that

$$Na = a(N-1), \qquad Na^\dagger = a^\dagger(N+1).$$

Now suppose we have a vector $|n\rangle$ $(n \in \mathbf{R})$ such that $N|n\rangle = n|n\rangle$. Then

$$Na|n\rangle = (n-1)a|n\rangle, \qquad Na^\dagger|n\rangle = (n+1)a^\dagger|n\rangle.$$

That is, $a|n\rangle$ and $a^\dagger|n\rangle$, if they are not zero, are also eigenvectors of N. Also, we find that the square of the norm of $a|n\rangle$ is

$$\langle n|a^\dagger a|n\rangle = n\langle n|n\rangle.$$

Therefore, n must be nonnegative. Furthermore, if $n = 0$, then $a|n\rangle = 0$, and conversely (whereas $a^\dagger|n\rangle \neq 0$ always). By induction, n must be a nonnegative integer, since the sequence of eigenvalues corresponding to the vectors of the form $a \cdots a|n\rangle$ is not allowed to jump over 0. Starting from $|0\rangle$, we can construct the normalized eigenvectors

$$|n\rangle = \frac{1}{\sqrt{n!}} \left(a^\dagger\right)^n |0\rangle.$$

(Do not confuse $|0\rangle$ with 0, the zero vector in the Hilbert space!) The action of the creation and annihilation operators on these basis vectors is

$$a^\dagger|n\rangle = \sqrt{n+1}\,|n+1\rangle, \qquad a|n\rangle = \sqrt{n}\,|n-1\rangle.$$

We have found a Hilbert space (the closed span of the orthonormal basis $\{|n\rangle\}$) in which the operators x, p, and H act with their correct commutation relations. Each of these operators is actually self-adjoint and hence has a spectral representation; in the representation where x acts diagonally, $|n\rangle$ is identified with the Hermite function $\psi_n(x)$ previously discussed. The spectral representation of H is, of course, the one we are looking at, since H is manifestly diagonal here. One thinks of $|n\rangle$ as representing a state in which n "excitations" or "quanta" are present. N counts such excitations, $a^\dagger$ creates them, and a destroys them. Apart from the constant term $\frac{1}{2}\omega$, the energy (eigenvalue of H) is proportional to n. *This construction is the prototype for the introduction of particles into quantum field theory.* Particles are merely excitations of a field.

If the Hilbert space spanned by $\{|n\rangle\}$ is not the entire Hilbert space of the system, then any eigenvector of H linearly independent from it would, by the same argument, yield another copy of the whole sequence of eigenvectors. This would mean that the representation of x and p is not irreducible; there are additional, independent observables in the theory. For example, a spin-$\frac{1}{2}$ oscillator's Hilbert space is the direct sum of two copies of this elementary oscillator space, with basis $\{|n, \pm\rangle\}$; and a two-dimensional oscillator's space is the direct product, with basis $\{|n_1, n_2\rangle\}$.

Of course, there are some gaps in this argument. How (without appealing to the Stone–Von Neumann theorem) does one exclude the possibility of a representation in which N has continuous spectrum? Or the possibility that some steps in the argument are meaningless because a vector is not in the domain of the unbounded operator that is applied to it?

Nevertheless, this example illustrates that the formal algebra of commutation relations can carry one a long way. It even separates the main features of the model system from irrelevant details of special-function theory.

Example 3: **The repulsive oscillator.** The Hamiltonian

$$H = \tfrac{1}{2}p^2 - \tfrac{1}{2}\omega^2 x^2$$

classically describes an exponentially runaway particle. Titchmarsh 1962, Theorem 5.10, shows that H is essentially self-adjoint (i.e., defines a self-adjoint operator in $\mathcal{L}^2(\mathbf{R})$ without the aid of extra boundary conditions at infinity), and that $\sigma(H) = (-\infty, \infty)$. (The spectrum is continuous and is unbounded below.) The figure shows the potential and a typical eigenfunction (a parabolic cylinder function plotted from data in the National Bureau of Standards Handbook (Abramowitz & Stegun 1968)). Note the transitions from exponential to oscillatory behavior where the potential crosses the energy of the eigenfunction (a small negative number in this case).

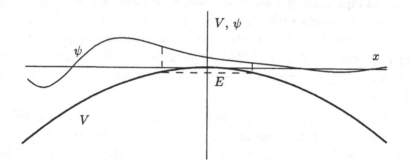

Models such as this, with arbitrarily negative energy, are in physical disrepute. The idea is that the slightest perturbation of the dynamics would introduce a coupling between the repulsive oscillator and the other, normal degrees of freedom of the world. If such an object existed, therefore, there would be a dynamical instability in which the energy of the runaway particle tumbled to an arbitrarily low value while the energy of the rest of the world grew without bound. Thus, the predictions of such a theory could not be physically plausible. If an H of this sort arose by linearization of some physically realistic theory, one would expect that the neglected nonlinear interactions actually halt the runaway motion and stabilize the system. The linearized theory would therefore be a bad approximation to the full theory. Nevertheless, field theories with inverted-oscillator potentials are amusing to contemplate and raise some important issues about quantization procedures and the relation between fields and particles (see the appendix).

THE CONNECTION BETWEEN THE TWO PICTURES

If H is independent of t, the solution of the Schrödinger equation, $i\,\partial\Psi/\partial t = H\Psi$, can be written

$$\Psi(t,x) = U(t)\Psi(0,x), \qquad U(t) \equiv e^{-itH}.$$

One can *redefine* the state vectors and operators by a time-dependent similarity transformation:

$$\tilde{\Psi} \equiv U(t)^{-1}\Psi(t), \qquad \tilde{A}(t) \equiv U(t)^{-1}AU(t).$$

Then the new state vector is independent of time, while the observable operators evolve according to the Heisenberg picture. All the matrix elements $\langle\Psi|A|\Phi\rangle$, hence all physical predictions of the theory, are unchanged.

If H depends on t, then the quantum dynamics is given by a two-parameter family of operators, $U(t_2, t_1)$, rather than the one-parameter Lie group, $U(t_2 - t_1) = e^{-i(t_2 - t_1)H}$. (In the Schrödinger picture, for example, $U(t_2, t_1)$ maps initial data at t_1 to the wave function at time t_2.) In quantum theories with finitely many degrees of freedom, one can still use these operators to move back and forth between Heisenberg and Schrödinger pictures. In quantum field theory, however, especially with a time-dependent dynamical law, the very *existence* of a U (or an H, or a Schrödinger picture) is not a foregone conclusion.

THE CLASSICAL LIMIT

Under what circumstances do we get classical behavior from a (canonical) quantum system? Let us consider a particle of mass m in one dimension. For the details see Messiah 1961, Chap. 6; see also Maslov & Fedoriuk 1981 for a different aspect of the question. The essential conclusions can be carried over to field systems (at least to bosonic ones).

The important question is: What are the ratios of the characteristic lengths of the situation at hand?

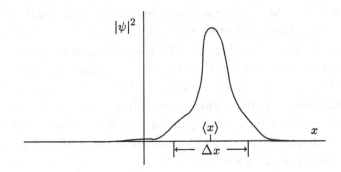

(1)   For a classical description of the position to be meaningful, the wave function $\psi(x)$ must be sharply peaked about its mean or expectation value, $\langle x \rangle$. That is,

$$\Delta x \equiv \langle (x - \langle x \rangle)^2 \rangle^{\frac{1}{2}}$$

must be small compared to lengths of experimental significance.

(2) If we demand *localization in phase space*, not just in configuration space, then we must also have Δp small compared to momenta of experimental significance. Here, however, we must accept the implications of the celebrated *uncertainty principle*, which in this context is a theorem of classical Fourier analysis in $\mathcal{L}^2(\mathbf{R})$:

$$(\Delta x)(\Delta p) \geq \frac{1}{2} \quad \text{for any } \psi.$$

Therefore, the product of the "experimentally significant" units of x and p must be sufficiently large for the classical picture of the system to be valid.

(3)   Let's suppose that we don't care about measuring p for its own sake. We still want a classical description of x over a significant time interval. From (1.6),

$$\Delta x(t) = \Delta x(0) + \frac{\Delta p}{m} t.$$

So to keep Δx small we must keep $\frac{t}{m \Delta x}$ small. (For the harmonic oscillator, from (1.5), we must keep $(m\omega \Delta x)^{-1}$ small.) In particular, this means that potentially classical behavior is associated with *large* m. This classical limit *of systems*, $m \to \infty$, is nonuniform in t — and also nonuniform in the state ψ, as described by Δx and Δp.

Parallel to the mathematician's sensitivity to nonuniform convergence is the physicist's dictum that only limits of *dimensionless parameters* have intrinsic significance. The classical limit of quantum theory is often said to be the limit $\hbar \to 0$, but $\hbar$ is a red herring. One can always choose units where $\hbar$ equals 1. In fact, to do otherwise distracts attention from the genuine small parameters in the problem at hand — which are ratios of $\hbar$ to other quantities with the dimensions of action (length $\times$ momentum, or energy $\times$ time) — and from the accompanying nonuniformities in the limit. (Nevertheless, taking $\hbar$ to zero is a convenient figure of speech, and expanding things in power series in $\hbar$ can be a useful bookkeeping device in practical calculations.)

Exercise 1: In the paragraphs numbered (2) and (3) above, what are the quantities with dimensions of action which are large compared to (the invisible) $\hbar$?

RESPECTIVE ADVANTAGES OF THE TWO PICTURES

In elementary quantum mechanics, the Schrödinger picture is preferred, for several reasons:

(a) The wave function for time t has a direct interpretation, independent of t, in terms of measurements made at t.

(b) The Schrödinger equation is always linear, while the Heisenberg equation is generally nonlinear. (Of course, we have looked only at the exceptions, the free particle and the oscillator.)

(c) Explicit construction of the operators which satisfy the Heisenberg equation is usually tantamount to solving the Schrödinger equation anyway. (One must deal with **p** and H as partial differential operators in $\mathcal{L}^2$ and find $U(t)$.) The harmonic oscillator, with its purely algebraic solution, is a rare exception.

However, in quantum field theory the Heisenberg picture is preferred.

(a) The Schrödinger formalism gives *time* a privileged role. The Heisenberg point of view permits t and the spatial coordinates to be treated on the same footing, hence permits a geometrically covariant formulation in keeping with the spirit of relativity theory.

(b) Attempting to do Schrödinger quantum mechanics with field systems would lead one into considering infinite-dimensional "differential operators", which are rather problematical objects. It is often easier to work with the formal algebra of observables and its Heisenberg equations of motion, and to hunt for physically acceptable representations *after* solving these. This is especially true when the field equations are *linear*, as they will be throughout the present work.

(c) As previously remarked, it is far from clear that a Schrödinger formulation should even make sense for a field system — especially with explicitly time-dependent field equations — because of the difficulty of constructing a Hamiltonian operator.

In summary, we prefer to work with that picture in which the classical theory of linear partial differential equations can be applied. In (possibly nonlinear) quantum mechanics, we solve the Schrödinger equation for the wave function; in linear quantum field theory, we solve the Heisenberg equation for the field operator. Let me emphasize that in a quantum field theory the field is *not* a wave function giving probabilities; despite the applicability of related mathematical techniques, a field equation such as the *Klein–Gordon equation,*

$$\frac{\partial^2 \phi}{\partial t^2} - \nabla^2 \phi + m^2 \phi = 0 \qquad \left(\phi = \phi(t, \mathbf{x}) \right),$$

is physically analogous not to the Schrödinger equation,

$$i \frac{\partial \psi}{\partial t} = -\nabla^2 \psi + V(\mathbf{x})\psi \qquad \left(\psi = \psi(t, \mathbf{x}) \right),$$

but to the Heisenberg (or classical) equation (1.1),

$$\frac{d^2 \mathbf{x}}{dt^2} = -\nabla V(\mathbf{x}) \qquad \left(\mathbf{x} = \mathbf{x}(t) \right).$$

In time-independent cases, this linear PDE theory becomes, after separation of variables, the theory of expansions in eigenfunctions of elliptic partial differential operators. Whether one starts from a Schrödinger equation or from the hyperbolic equation of a relativistic field, the elliptic eigenvalue equation belongs to the same class. The mathematics of spectral theory and eigenfunction expansions is the same, regardless of the physical context. Indeed, insight gained from one context may be applied to another. For example, the Schrödinger time evolution is unitary, while the hyperbolic time evolution is characterized by a finite propagation speed; each of these features has been used as a technical tool to gain information about the elliptic operator concerned, information which can then be used in the other context. I shall devote a chapter to this elliptic spectral theory before turning to field quantization.

Chapter 2
Self-Adjoint, Elliptic Differential Operators and Eigenfunction Expansions

The familiar Fourier series and Fourier transform are expansions of arbitrary functions in terms of eigenfunctions of the Laplace operator, $-\nabla^2$. The aim of this chapter is to describe the generalization to differential operators with variable coefficients. Such operators arise when one of the fields of nature, such as the gravitational field, is treated as "external". That is, that field is assumed to be known (as a function of space and time) and appears in the equations of motion of other fields; those equations may thereby be rendered linear.

DEFINITIONS

Following modern practice in the theory of partial differential (and pseudodifferential) operators, let us use the following notation for partial derivatives:

$$D_j \equiv -i\partial_j \equiv -i\,\frac{\partial}{\partial x^j}\,.$$

Note that D_j is the operator that replaces p_j in elementary quantization (Chapter 1). In the mathematical literature, the letter ξ is usually used instead of p in this context.

The most general second-order linear partial differential operator on scalar functions is

$$H \equiv g^{jk}(x)\,D_j D_k + b^j(x)\,D_j + c(x). \tag{2.1}$$

For the present, we assume

(1) $x \in \Omega$, which is an open, connected subset of $\mathbf{R}^d$. In (2.1), j and k are summed from 1 to d. (Ultimately, we are interested in manifolds, which may be pieced together from many such coordinate charts.)

(2) g, b, and c are smooth functions. (I.e., each component is an element in $C^\infty(\Omega)$, the space of functions which can be differentiated to arbitrarily high order.) They may become singular at the

boundary, $\partial\Omega$. (Most of my assertions remain true, perhaps with minor modifications, for less smooth coefficients, but I don't want to clutter the discussion with technicalities at this point.)

(3) The components of g are real-valued (whereas b and c may be complex-valued).

The matrix-valued function $\{g^{jk}(x)\}$ defines a metric tensor on Ω (or on the manifold for which Ω is a coordinate chart). These are the *contravariant components* of the metric, not the usual covariant components.

H is *elliptic* if $\{g^{jk}(x)\}$ is a positive definite (or negative definite) matrix for each x. More generally, an nth-order operator

$$H = \sum_j a^{j_1 j_2 \cdots j_n}(x) D_{j_1} \cdots D_{j_n} + \text{lower-order terms}$$

is *elliptic* if and only if $a^{j_1 \cdots j_n} p_{j_1} \cdots p_{j_n}$ is nonzero for every *real*, nonzero $\mathbf{p}$ in $\mathbf{R}^d$.

A fundamental property of elliptic operators is that the solutions of elliptic equations are at least as smooth as their coefficients. Precise formulations of this principle are *elliptic regularity theorems*. The simplest is the following [Folland 1976, Sec. 6B; Gilkey 1974, p. 30]:

Theorem: If H is elliptic, the functions f, g, b, and c are in $C^\infty(\Omega)$, and $Hu = f$, then $u \in C^\infty(\Omega)$. In particular, $Hu = Eu$ ($E \in \mathbf{R}$) implies $u \in C^\infty$.

In the latter situation, u is an *eigenfunction* of H. This term (unlike "eigenvector") does not imply square-integrability of u.

Let μ_ρ be a measure defined by a smooth, positive function $\rho(x)$ on Ω. That is,

$$d\mu_\rho \equiv \rho(x)\,d^d x.$$

Define the Hilbert space $\mathcal{L}^2_\rho(\Omega)$ by the inner product

$$\langle \psi, \phi \rangle \equiv \int_\Omega \psi(x)^* \phi(x)\,d\mu_\rho.$$

Consider the operator H restricted to the domain $C_0^\infty(\Omega)$. (A smooth function belongs to $C_0^\infty(\Omega)$ if it is nonzero only inside a closed, bounded region contained in the *interior* of Ω). H will be called *Hermitian* if

$$\langle \psi, H\phi \rangle = \langle H\psi, \phi \rangle \quad \text{for all } \psi,\, \phi \in C_0^\infty.$$

(More generally, a Hilbert-space operator with a specified domain is *Hermitian* if this equation is satisfied for all vectors ψ, ϕ in the domain. I should warn the nonexperts that "domain" here refers not to Ω, but to the set of vectors (functions) in the Hilbert space on which H is regarded as defined; typically a choice of domain corresponds to a particular choice of *boundary conditions* in a classical PDE problem.) For differential operators with smooth coefficients, a condition equivalent [e.g., Gårding 1954] to Hermiticity is that H be *formally self-adjoint*, which means that

$$\int_\Omega \psi^* H\phi\, \rho\, d^d x = \int_\Omega (H\psi)^* \phi\, \rho\, d^d x$$

by classical integration by parts, *all boundary terms being ignored*. The "0" in C_0^∞ is a brute-force guarantee that all integrals converge and all boundary terms vanish.

Example: Let $\rho = 1$, $g^{jk} = \delta^{jk}$, so that

$$H = -\nabla^2 + b^j D_j + c.$$

Then H (with domain C_0^∞) is Hermitian if and only if b^j is real-valued and $i\,\mathrm{Im}\,c = \frac{1}{2}D_j b^j$.

Exercise 2:

(a) Find analogous conditions for general ρ and g.

(b) Show that an operator satisfying those conditions can be written as

$$H\psi = \frac{1}{\rho}(D_j + A_j)\left[\rho g^{jk}(D_k + A_k)\psi\right] + V\psi,$$

where $A_j(x)$ and $V(x)$ are real-valued. (Some readers will recognize **A** as a connection on a vector bundle. See Chapter 8.)

Introduce coordinate-free notations if you feel so inclined. (Answers appear at the end of the chapter.)

The Hermitian operators which possess eigenfunction expansions satisfy an additional technical condition:

Definition: An Hermitian H is *essentially self-adjoint* if the ranges $\mathrm{ran}\,(H - i)$ and $\mathrm{ran}\,(H + i)$ are both *dense* in $\mathcal{L}_\rho^2$; it is *self-adjoint* if they are *equal* to $\mathcal{L}_\rho^2$.

Remarks:

(a) An essentially self-adjoint operator has a *unique* self-adjoint extension to a larger operator domain, and conversely.

(b) Any number in the upper half-plane could be used in place of i in the definition.

(c) If one of the classical partial differential equations $H\psi = \pm i\psi$ has a solution in $\mathcal{L}_\rho^2(\Omega)$, then H restricted to $C_0^\infty(\Omega)$ is *not* essentially self-adjoint; this is a frequent test in practice for self-adjointness of elliptic differential operators.

SELF-ADJOINT EXTENSIONS: THE CLASSIC EXAMPLES

If H is Hermitian, it may be possible to make it (essentially) self-adjoint by enlarging its domain. H may have *many* self-adjoint extensions, or (more rarely) *none*. In the former case, each such extension is obtained by relaxing the condition "$C_0^\infty(\Omega)$" to "$C^\infty(\Omega)$ satisfying certain homogeneous boundary conditions".

For example, consider $\Omega = (0,1)$, $H = -\dfrac{d^2}{dx^2}$. If $H\psi = E\psi$, then

$$\psi = c_1 e^{ikx} + c_2 e^{-ikx}, \qquad k \equiv \sqrt{E} \in \mathbf{C}.$$

All the eigenfunctions are in $\mathcal{L}^2(\Omega)$ (and in $C^\infty(\Omega)$). *None*, however, are in $C_0^\infty(\Omega)$. H is not Hermitian on the domain C^∞ (which is too large), and it is not essentially self-adjoint on C_0^∞ (which is too small).

Recall that this eigenvalue problem arises from separating variables in PDE problems such as

$$\frac{\partial u}{\partial t} = \frac{\partial^2 u}{\partial x^2} \equiv -Hu, \qquad u(0,x) = f(x) \in \mathcal{L}^2(0,1) \qquad (2.2)$$

(the Cauchy problem for the heat equation), or

$$\frac{\partial^2 u}{\partial t^2} = \frac{\partial^2 u}{\partial x^2}, \qquad u(0,x) = f(x), \qquad \frac{\partial u}{\partial t}(0,x) = g(x) \qquad (2.3)$$

(the Cauchy problem for the wave equation). To make these problems well-posed, we must tell where the heat or the wave goes when it hits the boundary ($x = 0$ or 1). Different boundary conditions correspond to different physical situations. Thus the need to choose among self-adjoint extensions just amounts to the obligation in classical applied mathematics to make sure that the problem has been completely stated. More specifically, for the heat problem (2.2) we have the standard choices of boundary conditions:

Dirichlet: $u(t,0) = 0$ end at fixed temperature

Neumann: $\dfrac{\partial u}{\partial x}(t,0) = 0$ insulated end

Robin: $\dfrac{\partial u}{\partial x} = \kappa u$ convective cooling

Periodic: $u(t,0) = u(t,1)$, $\dfrac{\partial u}{\partial x}(t,0) = \dfrac{\partial u}{\partial x}(t,1)$

In the last case, Ω represents a space which is physically a ring (i.e., 0 and 1 represent the same point); this is our first concrete, nontrivial manifold.

Given a boundary condition at each endpoint (or, in the periodic case, two boundary conditions relating the two endpoints), we get a discrete spectrum of allowed solutions. For instance, in the Dirichlet case, the allowed eigenvalues are $E_n \equiv (n\pi)^2$ (n a positive integer), with eigenfunctions $\psi_n \equiv \sin n\pi x$. An arbitrary square-integrable function on the interval can be expanded in the corresponding Fourier series, and the resulting factored solutions superposed to form the general solution of the initial-value problem (2.2):

$$u(t, x) = \sum_{n=1}^{\infty} e^{-E_n t} f_n \sin n\pi x,$$

$$f_n = 2 \int_0^1 \sin n\pi x \, f(x) \, dx \equiv \frac{\langle \psi_n, f \rangle}{\|\psi_n\|^2},$$

since $\|\psi_n\|^2 = \int_0^1 \sin^2 n\pi x \, dx = \frac{1}{2}$. In other words, $\{\sqrt{2} \sin n\pi x\}$ is an orthonormal basis for $\mathcal{L}^2(0,1)$, and H is thereby defined on the domain of functions $f = \sum_n f_n \psi_n$ such that $\sum_n |n^2 f_n|^2 < \infty$. (These are the same as those functions that are sufficiently smooth — almost in C^2 — and vanish at the endpoints.) We can write $u(t, \cdot) \equiv e^{-tH} f$.

Similarly, for any function $G(\lambda)$ ($\lambda \in \mathbf{R}$), one defines

$$[G(H)f](x) \equiv \sum_{n=1}^{\infty} G(E_n) f_n \sin n\pi x.$$

For example, the solution of the wave problem (2.3) is

$$[\cos(t\sqrt{H})]f + \frac{\sin(t\sqrt{H})}{\sqrt{H}}\,g. \tag{2.4}$$

The solution of the initial-value problem for the Schrödinger equation is $e^{-itH}f$.

Exercise 3: Convince yourself of (2.4).

The point here is that H is not completely defined until we state the boundary conditions. When the domain of H is that defined by Dirichlet boundary conditions, the formula (2.4) represents a different operator from when the domain is defined by periodic conditions; this is the difference between waves bouncing off the ends of a vibrating string and waves travelling around a homogeneous ring.

We have left a loose end: For a given Hermitian operator, how do we know it has any self-adjoint extensions at all? I'll state two well-known theorems, which follow fairly directly from the definition of self-adjointness (cf. Reed & Simon 1975, Theorem X.1 and following discussion).

Definition: H is *semibounded* (or *bounded below*) if there is an $a \in \mathbf{R}$ such that

$$\langle \psi, H\psi \rangle \geq a\langle \psi, \psi \rangle \quad \text{for all } \psi \in \text{dom } H.$$

Theorem: If H is Hermitian and semibounded, then it has self-adjoint extensions.

Observations: It is hard for an elliptic Hermitian operator on scalar functions to fail to be semibounded. In the context of Exercise 2(b), the first term in $\langle \psi, H\psi \rangle$ is manifestly positive, since integration by parts puts it into the form

$$\int_{\Omega} \|(\mathbf{D} + \mathbf{A})\psi\|_g^2\, \rho\, d^d x.$$

(The norm is that defined on vector-valued functions by the contravariant metric g.) Therefore, if $V(x)$ is bounded below (as a function), H is semibounded (as an operator).

Theorem: An Hermitian differential operator with *real coefficients* has self-adjoint extensions.

In the notation of the solution to Exercise 2, this means that the values of **b** are pure imaginary and those of c are real; equivalently, that the connection form **A** is 0. Putting the two theorems together, we see that trouble could arise for a second-order elliptic operator only if $\mathbf{A} \neq 0$ *and* V were unbounded below. (Recall that $V = \text{Re } c - \|\mathbf{A}\|_g^2$.)

Another ramification of the eigenvalue problem is the *continuous spectrum*. For example, consider $\Omega = \mathbf{R}$, $H = -\dfrac{d^2}{dx^2}$. Again $H\psi = E\psi$ implies

$$\psi = c_1 e^{ikx} + c_2 e^{-ikx}, \qquad k \equiv \sqrt{E}.$$

But this time *no* solutions are in $\mathcal{L}^2(\mathbf{R})$ for *any* E. Therefore, H is already self-adjoint; no boundary conditions are needed. But on the other hand, there are no eigenvectors! Nevertheless, an arbitrary vector can be expanded as an integral over the nonnormalizable eigenfunctions (the Fourier transform): Define

$$\psi_k(x) \equiv (2\pi)^{-\frac{1}{2}} e^{ikx},$$

$$\hat{f}(k) \equiv \int_{-\infty}^{\infty} \psi_k(x)^* f(x)\, dx.$$

Then

$$f(x) = \int_{-\infty}^{\infty} \psi_k(x)\, \hat{f}(k)\, dk, \qquad (2.5)$$

and

$$\left[e^{-tH} f\right](x) = \int_{-\infty}^{\infty} \psi_k(x)\, e^{-tk^2}\, \hat{f}(k)\, dk,$$

and so on. This raises two new questions:

1. Where did we get the factor $(2\pi)^{-\frac{1}{2}}$? Why could not (2.5) have been something like

$$f(x) = \int_{-\infty}^{\infty} 40\sqrt{k^2 + \pi^2}\, e^{ikx}\, \hat{f}(k)\, dk\,?$$

2. How did we know to integrate over all real k, and *only* over real k? Obviously square-integrability is not the criterion which distinguishes e^{ikx} with k real from e^{ikx} with k complex. (Note, in

this connection, that a pure imaginary k corresponds to a real, but negative, E.)

In the example at hand, the answers to these questions are facts of the classical theory of Fourier integrals [e.g., Titchmarsh 1937]. The point, however, is that we must anticipate encountering similar issues in the general case (to which we now return). We shall need a general theory to cover them.

THE SPECTRAL THEOREM

Definition: $z \in \mathbf{C}$ is in the *spectrum* of H, $\sigma(H)$, if the following condition is *false*:

$(H - z)^{-1}$ exists and is a bounded operator (with domain equal to the whole Hilbert space, $\mathcal{H}$).

Observations: If z is an eigenvalue of H (i.e., $H\psi = z\psi$ for a nonzero $\psi \in \mathcal{H}$), then $(H-z)^{-1}$ doesn't exist; this is the *point spectrum* of H. For *self-adjoint* H (our principal concern), there is only one other possibility: $H - z$ is injective (one-to-one) but not surjective (onto), its range (which is dom $(A-z)^{-1}$) is a dense subspace of $\mathcal{H}$, and $(A-z)^{-1}$ is an unbounded operator. Then one says that z is in the *continuous spectrum* of H. In this case z is an *approximate eigenvalue* of H (and conversely):

For every $\epsilon > 0$ there is a $\psi \in \mathcal{H}$ such that $\|\psi\| = 1$ and $\|(A - z)\psi\| < \epsilon$.

(Approximations to the Dirac delta function, such as $\frac{\epsilon}{\pi} \frac{1}{x^2+\epsilon^2}$, are approximate eigenfunctions of the position operator. Plane waves "cut off" at infinity are approximate eigenfuctions for the momentum operator.)

If H is Hermitian, $\sigma(H)$ is real. A typical $\sigma(H)$ with enough structure to be interesting is the spectrum of the hydrogen atom, as calculated in any textbook of quantum mechanics, such as Messiah 1961. All nonnegative energies make up the continuous spectrum, corresponding to the possible quantum states of a free electron *scattering* off the proton. In addition, there are infinitely many negative numbers in the point spectrum, corresponding to the states in which the electron is *bound* in the electric field of the proton. (These energy levels are bounded below and have 0 as an accumulation point above.) A "sum" over the spectrum, such as occurs in the theorems to be stated presently, therefore consists of an integration over the continuous spectrum plus a true sum over the point spectrum. Such a sum is more properly described as an integration with respect to a Stieltjes measure, $\int \ldots d\mu$, where $\mu(\lambda)$ is an increasing function which is continuous for $\lambda \geq 0$, while for $\lambda < 0$ it is piecewise constant with finite jumps at the eigenvalues. (Readers unfamiliar with Stieltjes measures may think of $d\mu$ as $\mu'(\lambda)\, d\lambda$, where $\mu'(\lambda)$ is a sum of delta functions on the negative real axis and a positive, continuous function on the positive axis.)

Spectral theorem (function-space version): Let H be a self-adjoint linear operator in a Hilbert space $\mathcal{H}$ (i.e., defined on a dense subspace of $\mathcal{H}$ and taking values in $\mathcal{H}$). Then $\mathcal{H}$ is isomorphic to a Hilbert space of functions, $\int^{\oplus} \mathcal{H}_\lambda \equiv \mathcal{L}^2\big(\sigma(H); \mathcal{H}_\lambda\big)$, where for each $\lambda \in \sigma(H)$, $\mathcal{H}_\lambda$ is a Hilbert space, whose dimension may depend on λ; if ϕ is in $\mathcal{L}^2\big(\sigma(H); \mathcal{H}_\lambda\big)$, then for each $\lambda \in \sigma(H)$, the *value* $\phi(\lambda)$ is a vector in $\mathcal{H}_\lambda$, and the norm and scalar product in $\mathcal{L}^2\big(\sigma(H); \mathcal{H}_\lambda\big)$ are determined by the formula

$$\|\phi\|^2 = \int_{\sigma(H)} \|\phi(\lambda)\|^2_{\mathcal{H}_\lambda}\, d\mu(\lambda),$$

where $\|\cdots\|_{\mathcal{H}_\lambda}$ is the norm in $\mathcal{H}_\lambda$ and μ is a measure on $\sigma(H)$. ($\|\cdots\|_{\mathcal{H}_\lambda}$ and μ are not unique, in the sense that an arbitrary positive factor can be transferred from one to the other.) *Under the isomorphism, H corresponds to the operator of multiplication by λ.* That is, if U is the norm-preserving bijection mapping $\int^{\oplus} \mathcal{H}_\lambda$ onto $\mathcal{H}$, then

$$H = UDU^{-1},$$

where D is defined by

$$[D\phi](\lambda) \equiv \lambda\, \phi(\lambda) \qquad \text{for all } \lambda \in \sigma(H).$$

I rush to reassure the physicist reader that this theorem is merely (1) the natural generalization to infinite dimensions of the elementary theorem describing the diagonalization of a symmetric matrix, and (2) a formal statement of what physicists do all the time when expanding wave functions in terms of eigenstates of an observable. The formalism needs to be slightly complicated to encompass the possibilities of both continuous spectrum and spectral multiplicity (degeneracy of eigenvalues and the corresponding phenomenon for continuous spectrum).

Unfortunately, the spectral theorem in this abstract generality does not tell us how to calculate the transforms U and U^{-1}. For elliptic differential operators, however, we can go further:

Corollary: If, moreover, $\mathcal{H} = \mathcal{L}_\rho^2(\Omega)$ and H is an elliptic operator with C^∞ coefficients, then the map U^{-1} (more precisely, the map

$$U\phi \mapsto \langle e, \phi(\lambda)\rangle_{\mathcal{H}_\lambda}$$

for fixed $\lambda \in \sigma(H)$ and fixed $e \in \mathcal{H}_\lambda$) is given (almost everywhere) by an integral transform

$$\langle e, \phi(\lambda)\rangle = \int_\Omega e(x)^* [U\phi](x)\, \rho(x)\, d^d x, \qquad (2.6)$$

where $e(x)$ is some solution of the classical differential equation $He = \lambda e$. ("Classical" in this context means that e is regarded as a literally differentiable, and not necessarily square-integrable, function, rather than a vector in the domain of H within the Hilbert space.) Let $\{e_{\lambda j}\}$ $(j = 1, \dots, \dim \mathcal{H}_\lambda \leq \infty)$ be an orthonormal basis for $\mathcal{H}_\lambda$, and $\phi_j(\lambda) \equiv \langle e_{\lambda j}, \phi(\lambda)\rangle_{\mathcal{H}_\lambda}$. Then the inverse transform is

$$[U\phi](x) = \int_{\sigma(H)} \sum_j e_{\lambda j}(x)\, \phi_j(\lambda)\, d\mu(\lambda). \qquad (2.7)$$

If the basis is not orthonormal, the formula will be of the more general form

$$[U\phi](x) = \int_{\sigma(H)} \sum_{j,k} e_{\lambda j}(x)\, \phi_k(\lambda)\, d\mu^{jk}(\lambda) \qquad (2.8)$$

for some matrix-valued measure μ^{jk}, and

$$\int_\Omega |U\phi|^2\, \rho\, dx \equiv \|U\phi\|^2$$

$$= \|\phi\|^2 = \int_{\sigma(H)} \sum_{j,k} \phi_j(\lambda)^*\, \phi_k(\lambda)\, d\mu^{jk}(\lambda).$$

(This is the *Parseval equation* for the unitary transformation U. It contains the same information which physicists are accustomed to express by "orthonormality and completeness relations" for the generalized eigenfunctions $e_{\lambda j}$. Examples of the latter will be offered below.)

Remarks:

(a) I will not state precise theorems about the nature of the convergence of the integrals (2.6–8); it depends on how nice the function $[U\phi](x)$ or $\phi_j(\lambda)$ is, just as in classical Fourier analysis. (See also (d).)

(b) If $\mathcal{H}$ were finite-dimensional, the U in the spectral theorem would be [represented by] a matrix whose columns are the eigenvectors of the Hermitian matrix [representing] H. (This is the elementary diagonalization theorem mentioned previously.) Formula (2.7) can be regarded as a generalization of this; x plays the role of the row index (labelling the components of each eigenvector), while the ordered pair (λ, j) plays the role of the column index (labelling the eigenvectors).

(c) Unfortunately, the corollary still doesn't tell us how to calculate $\langle e, e' \rangle_{\mathcal{H}_\lambda}$, hence how to construct an orthonormal basis. An equivalent remark is that the corollary doesn't tell us explicitly what the functions μ^{jk} in (2.8) are.

(d) The elliptic regularity principle guarantees that solutions of the differential equation $He = \lambda e$ are smooth functions. (They are eigen*functions* of H but not necessarily eigen*vectors*.) In more general situations (e.g., for a hyperbolic H) the "generalized eigenvectors" $e_{\lambda j}$ will be continuous linear functionals (distributions) defined on some subspace of $\mathcal{L}^2$ with a stronger topology. (For example, $e_{\lambda j}$ might belong to $\mathcal{D}' \equiv (C_0^\infty)'$, the space of distributions defined on smooth test functions of compact support.) They will satisfy $He = \lambda e$ *weakly*. (That is, the equation holds when both sides are applied to a test function and the operator is moved onto the test function by formal integration by parts.) In that case, (2.6–8) *a priori* require the expanded function to lie in the test-function space.

(e) Presumably, whenever the definition of H as a self-adjoint operator includes boundary conditions, the $e_{\lambda j}(x)$ will satisfy the boundary conditions as well as the classical differential equation. This assertion is not made clearly in the literature of the subject, however. (See, however, Gel'fand & Shilov 1967, Sec. IV.5.3.)

(f) If H is a quantum observable and $\|\phi\| = 1$, then

$$\int_\Delta \|\phi(\lambda)\|_{\mathcal{H}_\lambda}^2 \, d\mu(\lambda)$$

is the probability of observing H to have a value in $\Delta \subset \sigma(H)$.

(g) References for the corollary include Gårding 1954, Mautner 1953, Beals 1967, Maurin 1967 (Chap. 17), Maurin 1968 (Chap. 2), and Berezansky 1968.

THE PROBLEM OF EIGENFUNCTION EXPANSION

In any concrete problem the physicist is confronted with the task of "finding a complete set of modes" — that is, constructing explicitly the apparatus described in the elliptic corollary to the spectral theorem. There are three (overlapping) questions here:

1. What kind of boundary conditions, if any, are needed to make H self-adjoint? Which choice corresponds to the physics of the problem? (If too few (or too weak) boundary conditions are imposed, H will not be Hermitian on its oversized domain; then the eigenfunctions will be redundant and nonorthogonal. If the boundary conditions are too many (or too strong), H will not be essentially self-adjoint; the eigenfunction expansion will not be uniquely specified.)

2. Which numbers are in the spectrum? (These values can be further classified into point spectrum, σ_p, and continuous spectrum, σ_c, with further ramifications into absolutely continuous spectrum, singular spectrum, limit points of eigenvalues, etc.) What are their multiplicities $(\dim \mathcal{H}_\lambda)$? What are the corresponding eigenfunctions?

3. Given a "basis" of eigenfunctions, $\{e_{\lambda j}\}$, what are the weight factors, $\mu^{jk}(\lambda)$, in the inversion formula, (2.8)? Equivalently, what is the Parseval identity, or what are the orthogonality-completeness relations?

Remarks on Question 1: For Schrödinger operators $(H = -\nabla^2 + V)$ with $V(\mathbf{x})$ bounded below, one has the rules of thumb (not precise theorems):

(1) No boundary conditions are needed at infinity.

(2) "Regular" (or Robin-type) boundary conditions are appropriate at any boundary of Ω:

$$\frac{\partial \psi}{\partial n} = \kappa(\mathbf{x})\psi.$$

(The Neumann boundary condition is the special case $\kappa = 0$; the Dirichlet condition, $\psi = 0$, can be regarded as a singular limiting case, $\kappa = \infty$.)

If V is unbounded below near ∞ or the boundary of Ω, then more subtle ("singular") boundary conditions may be needed. (In one dimension, this is the "limit-circle case", of which more below.)

Remarks on Question 2: If V approaches $+\infty$ as $\mathbf{x} \to \infty$, H will have a pure point spectrum [Reed & Simon 1978, Thm. XIII.67]. (The key fact here is that H has compact resolvents.) A rule of thumb for anticipating the nature of the spectrum is to look at the behavior of a classical particle having H as Hamiltonian:

(1) If the particle cannot escape to infinity, then there will be only bound states (pure point spectrum).

(2) If the particle escapes as $t \to \infty$, then there will be a continuum in the spectrum.

(3) If the particle reaches infinity within a *finite* time, then boundary conditions will be needed at spatial infinity (to specify where the particle, or the probability current in the quantum theory, will go once it reaches there). In dimension 1 this leads to $\int^{\infty} |V(x)|^{-\frac{1}{2}} dx = \infty$ as the criterion for essential self-adjointness (lack of need for boundary conditions). If boundary conditions *are* needed, their effect may be to make the spectrum discrete again!

There very definitely are exceptions to these principles, and I refer to Reed & Simon 1975, Appendix X.1, for a thorough discussion.

In curved-space problems (or theories with external gauge fields) the operator H is not simply of the form $H = -\nabla^2 + V$; the Laplacian is replaced by a spatially inhomogeneous operator whose coefficients represent the gravitational (or gauge) field. The lore concerning questions 1 and 2 for such operators is not as complete and well known as for the Schrödinger operators, but some facts are known and will be presented in due course.

Remarks on Question 3: Complete results are known for special categories of operators.

1. Pure point spectrum: All that is needed is to orthonormalize the eigenvectors:

$$\langle \psi_j, \psi_k \rangle = \delta_{jk}; \qquad H\psi_j = \lambda_j \psi_j.$$

Then the coefficients in the expansion of an arbitrary vector can be calculated in the usual way:

$$\phi(j) \equiv \langle \psi_j, \phi \rangle, \qquad \psi = \sum_j \phi(j)\psi_j \,;$$

$$\sum_j |\phi(j)|^2 = \|\psi\|^2 \,; \qquad [U^{-1}HU\phi](j) = \lambda_j \phi_j \,.$$

(Here I have avoided the notational complications associated with spectral multiplicity by writing $\phi(j)$ (a scalar-valued function) instead of $\phi(\lambda_j)$ (which might need to be vector-valued). Such pragmatic revisions of notation are common and unavoidable, and will be performed without comment when there is little chance of confusion.)

2. One-dimensional scattering theory: See Messiah 1961, Chapter 3, for a complete treatment. We consider $H = -\nabla^2 + V(x)$ still.

a) *One-sided*: Let V be in $C^\infty(0, \infty)$, $V(x) \to 0$ sufficiently fast as $x \to \infty$. Impose a regular Sturm–Liouville boundary condition,

$$\psi'(0) = \kappa\psi(0) \qquad (\text{or } \psi(0) = 0).$$

Then $\sigma_c = [0, \infty)$ with multiplicity 1. There may also be a point spectrum $\sigma_p \subset (-\infty, 0]$; its contribution to the eigenfunction expansion may be handled as just described.

For every $k \in [0, \infty)$, where $E_k \equiv k^2$ is the pertinent element of σ_c, there is an eigenfunction (in the sense of the Corollary), ψ_k, with the asymptotic behavior

$$\psi_k(x) \sim N_k \sin(kx + \delta_k) \quad \text{as } x \to \infty,$$

where N_k is a normalization constant. Physically, the Schrödinger solution

$$\psi(t, x) \equiv \int_0^\infty \hat{f}(k)\, e^{-iE_k t}\, \psi_k(x)\, dk, \qquad (2.9)$$

with $\hat{f}$ peaked around k_0, is a pulse which travels in at speed $\simeq k_0/m$ and bounces back from the boundary. (m is the mass, which would be equal to $\frac{1}{2}$ in the normalization adopted here.)

The crucial fact about such problems is that *the normalization is the same as that for the Fourier sine transform* (that is, the special case $V(x) \equiv 0$, $\psi(0) = 0$). Namely, if we take $N_k = 1$ and define

$$\hat{\phi}(k) \equiv \int_0^\infty \psi_k(x)^* \,\phi(x)\, dx,$$

then

$$\phi(x) = \frac{2}{\pi} \int_0^\infty \hat{\phi}(k)\, \psi_k(x)\, dk + \text{bound-state contributions.}$$

(The last term is the projection of ϕ onto the subspace spanned by the true eigenvectors, if any — which correspond to certain nonpositive values of E.) In other words, $d\mu(E) = \frac{2}{\pi} d\sqrt{E}$. Still another way of expressing this fact of normalization, more in keeping with the quantum-mechanics textbooks, consists of the *orthonormality relation*,

$$\int_0^\infty \psi_k(x)^*\, \psi_{k'}(x)\, dx = \frac{\pi}{2}\, \delta(k - k'),$$

together with the *completeness relation*,

$$\int_0^\infty \psi_k(x)\, \psi_k(x')^*\, dk + \text{bound-state contributions} = \frac{\pi}{2}\, \delta(x - x').$$

(These relations respectively express the conditions $U^*U = 1$ and $UU^* = 1$, which together state that the spectral transform U is a unitary operator.) Finally, the Parseval relation is

$$\|\phi\|^2 \equiv \int_0^\infty |\phi(x)|^2\, dx = \int_0^\infty |\hat{\phi}(k)|^2\, \frac{2}{\pi}\, dk + \text{bound-state terms.}$$

It is customary, however, to change the normalization convention to $N_k \equiv \sqrt{\frac{2}{\pi}}$. This eliminates the factor $\frac{\pi}{2}$ from the basic formulas just given. (More precisely, it hides the factor in the formula for the eigenfunction; its square root reappears whenever an explicit integration has to be done.)

Kodaira 1949 gives a proof of the "crucial fact". A physical argument making it plausible is as follows: The pulse (2.9) will eventually move into a region where $V(x)$ is almost zero, so that the asymptotic expression for ψ_k is almost exact. (Any square-integrable wave function can be regarded as such a pulse, perhaps very broad. Although the wave function changes its shape with time, it can be verified that at large enough t, most (in the sense of the $\mathcal{L}^2$ norm) of the function will be in the asymptotic region.) Therefore, the eigenfunction expansion (for a fixed ψ) reduces approximately to an ordinary Fourier transform, with arbitrary accuracy if t is taken sufficiently large. But the dynamical operator e^{-iHt} is unitary and acts diagonally with respect to k;

so this Fourier decomposition may be carried back, at the cost of a k-dependent phase factor, into the regime where the pulse sits in the region where the potential is large.

b) *Two-sided*: Let $V \in C^{\infty}(-\infty, \infty)$ with $V \to 0$ sufficiently fast as $|x| \to \infty$. This case is actually of considerable interest to us, because it arises upon separation of variables in the radially symmetric gravitational potential around a black hole [e.g., DeWitt 1975]. (In contrast, radially symmetric three-dimensional potentials in ordinary atomic or nuclear physics give rise to one-sided one-dimensional scattering problems similar to that just discussed.)

In this case, the continuous spectrum is $[0, \infty)$ with multiplicity 2. *The normalization problem can be reduced to that for the full Fourier transform* (again corresponding to $V(x) \equiv 0$ on the relevant interval, $(-\infty, \infty)$). For every $k \in (0, \infty)$ there is a Fourier-normalized eigenfunction

$$\sqrt{2\pi}\,\psi_k^{\mathrm{in}}(x) \sim \begin{cases} e^{ikx} + R_k e^{-ikx} & \text{as } x \to -\infty, \\ T_k e^{ikx} & \text{as } x \to +\infty. \end{cases}$$

The reflection and transmission coefficients, R_k and T_k, depend upon V, and their computation is a problem in classical asymptotic theory for ordinary differential equations. These functions, when combined into normalizable solutions of the Schrödinger equation, describe pulses which impinge on the potential from the left, then break into transmitted and reflected parts.

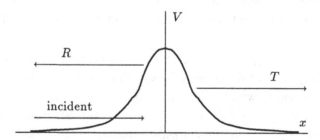

These are only half of the necessary basic eigenfunctions. For every $k \in (-\infty, 0)$ there is an eigenfunction with the behavior

$$\sqrt{2\pi}\,\psi_k^{\mathrm{in}}(x) \sim \begin{cases} T_k e^{ikx} & \text{as } x \to -\infty, \\ e^{ikx} + R_k e^{-ikx} & \text{as } x \to +\infty. \end{cases}$$

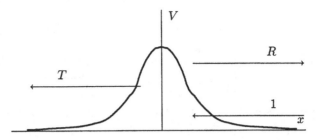

These represent motion of the complementary type, shown in the second figure.

As usual, the spectral variable is $E = k^2$. The spectral transform is

$$\hat{\phi}^{in}(k) \equiv \int_{-\infty}^{\infty} \psi_k^{in}(x)^* \, \phi(x) \, dx,$$

$$\phi(x) = \int_{-\infty}^{\infty} \psi_k^{in}(x) \, \hat{\phi}(k) \, dk + \text{bound-state terms}.$$

There is an alternative basis, $\{\psi_k^{out}\}$, for which the pulses cohere at late time instead of early time, as shown in the third figure. The reader should be able to write down the correct asymptotic forms by analogy with the other case. I emphasize that a complete basis set must include both signs of k, but only one of the two choices, *in* or *out*.

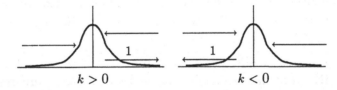

$$k > 0 \qquad\qquad k < 0$$

3. Three-dimensional scattering theory: Let $V(\mathbf{x}) \to 0$ sufficiently fast as $r \equiv |\mathbf{x}| \to \infty$. Then $\sigma_c = [0, \infty)$ with multiplicity ∞. For every $\mathbf{k} \in \mathbf{R}^3$, there is a Fourier-normalized eigenfunction

$$\psi_{\mathbf{k}}^{in}(\mathbf{x}) \sim (2\pi)^{-\frac{3}{2}} \left[e^{i\mathbf{k}\cdot\mathbf{x}} + \frac{e^{ikr}}{r} \, f_{\mathbf{k}}(\theta, \phi) \right].$$

That is, the solution consists asymptotically of an incoming plane wave and an outgoing spherical wave. Note that the one-dimensional formalism just described is the natural degeneration of this basis to one dimension. Naturally, there is also an *out* basis, whose spherical waves carry the factor e^{-ikr} so that the spherical waves are incoming when the time dependence is attached.

The formalism of three-dimensional scattering theory is presented thoroughly in many quantum-mechanics textbooks. For a rigorous treatment of the associated eigenfunction-expansion theory, see Ikebe 1960.

4. The general one-dimensional case: For the most general self-adjoint second-order linear ordinary differential operator, there is a complete eigenfunction-expansion theory, which illuminates all three questions. This is of interest not only as a source of solvable models in two-dimensional space-time, but also because many important models in physical, four-dimensional space-time can be reduced by separation of variables to one-dimensional spectral problems. This subject is not as well known among either mathematicians or physicists as I believe it should be. Therefore, I shall expound it in some detail.

The Weyl–Titchmarsh–Kodaira theory

References for this topic include Titchmarsh 1962, Kodaira 1949, Maurin 1967 and 1968, Richtmyer 1978 (Chap. 10), and Dunford & Schwartz 1963 (Chap. 13, especially Sec. 13.5).

We consider a differential operator defined on an interval $\Omega = (a, b) \subseteq \mathbf{R}$. Note that a one-dimensional second-order differential operator,

$$H = g(x)D^2 + b(x)D + c(x),$$

is always elliptic, provided that $g(x) > 0$. If it is Hermitian in $\mathcal{L}^2_\rho$, Exercise 2 says that it takes the form

$$H = \frac{1}{\rho(x)}\left(D + A(x)\right)\left[\rho(x)g(x)\left(D + A(x)\right)\cdot\right] + V(x).$$

To avoid technicalities I shall assume that the coefficients are of class C^2 in Ω (not necessarily at the endpoints), although this hypothesis is much stronger than necessary.

Rather than characterize the solutions by their behavior at infinity, we shall prescribe initial data at some distinguished point $x_0 \in (a, b)$.

Clearly this choice is somewhat arbitrary; but it turns out to be quite natural later, when we shall want to calculate and renormalize physical quantities, such as energy density, which are functions defined on Ω, by a process requiring series expansions about the argument point: It is convenient then to pick x_0 to be that point.

For each $z \in \mathbf{C}$, consider the pair of linearly independent solutions $\{\psi_{zj}\}_{j=0}^1$ of $H\psi = z\psi$ satisfying

$$\psi_{z0}(x_0) = 1, \qquad \rho g \psi'_{z0}(x_0) = 0,$$
$$\psi_{z1}(x_0) = 0, \qquad \rho g \psi'_{z1}(x_0) = 1.$$

We shall construct the *Green function* (the integral kernel of the *resolvent operator* $(H - z)^{-1}$) in terms of the ψ_{zj}.

At each endpoint of Ω there is a distinguished one-dimensional subspace of solutions of $H\psi = z\psi$. Consider, for example, the right endpoint, b. A theorem of Weyl describes the two possibilities:

1. **Limit-point case:** Whenever $\operatorname{Im} z \neq 0$, there is a unique (up to normalization) solution $\phi_{zb}(x)$ such that

$$\int_{x_0}^b |\phi_{zb}(x)|^2 \rho\, dx < \infty.$$

   (Any point in the interior of the interval could appear in the role of x_0 here. The point is that ϕ_{zb} is square-integrable in the immediate vicinity of b, or, for short, *square-integrable at b*.) This case will occur when H is essentially self-adjoint (on C_0^∞) without benefit of boundary conditions (or, more generally, when no boundary condition at b is needed). Incidentally, when z is real, the number of independent solutions that are square-integrable at b may be either 0 or 1 in this case.

2. **Limit-circle case:** *All* solutions are square-integrable at b. (If this is true of one z, it is true of all, real or complex.) This is the case when a boundary condition at b is needed to make H self-adjoint. (In particular, endpoints admitting regular Sturm–Liouville boundary conditions fall in this category.) One defines ϕ_{zb} (up to normalization) to be the solution satisfying the boundary condition. (The special case of periodic boundary conditions requires separate treatment.)

In each case it must be possible to write

$$\phi_{zb}(x) = c_0(z)\psi_{z0}(x) + c_1(z)\psi_{z1}(x),$$

for some coefficients c_j which are defined only up to a common factor. Examining the behavior of solutions at the left endpoint, one defines $\phi_{za}(x)$ similarly.

We turn to the Green function. Suppose that the solution of

$$(H - z)u = f$$

is given by

$$u(x) = \int_a^b G_z(x, y)\, f(y)\, \rho(y)\, dy.$$

It can be shown that this is equivalent to:

1) $(H_{(x)} - z)G_z(x, y) = \dfrac{\delta(x - y)}{\rho(y)}\,.$

2) As a function of x, $G_z(x, y)$ is square-integrable and satisfies the boundary conditions, if any.

From this in turn, standard calculations (see, e.g., Weinberger 1965) yield

$$G_z(x, y) = \frac{\phi_{za}(x_<)\,\phi_{zb}(x_>)}{(\rho g W_z)\big|_y}\,, \qquad (2.10)$$

$$x_< \equiv \min(x, y), \qquad x_> \equiv \max(x, y), \qquad W_z \equiv \phi_{za}\phi'_{zb} - \phi_{zb}\phi'_{za}\,.$$

Note that the Wronskian, W_z, compensates for the arbitrariness in the normalization and phase of the ϕ's. Also, $W_z = 0$ if and only if ϕ_{za} and ϕ_{zb} are proportional, in which case z is an eigenvalue of H, since we then have a single eigenfunction which satisfies the boundary condition or square-integrability condition at both endpoints. In particular, if z is not real, then W_z does not vanish.

Exercise 4: Fill in the details of the derivation of (2.10).

Let us now make a few observations with regard to Questions 1 and 2.

(1) The origin of boundary conditions is now plain: They are needed to make ϕ_{za} and ϕ_{zb} unique (up to normalization), so that $(H - z)^{-1}$ will be well-defined (whenever $\operatorname{Im} z \neq 0$) by the construction of G_z, and hence H will be self-adjoint.

(2) What can happen when z is real? There are three possibilities for each such z:

    a) The construction of G_z goes through without incident. Then z is not in the spectrum of H, since the resolvent operator exists.

    b) The construction of G_z fails because $W_z = 0$. Then $z \in \sigma_{\mathrm{p}}(H)$ (z is an eigenvalue), as previously remarked.

    c) The construction of G_z fails because ϕ_{za} or ϕ_{zb} does not exist. This happens in the limit-point case when no solution of $H\phi = z\phi$ is square-integrable at the endpoint concerned. In this case $z \notin \sigma_{\mathrm{p}}(H)$, but nevertheless $(H - z)^{-1}$ doesn't exist; hence $z \in \sigma_{\mathrm{c}}(H)$.

When $b = \infty$ and $V \to 0$ there, the eigenfunctions typically have the asymptotic behavior

$$\psi_z(x) \sim e^{\pm i \sqrt{z}\, x}.$$

Thus —

(α) If $\sqrt{z}$ is complex, one solution, ϕ_{zb}, decays and is square-integrable at ∞, while the other solution is unbounded at ∞. (Thus all solutions linearly independent of ϕ_{zb} grow as $x \to \infty$.) In this case z either is in the point spectrum or is not in the spectrum at all, depending on whether ϕ_{za} is proportional to ϕ_{zb}.

(β) If $\sqrt{z}$ is real, then both solutions are bounded, but neither is square-integrable at ∞. In this case z is in the continuous spectrum, and hence at least one of the solutions appears in the eigenfunction expansion (possibly both of them, depending on what happens at the other endpoint).

These facts are the origin of the folk wisdom that $\sigma_{\mathrm{c}}(H)$ corresponds to solutions of the Schrödinger equation that are bounded but not square-integrable. We see that this is slightly misleading: In case (α) with $z \notin \sigma(H)$, the point is not that ϕ_{za} is growing, but rather that its partner, ϕ_{zb}, is decaying, so that a Green function can be constructed. The existence of decaying and of growing solutions just happens to be correlated for most commonly studied differential equations. We might note that if V is linear in x, the eigenfunctions (Airy functions) are *not* bounded, yet every real z is in the continuous spectrum.

We turn now to the big problem, determining the spectral measure functions μ_{jk} in (2.8). This can be done by comparing the spectral expansion of the operator $(H-z)^{-1}$ with the Green function (2.10) we have independently constructed. First, we need to boil the latter down a bit; I leave the details to three increasingly sophisticated exercises.

Exercise 5: Suppose that $\mathbf{A}(x) = 0$, so that the coefficients in H are real.

a) Show that $\rho g W_z$ is a constant (with respect to x). It follows that $G_z(y,x) = G_z(x,y)$.

b) Conclude that (2.10) implies

$$
G_z(x,y) = \begin{cases} \sum_{j,k=0}^{1} M_<^{jk}(z)\,\psi_{zj}(x)\psi_{zk}(y) & \text{for } x \le y, \\ \sum_{j,k=0}^{1} M_>^{jk}(z)\,\psi_{zj}(x)\psi_{zk}(y) & \text{for } x \ge y, \end{cases}
\tag{2.11}
$$

where $M_<^{jk}(z) = M_>^{kj}(z)$.

Exercise 6: Returning to the general case —

a) Show that $G_z(y,x)^* = G_{z^*}(x,y)$. *Hints:* (2.10) is irrelevant; this is the kernel form of an operator identity.

b) Show from (a) and (2.10) that

$$
G_z(x,y) = \begin{cases} \sum_{j,k=0}^{1} M_<^{jk}(z)\,\psi_{zj}(x)\psi_{z^*k}(y)^* & \text{for } x \le y, \\ \sum_{j,k=0}^{1} M_>^{jk}(z)\,\psi_{zj}(x)\psi_{z^*k}(y)^* & \text{for } x \ge y, \end{cases}
\tag{2.12}
$$

where $M_<^{jk}(z) = M_>^{kj}(z^*)^*$.

c) Verify the consistency of (2.11) with (2.12), and that of (2.10) with (2.12).

Exercise 7: Extend the formalism to the case of periodic boundary conditions and more general boundary conditions that *mix* the two endpoints. (See Fulling 1982, Remark 2.2.) If the boundary conditions involve complex numbers, they may cause the same sort of complications as the gauge field **A**.

The following properties of M can be demonstrated (see Maurin's books):

(1) $M^{jk}_{<,>}(z)$ are analytic except on the real axis.
(2) $M^{jk}_{<} - M^{jk}_{>}$ is entire.
(3) $M^{00}_{<} = M^{00}_{>} \equiv M^{00}$.

We expect an inversion formula of the form

$$[U\phi](x) = \int_{\sigma(H)} \sum_{j,k=0}^{1} \psi_{\lambda j}(x)\,\phi_k(\lambda)\,d\mu^{jk}(\lambda),$$

where

$$\phi_k(\lambda) = \int_a^b \psi_{\lambda k}(y)^* \, [U\phi](y)\,\rho(y)\,dy.$$

(We're using the local basis $\{\psi_{\lambda j}\}$ as the $\{e_{\lambda j}\}$ of the earlier general formalism.) Also, for any function $F(\lambda)$, we expect

$$[[F(H)](U\phi)](x) = \int_{\sigma(H)} \sum_{j,k} \psi_{\lambda j}(x)\,F(\lambda)\,\phi_k(\lambda)\,d\mu^{jk}(\lambda)$$

$$= \int_a^b K_F(x,y)\,[[U\phi](y)]\,\rho(y)\,dy,$$

where

$$K_F(x,y) \equiv \int_{\sigma(H)} \sum_{j,k=0}^{1} \psi_{\lambda j}(x)\,F(\lambda)\,\psi_{\lambda k}(y)^*\,d\mu^{jk}(\lambda).$$

(K_F will not exist as a function unless F is sufficiently nice, of course.) In particular,

$$G_z(x,y) = \int_{\sigma(H)} \sum_{j,k} (\lambda - z)^{-1}\psi_{\lambda j}(x)\psi_{\lambda k}(y)^*\,d\mu^{jk}(\lambda). \qquad (2.13)$$

Comparing with (2.12), we see that

$$M^{00}(z) = G_z(x_0,x_0) = \int_{\sigma(H)} (\lambda - z)^{-1}\,d\mu^{00}(\lambda). \qquad (2.14)$$

(It is tempting to say also that

$$(\rho g)^{j+k} \frac{\partial^{j+k}}{\partial x^j \partial y^k} G_z(x_0, x_0) = \int_{\sigma(H)} (\lambda - z)^{-1} d\mu^{jk}(\lambda),$$

but this is gibberish since G_z is not differentiable on the diagonal. It *is* continuous there, but it has a cusp. It turns out that μ^{jk} increases with λ too fast for the integral on the right-hand side to converge, unless $j = 0 = k$.)

Now integrate (2.14) around a narrow rectangular contour surrounding the segment of the real axis from λ_1 to λ_2. The width of the rectangle, 2ϵ, will be taken to 0 eventually. For simplicity, assume that λ_1 and λ_2 are not poles of M^{00}. Then

$$\oint G_z(x_0, x_0)\, dz = -2\pi i \int_{\lambda_1}^{\lambda_2} d\mu^{00}(\lambda)$$
$$= -2\pi i [\mu^{00}(\lambda_2) - \mu^{00}(\lambda_1)].$$

But also,

$$\oint G_z(x_0, x_0)\, dz = \oint M^{00}(z)\, dz$$
$$= \lim_{\epsilon \downarrow 0} \int_{\lambda_1}^{\lambda_2} [M^{00}(\lambda - i\epsilon) - M^{00}(\lambda + i\epsilon)]\, d\lambda.$$

Thus we have a formula for the principal spectral measure:

$$\mu^{00}([\lambda_1, \lambda_2]) = \frac{1}{2\pi i} \lim_{\epsilon \downarrow 0} \int_{\lambda_1}^{\lambda_2} [M^{00}(\lambda + i\epsilon) - M^{00}(\lambda - i\epsilon)]\, d\lambda.$$

But $M^{00}(\lambda - i\epsilon) = M^{00}(\lambda + i\epsilon)^*$ by Exercise 6(b), so this can be simplified to

$$\mu^{00}([\lambda_1, \lambda_2]) = \frac{1}{\pi} \lim_{\epsilon \downarrow 0} \int_{\lambda_1}^{\lambda_2} \operatorname{Im} M^{00}(\lambda + i\epsilon)\, d\lambda \qquad (2.15)$$

(with suitable adjustments if either of the λ_j is a pole). Formally, one may write

$$\frac{d\mu^{00}}{d\lambda} = \frac{1}{\pi} \lim_{\epsilon \downarrow 0} \operatorname{Im} M^{00}(\lambda + i\epsilon).$$

This derivative is understood to include delta functions at the values of λ where μ^{00} is not continuous; this occurs at poles of M^{00} (see below).

To obtain some information about the other components of μ, consider the integral (around the same contour) of (2.13) with $x \neq x_0 \neq y$:

$$\oint G_z(x,y) \, dz = -2\pi i \int_{\lambda_1}^{\lambda_2} \sum_{j,k} \psi_{\lambda j} \psi_{\lambda k}(y)^* \, d\mu^{jk}(\lambda)$$

$$\equiv -2\pi i E_{[\lambda_1,\lambda_2]}(x,y).$$

$E_{[\lambda_1,\lambda_2]}(x,y)$ is the integral kernel of the projection operator onto the part of the Hilbert space corresponding to the interval $[\lambda_1, \lambda_2]$ of the spectrum — that is, the image under U of the functions in $\mathcal{L}^2(\sigma(H); \mathcal{H}_\lambda)$ with support in the interval $[\lambda_1, \lambda_2]$. If H has a pure point spectrum, then

$$E_{[\lambda_1,\lambda_2]}(x,y) = \sum_{\lambda_1 \leq \lambda \leq \lambda_2} \psi_j(x) \, \psi_j(y)^*$$

(where the ψ_j are orthonormalized and we are still assuming that λ_1 and λ_2 are not eigenvalues themselves). Unlike G_z, $E_{[\lambda_1,\lambda_2]}(x,y)$ *is a smooth function* [Gårding 1954]. On the other hand, from (2.12) we have

$$\oint G_z(x,y) \, dz = \sum_{j,k} \oint M^{jk}(z) \, \psi_{zj}(x) \, \psi_{z*k}(y)^* \, dz,$$

where $M(z) \equiv \frac{1}{2}[M_<(z) + M_>(z)]$. (We could equally well take $M = M_<$ or $M_>$, since the function $(M_< - M_>)\psi_z(\psi_{z*})^*$ is entire and makes no contribution to the integral.) When the integrals over the two significant parts of the contour are written out, this becomes

$$\sum_{j,k} \lim_{\epsilon \downarrow 0} \int_{\lambda_2}^{\lambda_1} [M^{jk}(\lambda - i\epsilon) - M^{jk}(\lambda + i\epsilon)] \, \psi_{\lambda j}(x) \, \psi_{\lambda k}(y)^* \, d\lambda.$$

Comparing this with the previous result, we get a formula for $E_{[\lambda_1,\lambda_2]}$; when we strip off the basis functions on each side, it becomes

$$d\mu^{jk}(\lambda) = \frac{1}{2\pi i} \lim_{\epsilon \downarrow 0} [M^{jk}(\lambda + i\epsilon) - M^{jk}(\lambda - i\epsilon)] \, d\lambda. \qquad (2.16)$$

Since $M(z)^{\dagger} = M(z^*)$ by Exercise 6(b), (2.16) can be written in terms of the anti-Hermitian part of the matrix $M(\lambda + i\epsilon)$. If $\mathbf{A} = 0$, then Exercise 5(b) shows that M is symmetric, and it follows that

$$d\mu^{jk}(\lambda) = \frac{1}{\pi} \lim_{\epsilon \downarrow 0} \operatorname{Im} M^{jk}(\lambda + i\epsilon) \, d\lambda. \qquad (2.17)$$

Of course, the previous remark about poles and delta functions applies here as well.

More precisely, there are three types of behavior of the matrix M at the real axis, three corresponding types of behavior of the matrix μ, and thus a threefold classification of real numbers λ. First, μ is constant over any interval of the real axis on which M is continuous; such intervals are not in $\sigma(H)$. At any λ where M has a pole, μ has a jump discontinuity equal to the negative of the residue of M; these points are the eigenvalues $(\sigma_{\mathrm{p}}(H))$. Finally, in the continuous spectrum $(\sigma_{\mathrm{c}}(H))$ M displays a finite jump across a branch cut lying along the real axis, so that μ is continuous and increasing.

Remarks:

(1) (2.16) is the result we would have got earlier by naively ignoring the discontinuity in the derivatives of G_z. That discontinuity is due entirely to the analytic part of G_z (associated with $M_< - M_>$), so it drops out of the formula when the contour integration is performed.

(2) If one endpoint, say a, is *regular* (in the Sturm–Liouville sense), then it is possible and customary to use a in the role of x_0 and to choose basis solutions so that $\psi_{z0} \equiv \phi_{za}$ satisfies the Sturm–Liouville boundary condition defining H, rather than the usual unit initial condition. This diagonalizes $\operatorname{Im} M$ to

$$\begin{pmatrix} \operatorname{Im} m(z) & 0 \\ 0 & 0 \end{pmatrix}$$

and hence diagonalizes μ similarly. Consequently, the entire formalism can be developed in terms of scalar quantities in place of 2×2 matrices; one can start from the basic equation

$$\phi_{zb}(x) = m(z)\phi_{za}(x) + \psi_{z1}$$

and avoid much of the mess of the general case. This simplification reflects the fact that the multiplicity of the spectrum is 1 in this case;

multiplicity 2 requires limit-point behavior at both endpoints. Many examples can be found in Titchmarsh 1962 and in many other places, such as Dean & Fulling 1982.

(3) Other families of functions of H can be used instead of $(H - \lambda)^{-1}$ to obtain information about $E_\Delta(x,y)$ or $\mu^{jk}(\lambda)$. For instance, the kernel of the heat operator, e^{-tH}, is essentially the Laplace transform of E, while the kernel of $e^{-it\sqrt{H}}$, which gives a certain solution of the wave equation associated with H, is essentially the Fourier transform.

Solution to Exercise 2

In classical notation, the Hermiticity conditions (a) are

$$\operatorname{Im} b^j = -\frac{1}{\rho}\partial_k(\rho g^{kj}),$$

$$\operatorname{Im} c = -\frac{1}{2\rho}\partial_j(\rho\operatorname{Re} b^j).$$

The formulas (b) relating the original coefficients to the more geometrically meaningful ones are

$$\operatorname{Re} b^j = 2g^{jk}A_k, \qquad A_k = \tfrac{1}{2}g_{kj}\operatorname{Re} b^j;$$

$$\operatorname{Re} c = V + \tfrac{1}{4}g_{jk}(\operatorname{Re} b^j)(\operatorname{Re} b^k) = V + g^{jk}A_jA_k.$$

Regarding **b** as a contravariant and **A** as a covariant vector, we see that the relationship between them involves the isomorphism between the two types of vectors which exists whenever a Riemannian metric is present, and that the formulas relating V and c involve the norms of these vectors. The formulas (a) involve a divergence operation defined by the function ρ on vectors and tensors. I leave the choice of coordinate-free notations for these objects to the reader.

Chapter 3
Quantization of a Static, Scalar Field Theory

When the quantum mechanics of particles appeared, circa 1925, its own inventors immediately realized that it was not really adequate for physics, for several reasons:

(1) It's nonrelativistic. This shortcoming can't be overcome simply by using (special) relativistic classical particle mechanics in place of nonrelativistic mechanics as the source of the equations of motion or the Hamiltonian. Attempts to do so led to relativistic wave equations, the Klein–Gordon and Dirac equations, which were afflicted with negative probabilities or negative energies, respectively. Superficially these could be eliminated by discarding half the solutions *ad hoc*, but the resulting theories developed inconsistencies when interactions were included.

(2) The states describe fixed numbers of particles. Therefore, the theory can't describe processes in which particles are produced or destroyed. Such interactions are observed experimentally, as when a proton and an antiproton annihilate into an electron, a positron, and a number of pions and photons.

(3) In electromagnetism (and also in gravitation) the principal object in the classical theory is a field, not a particle. It was therefore expected that the quantization process could be extended to fields.

In a sense, all these problems are the same problem. Special relativity implies the equivalence of mass and energy, hence the possibility that new particles can be produced when the particles coming into an interaction have total kinetic energy in excess of the total rest mass of the prospective products. Electromagnetism turned out, after quantization, to have a particle structure after all; but the particles have zero rest mass, so it's especially easy for them to lose their individuality as they are produced and absorbed at little energy cost. The interesting states of the theory tend to have indefinite particle number and are better interpreted in terms of *field* concepts. Even for particles with rest mass greater than zero, the states of various particle number can be combined into the same Hilbert space in a formalism ("second quantization") where there is a field operator satisfying a relativistic equation. The physical problems with the latter arose because it was *misinterpreted* as the equation for the wave function of one particle (in analogy to the Schrödinger equation) rather than as the equation governing a quantized field.

In some circumstances either a particle or a field interpretation is tenable. Consider a field satisfying an equation

$$-\frac{\partial^2 \phi}{\partial t^2} = K\phi, \tag{3.1}$$

where K is an operator of the type considered in Chapter 2: We assume that K is a second-order elliptic linear differential operator in a variable x ranging over a region $\Omega \subset \mathbf{R}^d$, or over a manifold, and that K (plus boundary conditions, if necessary) defines a strictly positive, self-adjoint operator in a Hilbert space $\mathcal{L}^2_\rho(\Omega)$. (*Strictly positive* means that $\langle \psi, K\psi \rangle \geq C > 0$ for all normalized ψ.) K has no dependence on the time variable, t.

This time independence is the most important special feature of (3.1). Time dependence raises the possibility of particle creation, and hence the inappropriateness of a single-particle interpretation of the wave equation. However, (3.1) is not even the most general t-independent, second-order, linear, hyperbolic, scalar equation: it contains no terms proportional to $\partial\phi/\partial t$ or $\partial^2\phi/\partial t \partial x^j$. Physically, such terms would correspond respectively to an electrostatic potential and to the type of gravitational field characteristic of a rotating system. When they are present, a single-particle interpretation *may* not be tenable. (This is the *Klein paradox*, discussed in the appendix.) Nonpositivity of K can cause similar problems.

In developing a quantum theory based on (3.1) I shall adopt the "particle" point of view and "discover" the field, in analogy to the second quantization of nonrelativistic quantum mechanics. This perspective, along with related mathematical simplifications, allows this class of models to be much further developed than the general case. Moreover, the topic is of great historical importance in the development of the subject, and an understanding of the particle (Fock space) point of view is indispensible for reading the literature. However, I should emphasize once more that this approach is *not adequate* for quantum field theory in most curved-space situations. (Indeed, much of my own professional work has been involved with establishing that point.)

As a special case, (3.1) includes the equation of motion of the free scalar field in $\mathbf{R}^d$, for which

$$K = -\nabla^2 + m^2.$$

Also instructive are models with this K and with Ω a subregion of $\mathbf{R}^d$ with suitable boundary conditions. Such models raise some of the issues to be encountered when Ω is curved, without some of the complications.

POSITIVE-FREQUENCY SOLUTIONS

General spectral theory allows us to extend (2.4) to the general positive, self-adjoint operator K: The general solution of (3.1) is

$$\phi(t, \cdot) = \cos(t\sqrt{K})f + \frac{\sin(t\sqrt{K})}{\sqrt{K}}g, \qquad (3.2)$$

where

$$f(x) = \phi(0, x), \qquad g(x) = \frac{\partial\phi}{\partial t}(0, x).$$

Advanced remarks: "General solution" is ambiguous unless we state what function spaces ϕ, f, and g are supposed to belong to. In the context of (3.2), it's natural to take f, g, and $\phi(t, \cdot)$ to be in $\mathcal{L}_\rho^2$, since that's where spectral theory gives us a theorem without further work. In quantum *mechanics* this space is *uniquely* appropriate, because ϕ is the wave function of a particle and $|\phi(x)|^2$ has the interpretation of a probability density (see Chapter 1). In other applications, however, $\mathcal{L}_\rho^2$ may have merely a technical significance, and other choices of space may be useful in certain circumstances. For example, suppose that there is no boundary (in particular, that K is the Laplace–Beltrami operator (cf. (6.3)) on a *complete* Riemannian manifold Ω). Then the hyperbolicity of (3.1) implies that whenever f and g belong to $C_0^\infty(\Omega)$, then $f(t, \cdot)$ (t fixed) also has compact support; if K has smooth coefficients, then $\phi(t, x)$ will belong to $C^\infty(\Omega \times \mathbf{R})$. Since $C_0^\infty \subset \mathcal{L}_\rho^2$ (for reasonable ρ), (3.2) still applies in this setting. From this, by duality, we can give meaning to (3.1) and (3.2) with distributional data.

The classic reference for such facts about hyperbolic equations is the lecture notes of Leray 1953. See also Treves 1975, Chap. 2; Choquet-Bruhat 1968; Dimock 1980, Theorem I and Appendix. The literature on hyperbolic equations with boundary conditions is much less well developed, and some of my assertions about the quantum field theory of such systems will be correspondingly vague from a technical point of view.

Let us now, temporarily, make the additional assumption that K *is real.* That is, $\mathbf{A}$ in Exercise 2(b) is 0, and the boundary conditions are also real, so that the complex conjugate of any solution is also a solution.

I adopt the following conventions for the formalism of expansion in eigenfunctions of K:

(1) Introduce a nondegenerate "index" j to label the modes (independent eigenfunctions). That is, the pair (λ, j) in the notation of Chapter 2 becomes simply j, and λ becomes λ_j. Thus the spectral transform is

$$\phi(j) \equiv \int_{\Omega} \psi_j(x)^* \, [U\phi](x) \, \rho(x) \, d^d x, \qquad (3.3)$$

where $K\psi_j = \lambda_j \, \psi_j$.

In other words, we regard K as a function of some other operator or operators, J, with simple spectrum, and we work in the spectral representation of J. The range of j may be $\sigma(J) \subset \mathbf{R}$, or something more abstract: for instance, J might stand for one of the standard "complete commuting sets" of quantum mechanics, such as

$$\mathbf{P} \equiv -i\nabla \qquad\qquad (j \equiv \mathbf{p})$$

or

$$(|\mathbf{P}|, |\mathbf{L}|, L_z) \qquad\qquad (j = (|\mathbf{p}|, l, m)).$$

(2) Assume that the ψ_j have been chosen "orthogonal", so that the inversion formula takes the simple form

$$[U\phi](x) = \int \phi(j) \, \psi_j(x) \, d\mu(j) \qquad (3.4)$$

(in contrast to the complicated integration $\int \sum_{j,k} \cdots d\mu^{jk}(\lambda)$ in the general Weyl–Titchmarsh–Kodaira formalism, for instance). Thus the Hilbert space of the spectral representation, previously denoted $\mathcal{L}^2\big(\sigma(K); \mathcal{H}_\lambda\big)$ or $\int^{\oplus} \mathcal{H}_\lambda$, is now just $\mathcal{L}^2_\mu$.

(3) Define

$$\tilde{f}(j) \equiv [U^{-1}f](j) = \int_\Omega \psi_j(x)^* \, f(x) \, \rho(x) \, d^d x. \qquad (3.5)$$

(4) Define $\omega_j \equiv \sqrt{\lambda_j} \in \sigma(K^{\frac{1}{2}})$. Since we have assumed K to be positive, ω_j has a lower bound greater than 0. This assumption was made to guarantee that the operator $K^{-\frac{1}{2}}$ encountered in (3.2) is everywhere defined (or, in other words, so that in the spectral representation we can divide by ω_j with impunity). Unfortunately, that rules out the important case of the massless free field, for which the lower bound is equal to 0. So one eventually needs to do some extra work to accommodate *technically* the case where 0 is the greatest lower bound of the continuous spectrum (but is not an eigenvalue). Some of the issues that arise here are discussed in Fulling & Ruijsenaars 1987. For a massless field in a closed universe (generalizing the interval with periodic boundary conditions) we need to deal with the case where 0 is an eigenvalue; this requires *essential* changes in the formalism, which are discussed in the appendix.

(5) Finally, assume for simplicity that

$$\psi_j(x)^* = \psi_{j'}(x)$$

for some j' in the index set. Necessarily, $\omega_{j'}$ will equal ω_j. Note that in general $\psi_j(x)^*$ would be a linear combination (possibly infinite) of the $\psi_{j'}$ corresponding to the same ω. (For instance, in scattering theory, as briefly discussed in Chapter 2, $\psi_{\mathbf{k}}^{\text{out}} = \psi_{-\mathbf{k}}^{\text{in}}{}^*$, which is not equal to $\psi_{\mathbf{k}'}^{\text{in}}$ for any $\mathbf{k}'$.)

Since K is real, we are always free to choose ψ_j real, hence $j' = j$. The usual reason for not doing so is the existence of some first-order differential operator, say $i\,\partial/\partial u$, which generates a symmetry and which one wants to include in J. Then all the eigenfunctions will be proportional to factors of the form e^{iku}. (For example, k will be a component of the vector $\mathbf{p}$ if the symmetry is a translation, and k will be the "magnetic quantum number" m of the spherical harmonic Y_l^m if the symmetry is a rotation ($u \equiv \theta$).) Therefore, we allow in the general formalism for the possibility that $j' \neq j$.

The solution (3.2) is

$$\phi(t,x) = \int d\mu(j) \left[\tilde{f}(j)\psi_j(x)\cos(\omega_j t) + \tilde{g}(j)\psi_j(x)\frac{\sin(\omega_j t)}{\omega_j} \right].$$

We consider the general complex-valued solution, and we reexpress the trigonometric functions as exponentials, getting

$$\phi(t,x) = \int \frac{d\mu(j)}{\sqrt{2\omega_j}} \left[\tilde{\phi}_+(j)\,\psi_j(x)e^{-i\omega_j t} + \tilde{\phi}_-(j)\,\psi_j(x)^* e^{+i\omega_j t} \right], \quad (3.6)$$

where

$$\tilde{\phi}_+(j) \equiv \frac{1}{\sqrt{2}} \left[\sqrt{\omega_j}\,\tilde{f}(j) + \frac{i}{\sqrt{\omega_j}}\,\tilde{g}(j) \right],$$

$$\tilde{\phi}_-(j) \equiv \frac{1}{\sqrt{2}} \left[\sqrt{\omega_j}\,\tilde{f}(j') - \frac{i}{\sqrt{\omega_j}}\,\tilde{g}(j') \right]. \quad (3.7)$$

Because of the $\sqrt{\omega_j}$ factors in (3.7), the conditions $\tilde{f}$, $\tilde{g} \in \mathcal{L}^2_\mu$ aren't equivalent to $\tilde{\phi}_\pm \in \mathcal{L}^2_\mu$. Therefore, we *abandon* the earlier suggestion that the initial data should range precisely over $\mathcal{L}^2_\rho$.

Physical Axiom: The Hilbert space of possible quantum states of a single particle, in the theory governed by the relativistic wave equation (3.1), is the space of functions of the form

$$\phi(t,x) = \int \frac{d\mu(j)}{\sqrt{2\omega_j}}\,\tilde{\phi}_+(j)\,\psi_j(x)e^{-i\omega_j t} \quad (3.8)$$

with $\tilde{\phi}_+ \in \mathcal{L}^2_\mu$. The inner product on these states is induced by that of $\mathcal{L}^2_\mu$:

$$\|\phi\|^2 = \|\tilde{\phi}_+\|^2_\mu \equiv \int |\tilde{\phi}_+(j)|^2\, d\mu(j). \quad (3.9)$$

The solutions (3.8) are called *positive-frequency solutions* of (3.1). (The solutions $\tilde{\phi}_-(j)e^{+i\omega_j t}$, naturally, are said to have negative frequency. The terminology may seem to be backwards, but arises from the collision of two individually logical sign conventions.)

Remark: The restriction to positive frequencies can be restated in more modern mathematical language this way: The true wave equation of the system is not (3.1) after all, but rather the *first-order pseudo-differential equation*

$$i\frac{\partial\phi}{\partial t} = \sqrt{K}\,\phi. \quad (3.10)$$

The other assertion in the axiom is that physically relevant solutions of
(3.10) are those with initial data $\phi(0, x)$ in the domain of the operator
$K^{\frac{1}{4}}$ [i.e., U (multiplication by $\omega_j^{\frac{1}{2}}$) U^{-1}].

Since all infinite-dimensional, separable, complex Hilbert spaces
are isomorphic, the axiom has little content until one says something
about how the wave functions are used to predict results of observa-
tions. If K (or some other function of J) is an observable, then $|\tilde{\phi}_+(j)|^2$
provides a probability density for measurements of it, as usual.

For the free field, one can take $j = \mathbf{p}$ or $j = (|\mathbf{p}|, l, m)$, with $\omega_j = \sqrt{\mathbf{p}^2 + m^2}$, the energy of the particle. Thus energy, momentum, and
angular momentum are respectable observables in this theory. They are
the most important ones in practice for the free Klein–Gordon theory.
On the other hand, *position* as an observable is problematical, even for
the *free* Klein–Gordon theory:

THE NEWTON–WIGNER POSITION OPERATOR

Given $\tilde{\phi}_+ \in \mathcal{L}_\mu^2$, define

$$\phi_{\rm NW}(t, x) \equiv \int \tilde{\phi}_+(j)\, \psi_j(x) e^{-i\omega_j t}\, d\mu(j). \qquad (3.11)$$

Note that $\phi_{\rm NW} \neq \phi$; they are different functional representations of the
same abstract vector. Both happen to solve (3.1) (and (3.10)).

The inner product induced by (3.11) from (3.9) is just that of
$\mathcal{L}_\rho^2(\Omega)$:

$$\|\tilde{\phi}_+\|^2 = \|\phi_{\rm NW}\|_\rho^2 \equiv \int_\Omega |\phi_{\rm NW}(t, x)|^2\, \rho(x)\, d^d x$$

(independent of t!). Therefore, it makes mathematical sense to inter-
pret $|\phi_{\rm NW}(t, x)|^2$ as the probability density for observations of x at
time t. The corresponding self-adjoint operator

$$X: \phi_{\rm NW}(x) \mapsto x\, \phi_{\rm NW}(x)$$

is the *Newton–Wigner position operator*. In the free case ($K = -\nabla^2 + m^2$) it has been thoroughly studied. (See Newton & Wigner 1949;
Wightman & Schweber 1955; Wightman 1962; Segal & Goodman 1965;
Ruijsenaars 1981.) The conclusions reached are:

(1) X has surprising properties. In particular, a particle localized in a compact set Ω_0 at time t has a nonzero probability of being detected at any given point in $\mathbf{R}^d$ at $t + \delta$ — apparently travelling faster than light. (The explanation of the apparent mathematical conflict with the causality of propagation of solutions of the Klein–Gordon equation is this: A classical solution with localized initial data ($\phi(0, x)$ and $\frac{\partial \phi}{\partial t}(0, x)$ both of compact support) necessarily contains both positive and negative frequencies, so it is not a particle wave function in our theory.)

(2) X is unique, in the sense that any other definition of a position observable would have even more unacceptable properties.

These results indicate a breakdown of the single-particle picture even for the free field. The paradox seems less severe if we adopt a multiparticle point of view: If the states of the system include states of arbitrarily many particles, observation of a one-particle state localized in Ω_0 entails observation of the *absence* of particles everywhere else. Therefore, the later observation of one particle far away from Ω_0 need not be attributed to a transluminal signal from the original measurement.

In any case, no "ideal" position measurement device exists. Interactions are never of exactly finite range. In practice, energy, momentum, and angular momentum measurements are more common in high-energy physics.

Exercise 8:

(a) From (3.8) and (3.11), derive an integral formula for ϕ in terms of ϕ_{NW} (for the free massive field). (I.e., find an integral kernel for the operator $\frac{1}{\sqrt{2}} K^{-\frac{1}{4}}$. The answer is a certain Bessel function.)

(b) Try to do the same for the inverse transformation, $\sqrt{2}\, K^{+\frac{1}{4}} : \phi \mapsto \phi_{\mathrm{NW}}$. Observe that your integral diverges for most functions ϕ. Explain why; find a way to salvage the situation by reinterpreting the integral or modifying the formula.

(c) Prove the statement that a classically localized solution must contain frequencies of both signs.

Now let's find the inner product induced by (3.8) from (3.9). Invert (3.8):

$$\frac{\tilde{\phi}_+(j)}{\sqrt{2\omega_j}} = U^{-1}[\phi(t, \cdot)] = \int_\Omega \phi(t, x)\psi_j(x)^* \rho\, dx.$$

Similarly,

$$\frac{\partial \phi}{\partial t} = -i \int d\mu(j) \sqrt{\frac{\omega_j}{2}} \,\tilde{\phi}_+(j)\psi_j(x)e^{-i\omega_j t}$$

implies

$$\sqrt{\frac{\omega_j}{2}} \,\tilde{\phi}_+(j) = iU^{-1}\left[\frac{\partial \phi}{\partial t}\right] = i \int_\Omega \frac{\partial \phi}{\partial t}(t, x)\psi_j(x)^* \rho\, dx.$$

Thus

$$(\alpha) \qquad \|\phi\|^2 = \|\tilde{\phi}_+\|_\mu^2 = 2 \int d\mu(j) \frac{\tilde{\phi}_+(j)^*}{\sqrt{2\omega_j}} \sqrt{\frac{\omega_j}{2}} \,\tilde{\phi}_+(j)$$

$$= 2i \int d\mu(j) \int_\Omega \phi(t, x)^* \psi_j(x)\, \rho\, dx \int_\Omega \frac{\partial \phi}{\partial t}(t, y)\psi_j(y)^* \rho\, dy.$$

Now use the completeness relation for the eigenfunctions,

$$\int d\mu(j)\psi_j(x)\psi_j(y)^* = \frac{\delta(x - y)}{\rho(x)}$$

(the distribution kernel of the identity operator). One gets

$$(\beta) \qquad \|\phi\|^2 = 2i \int_\Omega \int_\Omega \phi(t, x)^* \frac{\partial \phi}{\partial t}(t, y)\, \rho\delta(x - y)\, dx\, dy$$

$$= 2i \int_\Omega \phi(t, x)^* \frac{\partial \phi}{\partial t}(t, x)\, \rho\, dx.$$

But the choice of the second factor as the locus of $\partial/\partial t$ was arbitrary; the quantity can be written more symmetrically as

$$\|\phi\|^2 = i \int_\Omega \left[\phi^* \frac{\partial \phi}{\partial t} - \left(\frac{\partial \phi}{\partial t}\right)^* \phi\right] \rho\, dx$$

$$\equiv i \int_\Omega \phi^* \overleftrightarrow{\partial}_t \phi\, \rho\, dx. \qquad (3.12)$$

Technical remark: The passage from (α) to (β) is *not* an application of Fubini's theorem. If ϕ and $\frac{\partial \phi}{\partial t}$ are both in $\mathcal{L}^2_\rho$, the equivalence of (α) and (β) is just Parseval's equality. Actually, we have $K^{\frac{1}{4}}\phi$ and

$K^{-\frac{1}{4}}\frac{\partial\phi}{\partial t}$ in $\mathcal{L}^2_\rho$ (since both are proportional to ϕ_{NW}). Their "pairing" is still defined, since they are in dual Sobolev spaces. So (β) is meaningful in this distributional sense *at least*.

Exercise 9: Show that for the general solution (3.6), the right-hand side of (3.12) is the *indefinite* sesquilinear form $\|\tilde{\phi}_+\|^2_\mu - \|\tilde{\phi}_-\|^2_\mu$. Is this true if one doesn't symmetrize the position of ∂_t ?

FOCK SPACE

Now that we have decided what is meant by a *state of a single particle* in the quantum theory based on the field equation (3.1), we ask what a *quantum state of a system consisting of two such particles* is.

This will require a short digression to review the concept of a *tensor product* of vector spaces. The vector spaces of interest to us are $\mathcal{L}^2$ spaces (spaces of square-integrable functions). For them, we have it on the authority of Paul Halmos (Halmos & Sunder 1978, p. 44) that it is respectable to ignore the intrinsic algebraic definition of tensor product in favor of the following:

Definition: If A and B are two measure spaces, then *the tensor product of $\mathcal{L}^2(A)$ and $\mathcal{L}^2(B)$ is*

$$(\sharp) \qquad\qquad \mathcal{L}^2(A) \otimes \mathcal{L}^2(B) \equiv \mathcal{L}^2(A \times B)$$

— that is, the space of square-integrable functions $\phi(x,y)$, where $x \in A$ and $y \in B$.

Remarks on the relationship of the tensor product to the direct sum: (1) If A and B are themselves vector spaces, then $A \times B$ is the same as $A \oplus B$. (2) The construction ($\sharp$) must not be confused with the similar one

$$(\flat) \qquad\qquad \mathcal{L}^2(A) \oplus \mathcal{L}^2(B) = \mathcal{L}^2(A \cup B),$$

where A and B are understood to be disjoint. (If $A \cup B \neq \emptyset$ to begin with, one can make them formally disjoint by attaching distinct "labels" to their respective elements. In that case the disjoint union is denoted $A \amalg B$ to avoid ambiguity.)

Example 1: If A is a set of 2 elements and B is a disjoint set of 3 elements, then $\mathcal{L}^2(A)$ is $\mathbf{C}^2$, the space of 2-component vectors, and

$\mathcal{L}^2(B)$ is $\mathbf{C}^3$; $A \cup B$ has 5 elements, and $\mathcal{L}^2(A \cup B)$ is the space of 5-component vectors; $A \times B$ has 6 elements (index pairs), and $\mathcal{L}^2(A \times B)$ is the space of 2×3 matrices.

Example 2: If $A = B = \mathbf{R}$, then the disjoint union $A \amalg B$ is two copies of $\mathbf{R}$ (a "railroad track", not a plane!), and an element of

$$\mathcal{L}^2(A \amalg B) \equiv \mathcal{L}^2(\mathbf{R}) \oplus \mathcal{L}^2(\mathbf{R}) \cong \mathcal{L}^2(\mathbf{R}; \mathbf{C}^2)$$

is a pair of complex-valued functions of a real variable, or, equivalently, a two-component function of a real variable. $A \times B = A \oplus B$ is the plane, $\mathbf{R}^2$, and an element of

$$\mathcal{L}^2(A \times B) \equiv \mathcal{L}^2(\mathbf{R}) \otimes \mathcal{L}^2(\mathbf{R})$$

is a complex-valued function of two real variables.

It is important to recall that an element $\phi(x, y)$ of $\mathcal{L}^2(A) \otimes \mathcal{L}^2(B)$ may not factor as $\phi_1(x)\phi_2(y)$. (The set of such products is not closed under addition.) In the context of Example 1, this simply means that not every matrix is of rank 1!

The repeated tensor product $\mathcal{L}^2(A) \otimes \mathcal{L}^2(A) \otimes \mathcal{L}^2(A)$ is denoted $\mathcal{L}^2(A)^{\otimes 3}$, and so on.

A first try at an answer to our question (*What is a two-particle state?*) is: Since a one-particle state is a function $\tilde{\phi}_+(j) \in \mathcal{L}^2_\mu$, a two-particle state is a function $\tilde{\phi}(j_1, j_2)$, so that $|\tilde{\phi}(j_1, j_2)|^2$ is the probability [density] of finding Particle 1 in state j_1 and Particle 2 in state j_2. That is, the state space of two particles is $\mathcal{L}^2_\mu \otimes \mathcal{L}^2_\mu$ **(FALSE)**.

The trouble with this is that quantum particles (of the same type) are indistinguishable; all that is meaningful is a probability [density] for finding *a* particle in state j_1 and *a* particle in state j_2. Therefore, one requires that $\tilde{\phi}(j_1, j_2) = \tilde{\phi}(j_2, j_1)$, and takes the state space to be the *symmetrized tensor product* $\mathcal{L}^2_\mu \odot \mathcal{L}^2_\mu \equiv (\mathcal{L}^2_\mu)^{\odot 2}$ consisting of all such symmetric elements of $(\mathcal{L}^2_\mu)^{\otimes 2}$. More generally, the state space of n particles is $(\mathcal{L}^2_\mu)^{\odot n}$, the totally symmetric subspace of the n-fold tensor product $(\mathcal{L}^2_\mu)^{\otimes n}$.

What has just been defined is *Bose statistics*. *Antisymmetric* wave functions (e.g., $\mathcal{L}^2_\mu \wedge \mathcal{L}^2_\mu$ for two particles) define *Fermi statistics*. Consistency of quantum field theory requires that scalar and tensor fields (integer-spin fields) obey Bose statistics, and that spinor (fractional-spin) fields obey Fermi statistics. Later (Chap. 7) we shall see evidence of this principle in the context of time-dependent external fields.

So far we have considered systems with a fixed number of particles. However, ultimately we want the *number of particles* itself to be an observable. This means that a state may be a linear combination of vectors from the various spaces $(\mathcal{L}_\mu^2)^{\odot n}$, including $n = 0$.

Definition: *Fock space*, $\mathcal{F} \equiv \overline{\mathcal{F}_0}$, is the Hilbert-space completion of

$$\mathcal{F}_0 \equiv \mathbf{C} \oplus \mathcal{L}_\mu^2 \oplus (\mathcal{L}_\mu^2)^{\odot 2} \oplus (\mathcal{L}_\mu^2)^{\odot 3} \oplus \cdots .$$

That is, an element of $\mathcal{F}$ is a sequence

$$\tilde{\Phi} = \{\, \tilde{\phi}_0 \,, \tilde{\phi}_1(j), \tilde{\phi}_2(j_1, j_2), \ldots, \tilde{\phi}_n(j_1, \ldots, j_n), \ldots \},$$

with $\tilde{\phi}_n \in (\mathcal{L}_\mu^2)^{\odot n}$ and

$$\|\tilde{\Phi}\|^2 \equiv \sum_{n=0}^{\infty} \|\tilde{\phi}_n\|^2 < \infty .$$

The vacuum: The vector $|0\rangle \equiv \{1, 0, 0, \ldots\}$ represents the state with no particles present.

Creation and annihilation operators: For $\tilde{u} \in \mathcal{L}_\mu^2$, define

$$a(\tilde{u})\tilde{\Phi} \equiv \{\ldots, \sqrt{n+1} \int d\mu(j)\, \tilde{u}(j)\tilde{\phi}_{n+1}(j, j_1, \ldots, j_n), \ldots \}, \quad (3.13a)$$

and

$$a^\dagger(\tilde{u})\tilde{\Phi} \equiv \{0, \tilde{u}(j_1)\tilde{\phi}_0, \ldots, \sqrt{n}\, \text{sym}[\tilde{u}(j_1)\tilde{\phi}_{n-1}(j_2, \ldots, j_n)], \ldots \}. \quad (3.13b)$$

Here "sym" is the operation of symmetrization of a function upon all its arguments j_i. It entails a division by $n!$, but this reduces to an explicit factor of merely n after the symmetry of $\tilde{\phi}_{n-1}$ is exploited; for instance,

$$\text{sym}[\tilde{u}(j_1)\tilde{\phi}_2(j_2, j_3)]$$
$$= \tfrac{1}{3}[\tilde{u}(j_1)\tilde{\phi}_2(j_2, j_3) + \tilde{u}(j_2)\tilde{\phi}_2(j_1, j_3) + \tilde{u}(j_3)\tilde{\phi}_2(j_1, j_2)].$$

Formulas (3.13) define unbounded operators. Thus, their right-hand sides do not always define elements of $\mathcal{F}$, but they do if $\tilde{\Phi}$ is, say, a terminating sequence (an element of $\mathcal{F}_0$). (In fact, $a(\tilde{u})$ and $a^\dagger(\tilde{u})$ are unbounded operators having $\mathcal{F}_0$ as a common *invariant* domain. They can be extended to larger domains by taking their closures.)

Exercise 10: Taking $\mathcal{F}_0$ as domain, show that the operator adjoint of $a^\dagger(\tilde{u})$ is an extension of $a(\tilde{u}^*)$ and vice versa. In other words, for $\tilde{\Phi}$ and $\tilde{\Psi}$ in $\mathcal{F}_0$,

$$\langle \tilde{\Phi}, a^\dagger(\tilde{u})\tilde{\Psi} \rangle = \langle a(\tilde{u}^*)\tilde{\Phi}, \tilde{\Psi} \rangle.$$

(The complex conjugation is needed to make the annihilation operator $a(\,\cdot\,)$ linear in $\tilde{u}$. Some authors deliberately leave it antilinear.)

The commutator: From (3.13) one calculates

$$(*) \quad a^\dagger(\tilde{v})a(\tilde{u})\tilde{\Phi}$$

$$= \{\ldots, \sqrt{n}\,\mathrm{sym}[\tilde{v}(j_0)\sqrt{n}\int d\mu(j)\,\tilde{u}(j)\tilde{\phi}_n(j,j_2,\ldots,j_n)], \ldots\}$$

and

$$a(\tilde{u})a^\dagger(\tilde{v})\tilde{\Phi}$$

$$= \{\ldots, \sqrt{n+1}\int d\mu(j)\,\tilde{u}(j)\sqrt{n+1}\,\mathop{\mathrm{sym}}_{j,j_1,\ldots,j_n}[\tilde{v}(j)\tilde{\phi}(j_1,\ldots,j_n)], \ldots\}$$

$$= \{\ldots, \langle \tilde{u}^*, \tilde{v}\rangle_\mu\,\tilde{\phi}_n(j_1,\ldots,j_n) + \text{same terms as in } a^\dagger a\tilde{\phi}, \ldots\}.$$

Therefore,

$$[a(\tilde{u}), a^\dagger(\tilde{v})] = \langle \tilde{u}^*, \tilde{v}\rangle_\mu. \qquad (3.14)$$

In particular, if $\|\tilde{u}\| = 1$, then $[a(\tilde{u}), a^\dagger(\tilde{u})] = 1$. Thus $[a(\tilde{u})$ and $a^\dagger(\tilde{u})$ act like the annihilation and creation operators of a harmonic oscillator (Chapter 1). Also, $[a(\tilde{u}), a(\tilde{v})] = 0$; and if $\langle \tilde{u}^*, \tilde{v}\rangle = 0$, then $[a(\tilde{u}), a^\dagger(\tilde{v})] = 0$, so that orthogonal test functions give rise to independent oscillators. In analogy with the single oscillator of Chapter 1, $a^\dagger(\tilde{u})a(\tilde{u}^*)$ must be the operator representing the observable "number of particles in state $\tilde{u}$"; indeed, from $(*)$ one can see that this operator picks out the part of each $\tilde{\phi}_n$ that contains $\tilde{u}$ and multiplies it by n.

Now let $\{e_\alpha\}_{\alpha=1}^\infty$ be an orthonormal basis for $\mathcal{L}_\mu^2$ (not necessarily consisting of eigenvectors of K). Then $\{e_{\alpha_1} \otimes \cdots \otimes e_{\alpha_n}\} \equiv$

$\{e_{\alpha_1}(j_1)\cdots e_{\alpha_n}(j_n)\}$ is an orthonormal basis for $(\mathcal{L}_\mu^2)^{\otimes n}$, and these vectors' symmetrizations, $e_{\alpha_1}\odot\cdots\odot e_{\alpha_n}$, form an orthogonal (not normalized) basis for $(\mathcal{L}_\mu^2)^{\odot n}$ after the duplications are omitted. Varying n, we get an orthogonal basis for $\mathcal{F}$. (Note that the most general product function, $\tilde{\phi}(j_1,j_2,\dots) = f(j_1)g(j_2)\cdots$, will be a sum of several terms when expressed in terms of this basis.)

If K, the operator in (3.1) defining our dynamics, has discrete (pure point) spectrum, then we can let $\alpha \equiv j$, $e_\alpha \equiv \psi_j$. Then

$$\tilde{\phi}_n(j_1,\dots,j_n) \cong \sum_{j_1,\dots,j_n} \tilde{\phi}_n(j_1,\dots,j_n)\, e_{j_1}\otimes e_{j_2}\otimes\cdots\otimes e_{j_n}.$$

Let $a_j \equiv a(e_j)$. Then

$$[a_j, a_k^\dagger] = \delta_{j,k}, \qquad [a_j, a_k] = 0. \tag{3.15}$$

We can characterize the vacuum by

$$a_j|0\rangle = 0 \quad \text{for all } j. \tag{3.16}$$

Furthermore, $a_{j_1}^\dagger\cdots a_{j_n}^\dagger|0\rangle$ is proportional to $e_{j_1}\odot\cdots\odot e_{j_n}$, and a calculation based on (3.13b), or on (3.14) and Exercise 10, shows that the norm of the former is a sum over permutations,

$$\sum_\sigma \delta_{j_1,j_{\sigma(1)}}\cdots\delta_{j_n,j_{\sigma(n)}} = \prod_{j=1}^\infty n_j!,$$

where n_j is the *occupation number* of state ψ_j: the number of index values equal to j in the list $\{j_1,\dots,j_n\}$. Some further combinatorics verifies that

$$\{0,0,\dots,\tilde{\phi}_n(j_1,\dots,j_n),0,\dots\}$$
$$= \frac{1}{\sqrt{n!}}\int d\mu(j_1)\cdots d\mu(j_n)\,\tilde{\phi}_n(j_1,\dots,j_n)\,a_{j_1}^\dagger\cdots a_{j_n}^\dagger|0\rangle. \tag{3.17}$$

(Here the integrations are actually summations, of course.) In this way all of $\mathcal{F}_0$ can be built up. We can write

$$a(\tilde{u}) = \int d\mu(j)\,\tilde{u}(j)a_j \tag{3.18}$$

and a similar formula for $a^\dagger$.

If K has continuous spectrum, we can retain (3.16), (3.17), and (3.18) in a distributional sense (which justifies the formal passage to the continuum limit in the less rigorous literature). The objects $a(\,\cdot\,)$ and $a^\dagger(\,\cdot\,)$ are *operator-valued distributions*: $\langle \tilde{\Phi}, a(\tilde{u})\tilde{\Psi}\rangle$ is in $\mathbf{C}$ for $\tilde{\Psi}, \tilde{\Phi} \in \mathcal{F}_0$ and $\tilde{u} \in \mathcal{L}_\mu^2$, and one can restrict the argument $\tilde{u}$ to smaller test function spaces and verify continuity in $\tilde{u}$ with respect to correspondingly stronger topologies. The analogue of (3.15) is

$$[a_j, a_k^\dagger] = \delta_\mu(j, k), \qquad [a_j, a_k] = 0,$$

where δ_μ is the Dirac delta function adapted to the measure μ (i.e., $\int f(k)\,\delta_\mu(j, k)\,d\mu(k) = f(j)$.)

The number operator: When $\sigma(K)$ is discrete and the eigenfunctions ψ_j are real, the operator $a_j^\dagger a_j$ represents the number of particles occupying the state ψ_j. Thus

$$N \equiv \int d\mu(j)\, a_j^\dagger a_j$$

is the operator for the *total* number of particles. This remains true in the general case; N can simply be defined by

$$N\tilde{\Phi} = \{\ldots, n\tilde{\phi}_n(j_1, \ldots, j_n), \ldots\},$$

which is formally consistent with the definitions (3.13). (Cf. (∗) with $\tilde{v} = \tilde{u} = \tilde{\psi}_j \equiv \delta_\mu(\,\cdot\, - j)$.) In general, the observable $a_j^\dagger a_j$ can be regarded as the *density* (with respect to μ) of particles at the point j in "energy space".

The Hamiltonian: Recall that our wave functions satisfy $\frac{\partial \phi}{\partial t} = K^{\frac{1}{2}}\phi$; hence their spectral transforms satisfy

$$i\frac{\partial \tilde{\phi}(t, j)}{\partial t} = \omega_j \tilde{\phi}(t, j), \qquad \tilde{\phi}(t, j) \equiv \tilde{\phi}_+(j) e^{-i\omega_j t}.$$

Therefore, the single-particle Hamiltonian is $K^{\frac{1}{2}}$ in position space and is multiplication by ω_j in energy space (the spectral representation of K). Consequently, the n-particle Hamiltonian is

$$\tilde{\phi}_n(j_1, \ldots, j_n) \mapsto (\omega_{j_1} + \cdots + \omega_{j_n})\, \tilde{\phi}_n(j_1, \ldots, j_n).$$

Thus, finally, the Hamiltonian in Fock space is

$$H \equiv \sum_{n=0}^{\infty} \left(\sum_{k=1}^{n} \omega_{j_k} \right) P_n \,,$$

where P_n is the projection onto the n-particle subspace. Equivalently, one may write

$$H \tilde{\Phi} \equiv \left\{ \ldots, \sum_{k=1}^{n} \omega_{j_k} \tilde{\phi}_n(j_1, \ldots, j_n), \ldots \right\},$$

or

$$H \equiv \int d\mu(j)\, \omega_j a_j^\dagger a_j.$$

Second quantization of arbitrary single-particle observables: In fact, any observable, O, of the single-particle theory has a Fock-space observable associated to it in this way:

$$\tilde{\Phi} \mapsto \left\{ \ldots, \sum_{k=1}^{n} O_{(k)} \tilde{\phi}_n(j_1, \ldots, j_n), \ldots \right\},$$

where $O_{(k)}$ indicates action on the kth argument of the multiparticle wave function. (O need not be diagonal in the energy representation.) In the mathematical literature of quantum field theory this operator is sometimes denoted $d\Gamma(O)$.

Furthermore, any unitary operator, U, on $\mathcal{L}_\mu^2$ induces a unitary operator on $(\mathcal{L}_\mu^2)^{\odot n}$ by

$$\tilde{\phi}_n(j_1, \ldots, j_n) \mapsto U_{(1)} U_{(2)} \cdots U_{(n)} \tilde{\phi}_n(j_1, \ldots, j_n).$$

(This is the infinite-dimensional analogue of the transformation law of an n-index tensor under change of basis.) Hence we have a unitary operator defined on $\mathcal{F}$, sometimes called $\Gamma(U)$. *Stone's theorem* relating unitary and self-adjoint operators on Hilbert space says that

$$U(t) = e^{itO}$$

is equivalent to

$$O = -i \left. \frac{dU}{dt} \right|_{t=0} = -i U^{-1} \left. \frac{dU}{dt} \right|_{\text{arbitrary } t} ;$$

the self-adjoint operator O is the *generator* of the unitary group $U(t)$ [Reed & Simon 1972, Sec. VIII.4]. If U and O are related in this way, then the generator of $\Gamma(U(t))$ is $d\Gamma(O)$:

$$\Gamma(U(t)) = e^{it\, d\Gamma(O)},$$

or

$$U_{(1)}(t) \otimes \cdots \otimes U_{(n)}(t) = \exp[it(O_{(1)} \otimes I \otimes \cdots \otimes I + I \otimes O_{(2)} \otimes \cdots + \ldots)].$$

Note that a *product* of U's corresponds to a *sum* of O's, in close analogy with the Leibnitz rule of elementary calculus.

FIELD OPERATORS

This exposition roughly follows the treatment in Sec. 3 of Fulling et al. 1981.

Test functions: Since the field operator will turn out to be a distribution, it is necessary to choose a class of nice functions to serve as permissible values of its argument. Let us (somewhat arbitrarily) consider the space $C_0^\infty(\Omega)$. If f is in this class, then $f \in \operatorname{dom} K$, hence $\in \operatorname{dom} K^p$ for $0 \leq p \leq 1$. If the coefficients in K are smooth, then C_0^∞ is an invariant domain and $f \in \operatorname{dom} K^p$ for all $p \geq 0$. (This generalizes the theorem that $f \in C^\infty(\mathbf{R})$ if and only if its Fourier transform is of rapid decrease. Here the rapid decrease is in the variable ω_j.)

For the time being, we are assuming that K is strictly positive. This implies that $C_0^\infty \subset \operatorname{dom} K^p$ for negative p also. (That is precisely the reason for making the assumption.)

x-space annihilation and creation operators: For $u \in \mathcal{L}_\rho^2(\Omega)$ (hence $\tilde{u} \in \mathcal{L}_\mu^2$) define

$$a^\dagger(u) \equiv a^\dagger(\tilde{u}), \qquad a(u) \equiv [a^\dagger(u^*)]^\dagger.$$

Recall that we defined $a(\tilde{u})$ as $[a^\dagger(\tilde{u}^*)]^\dagger$, in effect, in Exercise 10. If the eigenfunctions ψ_j are real, then $u^* = \tilde{u}^*$; hence $a(u) = a(\tilde{u})$. In general,

$$a(u) = [a^\dagger(\widetilde{u^*})]^\dagger = a(\widetilde{u^*}^*).$$

If u is real-valued, $a(u) = a(\tilde{u}^*)$.

This nuisance would go away if we defined the annihilation operator to be the object, *antilinear* in u, defined by $[a^\dagger(u)]^\dagger$, by $[a^\dagger(\tilde{u})]^\dagger$, or by

$$[a^\dagger(u)]^\dagger \Phi \equiv \left\{ \ldots, \sqrt{n+1} \int d\mu(j)\, u(j)^* \tilde{\phi}_{n+1}(j, j_1, \ldots, j_n), \ldots \right\}$$

(cf. (3.13a)). Alas, it is traditional and otherwise convenient to define a *linear* $a(\,\cdot\,)$. To do so we need a notion of complex conjugation, and that is not intrinsic to the Hilbert space but depends on the realization of the latter as an $\mathcal{L}^2$ space.

In this connection it should be noted that x-space is physically more fundamental than j-space. Therefore, we may regard the complex conjugation in x-space as "God-given", while that in j-space may be an artifact of the basis $\{\psi_j\}$ chosen. This is the reason why $a(u)$, rather than $a(\tilde{u})$, appears in the definition of a quantum field:

Definition: Let $f \in C_0^\infty(\Omega)$.

$$\phi(t, f) \equiv \frac{1}{\sqrt{2}} \left[a\left(K^{-\frac{1}{4}} e^{-it\sqrt{K}} f \right) + a^\dagger \left(K^{-\frac{1}{4}} e^{+it\sqrt{K}} f \right) \right]. \qquad (3.19)$$

I'll elucidate this definition by a sequence of observations:

1. The formal derivative of (3.19) is

$$\frac{d\phi}{dt}(t, f) = \frac{-i}{\sqrt{2}} \left[a\left(K^{+\frac{1}{4}} e^{-it\sqrt{K}} f \right) - a^\dagger \left(K^{+\frac{1}{4}} e^{+it\sqrt{K}} f \right) \right]. \qquad (3.20)$$

Since a and $a^\dagger$ acting on $\mathcal{F}_0$ can be shown to be strongly continuous in their function arguments, (3.20) is valid as a strong derivative. (The strong derivative of an operator-valued function is defined by $O'(t)\Phi \equiv \frac{d}{dt}[O(t)\Phi]$.)

Continuing to differentiate, we get

$$\frac{d^2\phi}{dt^2}(t, f) = -\frac{1}{\sqrt{2}} \left[a\left(K^{-\frac{1}{4}} e^{-it\sqrt{K}} (Kf) \right) + a^\dagger \left(K^{-\frac{1}{4}} e^{+it\sqrt{K}} (Kf) \right) \right]$$
$$= -\phi(t, Kf).$$

This shows that ϕ satisfies the field equation (3.1),

$$\frac{\partial^2 \phi}{\partial t^2} = -K\phi,$$

in the distributional sense. (Since K is self-adjoint and real, it is self-dual — equal to its own transpose. Therefore, applying K to f is the same thing as applying it to ϕ.)

2. If we write

$$a(\tilde{u}) = \int d\mu(j)\, a_j\, \tilde{u}(j),$$

etc., we get

$$\phi(t, f) = \int \frac{d\mu(j)}{\sqrt{2\omega_j}} \left[a_j e^{-it\omega_j} \widetilde{f^*}(j)^* + a_j^\dagger e^{+it\omega_j} \tilde{f}(j) \right].$$

But $\tilde{f}(j) = \int_\Omega \rho\, dx\, \psi_j(x)^* f(x)$, hence

$$\widetilde{f^*}(j)^* = \int_\Omega \rho\, dx\, \psi_j(x) f(x) = f(j').$$

Write $\phi(t, f) \equiv \int \rho\, dx\, \phi(t, x) f(x)$ and peel off the test function to get

$$\phi(t, x) = \int \frac{d\mu(j)}{\sqrt{2\omega_j}} \left[a_j e^{-it\omega_j} \psi_j(x) + a_j^\dagger e^{+it\omega_j} \psi_j(x)^* \right], \qquad (3.21)$$

the "unsmeared" version of (3.19).

This is precisely the general solution (3.6) of the field equation (3.1), with the arbitrary amplitude functions, $\tilde{\phi}_+(j)$ and $\tilde{\phi}_-(j)$, of the classical solution replaced by the operator-valued "functions" a_j and $a_j^\dagger$. Thus $\phi(t, x)$ is an operator-valued solution of the equation of motion — just what we would expect if we were doing quantum field theory in the Heisenberg picture!

We have presented $\phi(t, x)$ as a distribution in x and a function in t. *A fortiori*, it is a distribution in (x, t): For $F \in C_0^\infty(\mathbf{R} \times \Omega)$,

$$\phi(F) = \int_{\mathbf{R} \times \Omega} \phi(t, x) F(t, x)\, \rho(x)\, d^d x\, dt$$

$$\equiv \int_{\mathbf{R}} \phi\big(t, F(t, \cdot)\big)\, dt$$

is defined, and $\phi(t, x)$ satisfies (3.1) in the fully distributional sense:

$$\phi\left(\frac{\partial^2 F}{\partial t^2} + KF\right) = 0.$$

General quantized fields (satisfying nonlinear equations of motion) are expected to be distributions on space-time, but not necessarily distributions on space for each value of t. This additional singularity in the behavior of fields is related to something known in the physics literature as "infinite wave-function renormalization". A classic theorem demonstrating its existence was proved by Powers 1967.

In summary, $\phi(t, x)$ is an *operator-valued distribution*:

$$\langle \Phi, \phi(F)\Psi \rangle \in \mathbf{C} \quad \text{for } \Phi, \Psi \in \mathcal{F}_0 \text{ and } F \in C_0^\infty;$$

$\phi(F)$ is an operator (Hermitian, in the theory of a real scalar field); $\langle \Phi, \phi(t, x)\Psi \rangle$ is an ordinary distribution (real-valued in the real scalar theory if $\Phi = \Psi$).

3. *Initial data:* From (3.19) and (3.20),

$$\phi(0, f) = \frac{1}{\sqrt{2}} \left[a\left(K^{-\frac{1}{4}}f\right) + a^\dagger\left(K^{-\frac{1}{4}}f\right) \right], \qquad (3.22a)$$

$$\frac{\partial \phi}{\partial t}(0, f) = -\frac{i}{\sqrt{2}} \left[a\left(K^{+\frac{1}{4}}f\right) - a^\dagger\left(K^{+\frac{1}{4}}f\right) \right] \equiv \pi_\rho(0, f). \qquad (3.22b)$$

Suppose that f and g are real, so that $a(f) = a(\tilde{f}^*)$, etc. From (3.14) in the form $[a(\tilde{u}^*), a^\dagger(\tilde{v})] = \langle \tilde{u}, \tilde{v} \rangle_\mu$, and from $[a, a] = 0$, we calculate

$$[\phi(0, f), \phi(0, g)]$$
$$= \tfrac{1}{2}\left[a\left(K^{-\frac{1}{4}}f\right), a^\dagger\left(K^{-\frac{1}{4}}g\right) \right] + \tfrac{1}{2}\left[a^\dagger\left(K^{-\frac{1}{4}}f\right), a\left(K^{-\frac{1}{4}}g\right) \right]$$
$$= \tfrac{1}{2}\langle \tilde{f}, \tilde{g} \rangle_\mu - \tfrac{1}{2}\langle \tilde{g}, \tilde{f} \rangle_\mu$$
$$= 0.$$

Similarly, $[\pi, \pi] = 0$. Finally,

$$[\phi(0, f), \pi_\rho(0, g)]$$
$$= \tfrac{1}{2}\left[a\left(K^{-\frac{1}{4}}f\right), ia^\dagger\left(K^{\frac{1}{4}}g\right) \right] + \tfrac{1}{2}\left[a^\dagger\left(K^{-\frac{1}{4}}f\right), -ia\left(K^{\frac{1}{4}}g\right) \right]$$
$$= \tfrac{i}{2}\langle \tilde{f}, \tilde{g} \rangle_\mu + \tfrac{i}{2}\langle \tilde{g}, \tilde{f} \rangle_\mu$$
$$= i\langle \tilde{f}, \tilde{g} \rangle_\mu$$
$$= i\langle f, g \rangle_\rho.$$

(If f were complex, it would be replaced by f^* in the last step.) Strip off the test functions:

$$[\phi(0, x), \pi_\rho(0, y)] = i\delta_\rho(x, y) \equiv i\,\frac{\delta(x - y)}{\rho(y)}. \qquad (3.23a)$$

We have hereby derived the "covariant" form of the *canonical commutation relations*.

Define $\pi(0, y) \equiv \rho(y)\pi_\rho(0, y)$. Then

$$[\phi(0, x), \pi(0, y)] = i\delta(x - y). \tag{3.23b}$$

Supplemented by $[\phi, \phi] = 0 = [\pi, \pi]$, this may be called the "primordial" form of the canonical commutation relations — the form most directly analogous to the $[x_j, p_k] = i\delta_{jk}$ of the quantum mechanics of finitely many degrees of freedom, and the form which is meaningful in any theory whose equation of motion is derived from an action principle, without necessarily any extra geometrical structure (here, the density ρ):

Exercise 11: Construct a Lagrangian *density*,

$$\mathcal{L}(\phi(t, x), \nabla\phi(t, x), \partial_t\phi(t, x)),$$

such that

$$S \equiv \int_{t_1}^{t_2} L\, dt \equiv \int_{t_1}^{t_2} \int_K \mathcal{L}\, d^d x$$

is a *scalar*, and such that the resulting Euler–Lagrange equation is (3.1). (Here K is any compact subset of Ω. The key point is that ρ does not appear in the elements of integration, but in $\mathcal{L}$ itself.) Show that

$$\frac{\partial \mathcal{L}}{\partial[\dot{\phi}^*(t, x)]} = \pi(t, x).$$

Exercise 12: The Hamiltonian H defines a time-translation group on $\mathcal{F}$ by

$$V(t)\Phi \equiv e^{+itH}\Phi$$

$$= \left\{\ldots, \exp\left(it \sum_{k=1}^n \omega_{j_k}\right)\tilde{\phi}_n(j_1, \ldots, j_n), \ldots\right\}$$

$$\equiv e^{it\, d\Gamma(\sqrt{K})}\Phi \equiv \Gamma\left(e^{it\sqrt{K}}\right)\Phi.$$

(a) Show that $\phi(t, f) = V(t)\phi(0, f)V(t)^{-1}$.

(b) Why is it correct to bestow the name "time-translation symmetry group" on $V(t) \equiv e^{+itH}$, not on $U(t) \equiv e^{-itH}$, the operator family giving the time development of solutions of the Schrödinger equation?

4. It is instructive to start from (3.22) and reconstruct the time-dependent field (3.19). First we define some operators from $\mathcal{L}^2(\Omega)$ into itself which solve the classical Cauchy problem for our field equation:

$$G_t \equiv -\frac{1}{\sqrt{K}}\sin\left(t\sqrt{K}\right);$$

$$(G_0 f)(x) = 0, \qquad \frac{\partial}{\partial t}(G_0 f)(x) = -f(x).$$

$$C_t \equiv \cos\left(t\sqrt{K}\right) = -\frac{\partial G_t}{\partial t};$$

$$(C_0 f)(x) = f(x), \qquad \frac{\partial}{\partial t}(C_0 f)(x) = 0.$$

The solution $u(t, x)$ with data

$$u(0, x) = f(x), \qquad \frac{\partial u}{\partial t}(0, x) = g(x)$$

is

$$u(t, x) = (C_t f)(x) - (G_t g)(x).$$

(The distribution kernel of G will become important in later chapters.)

This principle should hold also for operator-valued data, since the differential equation is linear. That is,

$$\phi(t, x) = C_t \phi(0, x) - G_t \pi_\rho(0, x).$$

But C_t and G_t are self-dual (self-adjoint and real), so

$$\phi(t, x) = \phi(0, C_t f) - \pi_\rho(0, G_t f)$$

for all $f \in C_0^\infty$. Insert the definitions of C, G, and ϕ (i.e., (3.22)), and express the trigonometric functions in terms of $e^{\pm it\sqrt{K}}$. The result is (3.19), our starting definition.

We have seen here that the equation of motion plus the canonical commutation relations plus the Fock representation of the latter (through creation and annihilation operators acting on a vacuum) lead to a unique result for the quantum field operator. The same construction will work in non-Fock representations of the canonical relations to yield other operator-valued solutions of the field equation and the canonical commutator algebra. The possibility of such nonequivalent representations has already been mentioned in Chapter 1.

Remark: Strictly speaking, the theory of eigenfunction expansions is not necessary for the construction of the field operator. All of spectral theory that is needed is enough functional calculus to define operators such as $K^{\frac{1}{4}}$ and $\sin(tK^{\frac{1}{2}})$. Nevertheless, I believe that the introduction of the full-blown spectral representation is highly desirable, both in preparation for the detailed investigation of particular cases and in order to draw a clear connection with the usual treatment in the physics literature through a "sum over modes".

5. *Is particle density at a point meaningful?* Consider a one-particle state

$$\Phi = \{0, \tilde{u}_+(j), 0, \dots\}$$
$$\equiv a^\dagger(\tilde{u}_+)|0\rangle \equiv \int a_j^\dagger \tilde{u}_+(j)\, d\mu(j)|0\rangle.$$

For such a state we introduced the "covariant" configuration-space wave function (3.8),

$$\phi(t,x) = \int \frac{d\mu(j)}{\sqrt{2\omega_j}} \tilde{u}_+(j)\, \psi_j(x) e^{-i\omega_j t}.$$

Proposition 1: $u(t,x) = \langle 0|\phi(t,x)|\Phi\rangle.$ (3.24)

Proof: $\langle 0|\phi(t,x)|\Phi\rangle =$

$$\langle 0| \int \frac{d\mu(k)}{\sqrt{2\omega_k}} \left[a_k e^{-it\omega_k}\psi_k(x) + a_k^\dagger e^{it\omega_k}\psi_k(x)^* \right] \int d\mu(j)\, a_j^\dagger \tilde{u}_+(j)|0\rangle.$$

The fundamental tools for evaluating such vacuum expectation values of products of field operators are

$$[a_k, a_j^\dagger] = \delta_\mu(j,k), \qquad a_k|0\rangle = \langle 0|a_k^\dagger = 0.$$ (3.25)

It is easy to see that a nonvanishing contribution can occur only when every annihilation operator in a term can be "paired" with a creation operator standing to its right and all creation operators are absorbed in that way. In general, there will be a contribution for each such pairing of the factors in each term. (This is called *Wick's theorem.*) In the present simple example there is just one pairing in the first term and none in the second, so the expression reduces to

$$\langle 0| \int \frac{d\mu(k)}{\sqrt{2\omega_k}} \int d\mu(j)\, \delta_\mu(j,k) e^{-it\omega_k} \psi_k(x)\tilde{u}_+(j)|0\rangle$$

$$= \int \frac{d\mu(j)}{\sqrt{2\omega_j}} \tilde{u}_+(j)\psi_j(x) e^{-it\omega_j} = u(t,x).$$

An interpretation of (3.24) is attained by moving the (Hermitian) operator $\phi(t,x)$ to the other side of the inner product:

$$u(t,x) = \langle 0|\phi(t,x)|\Phi\rangle = \langle \phi(t,x)\, 0|\Phi\rangle.$$

That is, the single-particle wave function $u(t,x)$ is the inner product of the given quantum state of the field system with the vector formed by the action of the field operator $\phi(t,x)$ on the vacuum vector. (This is, of course, a "generalized" vector — not a true element of Hilbert space until "smeared" with a test function.) This vector, $\phi(t,x)|0\rangle$, belongs to the one-particle sector of the Fock space. Therefore, it is often said that $\phi(t,x)$ creates a particle from the vacuum "at the point (t,x)". This is a vestige of the role of the field operator in nonrelativistic second quantization, where the statement can be interpreted more literally. It should be noted that when acting on an n-particle state $(n > 1)$, ϕ both creates and annihilates, producing a sum of an $(n+1)$-particle state and an $(n-1)$-particle state.

A (formally) better "creation operator at a point" can be defined in the Newton–Wigner framework introduced earlier. Let

$$\phi_{\text{NW}}(x) \equiv \int d\mu(j)\, \psi_j(x) a_j,$$

$$\phi_{\text{NW}}^\dagger(x) \equiv \int d\mu(j)\, \psi_j(x)^* a_j^\dagger.$$

Thus for $u \in \mathcal{L}_\rho^2(\Omega)$,

$$\phi_{\text{NW}}^\dagger(u) \equiv \int_\Omega \phi_{\text{NW}}^\dagger(x) u(x)\, \rho\, dx = a^\dagger(\tilde{u}),$$

$$\phi_{\text{NW}}(u) \equiv \int_\Omega \phi_{\text{NW}}(x) u(x)\, \rho\, dx = a\big(\widetilde{u^*}\big).$$

Proposition 2: $\phi_{\mathrm{NW}}^\dagger(x)\phi_{\mathrm{NW}}(x)$ is the operator representing the number density of particles at x (in the Newton–Wigner interpretation).

Proof: Consider first a one-particle state, $\Phi = \{0, \tilde{u}_+(j_1), 0, \ldots\}$.

$$\langle\Phi|\phi_{\mathrm{NW}}^\dagger(x)\phi_{\mathrm{NW}}(x)|\Phi\rangle$$

$$= \langle 0|\int d\mu(k_1)\, a_{k_1}\, \tilde{u}_+(k_1)^* \int d\mu(k)\, \psi_k(x)^* a_k^\dagger$$

$$\times \int d\mu(j)\, \psi_j(x) a_j \int d\mu(j_1) a_{j_1}^\dagger\, \tilde{u}_+(j_1)|0\rangle.$$

This is a more substantive exercise in the use of Wick's theorem, although there is still only one way of pairing the annihilation and creation operators in the right order. Use of (3.25) yields

$$\int d\mu(k_1) \int d\mu(k)\, \delta_\mu(k_1, k)\tilde{u}_+(k_1)^*\, \psi_k(x)^*$$

$$\int d\mu(j) \int d\mu(j_1)\, \delta_\mu(j, j_1)\psi_j(x)\tilde{u}_+(j_1)^*$$

$$= \int d\mu(k)\, \tilde{u}_+(k)^*\, \psi_k(x)^* \int d\mu(j)\, \psi_j(x)\tilde{u}_+(j)$$

$$= u_{\mathrm{NW}}(x)^*\, u_{\mathrm{NW}}(x)$$

$$= |u_{\mathrm{NW}}(x)|^2,$$

which is the probability density for finding the particle at x (if one accepts the Newton–Wigner interpretation).

In the n-particle case a similar calculation gives

$$n \int \rho(x_2)\, d^d x_2 \cdots \int \rho(x_n)\, d^d x_n\, |u_{\mathrm{NW}}(x, x_2, \ldots, x_n)|^2.$$

Integrating over x we get n, as we should. Before integration, this quantity is the probability of finding a particular particle at x (regardless of where the others are) times n, the number of particles to be considered.

The field ϕ_{NW} has "diseases" traceable to those of u_{NW}: One can show that $[\phi_{\mathrm{NW}}(x), \phi_{\mathrm{NW}}^\dagger] = \delta(x - y)$, but if we look at the Heisenberg time dependence of ϕ_{NW} (or the Hermitian field $\phi_{\mathrm{NW}} + \phi_{\mathrm{NW}}^\dagger$), we find that $\left[\phi_{\mathrm{NW}}(x), \frac{\partial\phi_{\mathrm{NW}}}{\partial t}(y)\right]$ does not vanish even when $x \neq y$. Therefore,

$[\phi_{\mathrm{NW}}(t,x),\phi_{\mathrm{NW}}(s,y)] \neq 0$ even when the space-time points (t,x) and (s,y) have spacelike separation. Consequently, ϕ_{NW} is unacceptable as the basic object in a relativistic *field* theory, where the independence of observations not connectable by signals is axiomatic. In contrast, the field $\phi(t,x)$ does commute with itself at spacelike separations, as we'll see in the next chapter.

Chapter 4
Two-Point Functions

Our next goal is to display the relation between such objects as

$$\langle 0|\phi(t,x)\phi(s,y)|0\rangle$$

and certain integral kernels associated with the classical wave equation

$$\left(\frac{\partial^2}{\partial t^2} + K\right)\phi = 0.$$

Here (t,x) and (s,y) are two points in space-time $(\mathbf{R} \times \Omega)$. For both conceptual and typographical economy, we need to represent such things by single letters; since points in a manifold are *not* vectors, the standard vector notations are possibly misleading. Let us use $\underline{x} \equiv (t,x)$, $\underline{y} \equiv (s,y)$, with $d\underline{x} \equiv dt\, d^d x$, etc.

GREEN FUNCTIONS

Let's start by "formally" investigating the Green-function equation

$$\left(\partial_t^2 + K\right)_{(\underline{x})} G(\underline{x};\underline{y}) = \delta_\rho(\underline{x},\underline{y}) \equiv \frac{\delta(x-y)}{\rho(y)}\,\delta(t-s). \qquad (4.1)$$

G should be the integral kernel of $\left(\partial_t^2 + K\right)^{-1}$:

$$u(t,x) \equiv \int_\Omega \rho\, dy \int_{\mathbf{R}} ds\, G(t,x;s,y)\, f(s,y)$$

$$\Rightarrow \left(\partial_t^2 + K\right) u(t,x) = f(t,x).$$

This raises both a *fundamental issue*: What initial conditions are imposed to make $\left(\partial_t^2 + K\right)^{-1}$ well-defined? — and a *technical issue*: G may be a distribution, not a function. (In particular, its Fourier transform with respect to time may not exist classically.) Both these matters will be addressed in due course.

We take the formal Fourier transform in t and the spectral transform in x:

$$\int e^{+i\omega t}\, dt \int \psi_j(x)^*\, \rho\, dx \ (\,\cdot\,).$$

Then (4.1) becomes

$$(-\omega^2 + \omega_j^2)\tilde{G}(\omega, j; s, y) = \psi_j(y)^* e^{i\omega s}.$$

Thus

$$\tilde{G} = \frac{\psi_j(y)^* e^{i\omega s}}{\omega_j^2 - \omega^2} + \text{term with support on } \omega^2 = \omega_j^2.$$

Invert:

$$G(\underline{x}, \underline{y}) = -\frac{1}{2\pi} \int d\mu(j) \int d\omega \, \frac{1}{\omega^2 - \omega_j^2} e^{-i\omega(t-s)} \psi_j(x)\psi_j(y)^*$$

$$+ \text{ solution of homogeneous equation.} \quad (4.2)$$

Attempting to make sense of this formal expression, we evaluate the ω integral first. It should extend from $-\infty$ to ∞, but the poles at $\omega = \pm\omega_j$ make it ill-defined. One can respond by moving the poles off the real axis into the complex plane a small distance ϵ (or, equivalently, distorting the contour of integration in the opposite direction) with the announced intention of taking the limit $\epsilon \downarrow 0$ later; since the integrals will be evaluated by Cauchy's theorem, taking the limit actually turns out to be superfluous. The trouble is, however, that there is a choice of directions in which to move the poles. This ambiguity gives rise to what might be called the *Green function zoo*:

1. Let the contour run above the poles. Then when $t < s$, the contour can be closed in the upper half plane, and $G = 0$; when $t > s$, it must be closed in the lower half plane, giving rise to the *retarded Green function*:

$$G_{\text{ret}}(\underline{x}, \underline{y}) = \int d\mu(j)\, \frac{i}{2\omega_j} \left[e^{-i\omega_j(t-s)} - e^{+i\omega_j(t-s)} \right] \psi_j(x)\psi_j(y)^*$$

$$(\text{if } t > s)$$

$$= \int \frac{d\mu(j)}{\omega_j} \sin[\omega_j(t-s)]\psi_j(x)\psi_j(y)^* \, \theta(t-s).$$

2. Similarly, putting the contour below the poles yields the *advanced Green function*, which equals 0 if $t > s$:

$$G_{\text{adv}}(\underline{x}, \underline{y}) = -\int \frac{d\mu(j)}{\omega_j} \, \sin[\omega_j(t-s)]\psi_j(x)\psi_j(y)^* \, \theta(s-t).$$

3. Plow right through the poles, taking the *principal value* of each divergent integral:

$$\overline{G} \equiv \tfrac{1}{2}(G_{\text{ret}} + G_{\text{adv}}).$$

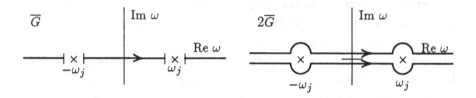

4. The *causal* (or *Feynman*) *propagator* is obtained by going under the left pole but over the right one. Since the square roots of $\omega_j^2 - i\epsilon$ are $\pm\left[\omega_j - \frac{i\epsilon}{2\omega_j} + \ldots\right]$, this can be written

$$G_{\text{F}}(\underline{x}, \underline{y}) \equiv \lim_{\epsilon \downarrow 0} \frac{-1}{2\pi} \int d\mu(j) \int d\omega \, \frac{e^{-i\omega(t-s)}}{\omega^2 - \omega_j^2 + i\epsilon} \, \psi_j(x)\psi_j(y)^*,$$

$$(4.3)$$

as well as

$$G_{\text{F}}(\underline{x}, \underline{y}) = i \int \frac{d\mu(j)}{2\omega_j} \, e^{-i\omega_j|t-s|} \, \psi_j(x)\psi_j(y)^*.$$

Obviously these are not the only logical possibilities, but they are the most useful and most commonly considered.

The difference between two Green functions (solutions of (4.1)) is a solution of the corresponding *homogeneous* wave equation and corresponds to a *closed* contour of ω integration in (4.2):

5. The right-hand pole contributes $-iG_+$, where G_+ is called the *Wightman function*:

$$G_+(\underline{x}, \underline{y}) = \int \frac{d\mu(j)}{2\omega_j} e^{-i\omega_j(t-s)} \psi_j(x)\psi_j(y)^*.$$

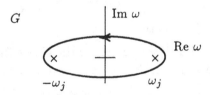

6. The integral around the left-hand pole is denoted $+iG_-$. Note that $G_- = G_+{}^*$, because $\sum_{\text{fixed } \omega_j} \psi_j(x)\psi_j(y)^*$, the spectral projection kernel, can equally well be written $\sum_{\text{fixed } \omega_j} \psi_j(x)^*\psi_j(y)$.

7. Go around both poles counterclockwise:

$$G \equiv G_{\text{adv}} - G_{\text{ret}}$$
$$= -i(G_+ - G_-)$$
$$= 2\,\mathrm{Im}\,G_+\,.$$

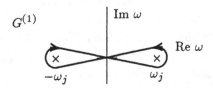

This function is called the *commutator*, for reasons to be made clear soon. The functions $G_\pm$ are the positive- and negative-frequency parts of G.

8. A contour that passes clockwise around the left pole and counterclockwise around the right one produces something denoted $-iG^{(1)}$, and variously named the *Hadamard function* or the *Schwinger function*:

$$G^{(1)} \equiv G_+ + G_-$$
$$= 2\,\mathrm{Re}\,G_+$$
$$= -2i(G_\mathrm{F} - \overline{G}).$$

We shall give it major attention in Chapter 9.

These homogeneous solutions are also sometimes loosely referred to as "Green functions".

Remarks:

(a) G_F equals $\overline{G} + \frac{i}{2}G^{(1)}$.

(b) G_ret, G_adv, $\overline{G}$, G, and $G^{(1)}$ are all real-valued; G_F, G_+, and G_- are complex.

(c)
$$G(\underline{x},\underline{y}) = -\int \frac{d\mu(j)}{\omega_j}\,\sin[\omega_j(t-s)]\psi_j(x)\psi_j(y)^*$$

$$= \text{kernel of}\ \ \frac{\sin\sqrt{K}(t-s)}{-\sqrt{K}}\,.$$

In Chapter 3 we used this function to solve the Cauchy problem, along with $-\partial_t G$, the kernel of $\cos\sqrt{K}(t-s)$. On the other hand,

$$G^{(1)}(\underline{x},\underline{y}) = \int \frac{d\mu(j)}{\omega_j}\,\cos[\omega_j(t-s)]\psi_j(x)\psi_j(y)^*$$

$$= \text{kernel of}\ \ \frac{\cos\sqrt{K}(t-s)}{\sqrt{K}}\,. \qquad (4.4)$$

This kernel is something new.

An important distinction among the Green functions appears when t is set equal to s. From the integral formulas one sees that

$$G(t,x;t,y) = 0,$$
$$-\partial_t G(t,x;s,y)\Big|_{s=t} = \int d\mu(j)\,\psi_j(x)\psi_j(y)^* = \delta(x-y),$$

but

$$G^{(1)}(t,x;t,y) = \int \frac{d\mu(j)}{\omega_j}\,\psi_j(x)\psi_j(y)^* = \text{kernel of } K^{-\frac12},$$

which is nonzero even for $x \neq y$.

As a consequence of its localized initial data, $G(\underline{x},\underline{y})$ as a function of $\underline{x}$ has support inside the light cone of $\underline{y}$. Therefore, G_ret is supported inside the future light cone and G_adv inside the past cone. It follows that $\overline{G}$ also lives inside the cone. In contrast, $G_\pm$, $G^{(1)}$, and G_F spread out over the whole space-time manifold.

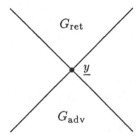

In fact, G_{ret} and G_{adv}, and hence G and $\overline{G}$, can be *characterized* by their support properties in space-time. The spectral representation plays no essential role in their definition. All that is needed is the basic theorem about solutions of hyperbolic partial differential equations: The solutions are uniquely defined by their Cauchy initial data, and they have finite propagation speed. (See, e.g., the appendix of Dimock 1980.) In contrast, G_+, G_-, $G^{(1)}$, and G_F have definitions which hinge on the decomposition of solutions into positive- and negative-frequency parts.

Green functions of the first type are defined in *time-dependent* field theories, provided that the system is *globally hyperbolic* — meaning, roughly, that the Cauchy problem is well-posed and the waves propagate with bounded speed. (See Chapter 6 for further discussion.) But there is no obvious, natural definition of the functions of the second type if the dynamics is not independent of time. This is a fundamental problem for the quantum theory of such systems; when we try to generalize the quantization of the free field in flat space to a general, time-dependent (although linear) dynamics, we realize that God made G_{ret}, G_{adv}, G, and $\overline{G}$, but all the others have been the work of man.

Much of the thrust of later parts of this book is toward making sense of time-dependent linear quantum field theories *without* defining precise counterparts of G_+, G_-, $G^{(1)}$, and G_F. Nevertheless, our renunciation of them must be qualified: Recall that G_F purports to be an integral kernel for $(\partial_t^2 + K)^{-1}$. Now $\partial_t^2 + K$ is a self-adjoint (albeit not elliptic) operator in $\mathcal{L}^2(\mathbf{R} \times \Omega)$, so $(\partial_t^2 + K - z)^{-1}$ should really make sense for $z \notin \sigma(\partial_t^2 + K)$ — in particular, for z complex. But G_F was defined, in effect, as

$$\lim_{\epsilon \downarrow 0} \left[\text{kernel of } (\partial_t^2 + K - i\epsilon)^{-1}\right]$$

[see (4.3)]. This suggests an intrinsic definition of G_F applicable to a general space-time (or other external potential). An interesting attempt to build up quantum field theory in curved space-time from this foundation has been made by Rumpf [Rumpf & Urbantke 1978; Rumpf 1976, 1980, 1981, 1983].

Two-point functions of a quantum field

Let's formally calculate

$$\langle 0|\phi(\underline{x})\phi(\underline{y})|0\rangle = \langle 0| \int \frac{d\mu(j)}{\sqrt{2\omega_j}} \, a_j\psi_j(x)e^{-i\omega_j t} \int \frac{d\mu(k)}{\sqrt{2\omega_k}} \, a_k^\dagger\psi_k(y)^* e^{i\omega_k s}|0\rangle$$

$$= \int \frac{d\mu(j)}{\sqrt{2\omega_j}} \, \psi_j(x)\psi_j(y)^* e^{-i\omega_j(t-s)}$$

$$\equiv G_+(\underline{x},\underline{y}). \tag{4.5}$$

Similarly,

$$G_-(\underline{x},\underline{y}) = \langle 0|\phi(\underline{y})\phi(\underline{x})|0\rangle.$$

Therefore,

$$iG(\underline{x},\underline{y}) = \langle 0|[\phi(\underline{x}), \phi(\underline{y})]|0\rangle,$$

and

$$G^{(1)}(\underline{x},\underline{y}) = \langle 0|\{\phi(\underline{x}), \phi(\underline{y})\}|0\rangle, \tag{4.6}$$

where the braces indicate the *anticommutator* of operators:

$$\{A, B\} \equiv AB + BA.$$

Now the field commutator, $[\phi(\underline{x}), \phi(\underline{y})]$, is a multiple of the identity operator, since $[a(f), a^\dagger(g)]$ is. Therefore (as may also be calculated directly),

$$[\phi(\underline{x}), \phi(\underline{y})] = iG(\underline{x},\underline{y}). \tag{4.7}$$

Indeed,

$$iG(\underline{x},\underline{y}) = \langle \Psi|[\phi(\underline{x}), \phi(\underline{y})]|\Psi\rangle$$

for any $\Psi \in \mathcal{F}_0$ with $\|\Psi\| = 1$, and

$$iG(\underline{x},\underline{y}) = \frac{\langle \Psi_1|[\phi(\underline{x}), \phi(\underline{y})]|\Psi_2\rangle}{\langle \Psi_1|\Psi_2\rangle}$$

for any nonorthogonal, normalized Ψ_1 and Ψ_2. In particular,

$$[\phi(\underline{x}), \phi(\underline{y})] = 0 \quad \text{if } \underline{x} \text{ and } \underline{y} \text{ have spacelike separation.}$$

Pressing further, one finds

$$G_{\text{ret}}(\underline{x}, \underline{y}) = i\theta(t - s)[\phi(\underline{x}), \phi(\underline{y})]$$

and similar expressions for G_{adv} and $\overline{G}$. Finally,

$$\begin{aligned} G_{\text{F}} &= i\theta(t - s)G_{+} + i\theta(s - t)G_{-} \\ &= i\langle 0|\mathcal{T}[\phi(\underline{x})\phi(\underline{y})]|0\rangle, \end{aligned} \qquad (4.8)$$

where

$$\mathcal{T}[\phi(\underline{x})\phi(\underline{y})] \equiv \begin{cases} \phi(\underline{x})\phi(\underline{y}) & \text{if } \underline{x} \text{ is in the future of } \underline{y}, \\ \phi(\underline{y})\phi(\underline{x}) & \text{if } \underline{x} \text{ is in the past of } \underline{y}. \end{cases}$$

This is called the *time-ordered product* of the two field operators; it arises naturally in perturbative calculations (systematized by the famous *Feynman diagrams*) — see any standard textbook on quantum field theory.

In keeping with the previous discussion, the four God-given Green functions turned out to be the expectation values of "c-number" operators (multiples of the identity), whereas the other four Green functions *depend on the state, $|0\rangle$.* In the time-independent theory we have a natural definition of the vacuum state, but in more general models this will not be so. In such situations, "choosing a vacuum" is tantamount to defining a particular Green function to be G_{F} (or a particular distributional solution of the homogeneous field equation to be $G^{(1)}$ or $G_{\pm}$). The latter strategy is slightly more general, since the distinguished Green function could be interpreted as coming from a *pair* of nonorthogonal distinguished states, as in

$$G_{\text{F}}(\underline{x}, \underline{y}) = \frac{\langle 0_{+}|\mathcal{T}[\phi(\underline{x})\phi(\underline{y})]|0_{-}\rangle}{\langle 0_{+}|0_{-}\rangle}.$$

This definition is, in fact, often adopted in models where there are a clearly defined "initial vacuum" and "final vacuum" (see Chapters 7 and 9). (Indeed, if the truth be told, it has often been adopted in situations where the definition of initial and final vacuum is mightily obscure.)

Of course, not *every* Green function has the necessary formal properties to be interpreted as the Feynman propagator or the anticommutator function associated with some state in Fock space. The characterization of those properties, and the construction of quantum field theory in curved space-time from that starting point, is an interesting and underdeveloped approach. It has a great deal in common, philosophically, with the "reconstruction" approach to nonlinear quantum field theory in flat space [Streater & Wightman 1964; Bogolubov et al. 1975].

A DISTRIBUTIONAL TREATMENT

The foregoing discussion indicates that if we know G_+ (or, equivalently, G and $G^{(1)}$), then we know all eight of the standard Green functions; for they are related by the formulas

$$G = 2 \operatorname{Im} G_+ , \qquad G^{(1)} = 2 \operatorname{Re} G_+ ,$$

$$G_- = G_+{}^*,$$

$$G_{\mathrm{ret}} = -G\,\theta(t - s), \qquad G_{\mathrm{adv}} = +G\,\theta(s - t), \qquad (4.9)$$

$$\overline{G} = \tfrac{1}{2}(G_{\mathrm{ret}} + G_{\mathrm{adv}}),$$

$$G_{\mathrm{F}} = \overline{G} + \frac{i}{2}\,G^{(1)} = iG_+\theta(t - s) - iG_-\theta(s - t).$$

Therefore, we concentrate on

$$G_+(\underline{x}, \underline{y}) = \int \frac{d\mu(j)}{2\omega_j}\, e^{-i\omega_j(t-s)}\, \psi_j(x)\psi_j(y)^* \qquad (4.10)$$

and ask whether anything can be said in general about evaluating the j integral.

We shall consider in detail the special case of the free massive field in four-dimensional space-time, for which

$$j = \mathbf{k} \in \mathbf{R}^3, \qquad \psi_j(\mathbf{x}) = (2\pi)^{-\frac{3}{2}} e^{i\mathbf{k}\cdot\mathbf{x}},$$

$$d\mu(j) = d^3k, \qquad \omega_j^2 = \mathbf{k}^2 + m^2.$$

[An alternative, widely used convention is worth noting:

$$d\mu(j) = \frac{d^3k}{2\omega_\mathbf{k}}, \qquad \psi_j = (2\pi)^{-\frac{3}{2}} \sqrt{2\omega_\mathbf{k}}\, e^{i\mathbf{k}\cdot\mathbf{x}}.$$

The advantage is that this turns the basic commutation relation into

$$[a_\mathbf{j}, a_\mathbf{k}^\dagger] = \delta_\mu(\mathbf{j}, \mathbf{k}) = 2\omega_j\delta(\mathbf{j} - \mathbf{k}),$$

which is Lorentz-invariant. (μ is an invariant measure on a hyperboloid parametrizing a representation of the symmetry group of flat space-time.) There is no need to choose the spectral measure to be the Lebesgue measure, even when the spectrum is absolutely continuous, if another convention has some advantage.] For this system one has

$$G_+(\underline{x}, \underline{y}) = \int \frac{d^3 k}{2\sqrt{\mathbf{k}^2 + m^2}} \, e^{i\mathbf{k}\cdot(\mathbf{x}-\mathbf{y}) - i\omega_{\mathbf{k}}(t-s)}. \qquad (4.11)$$

The integral is not obviously convergent, and it is certainly not absolutely convergent. We shall see that whenever $\underline{x} - \underline{y}$ is not a null vector (i.e., of zero length in the indefinite metric of space-time), the integral does nevertheless define a function. Where the argument vector is null, the function has a singularity.

We first review what it means to integrate a distribution with respect to a parameter:

Definition: If $\phi_\omega(\,\cdot\,)$ is a family of distributions depending on a parameter ω, then

$$\left[\int \phi_\omega \, d\mu(\omega) \right] (f) \equiv \int \phi_\omega(f) \, d\mu(\omega).$$

Thus $\int \phi_\omega(x) \, d\mu(\omega)$ is defined as a distribution; in the integral notation for distributions, we have

$$\int \int \phi_\omega(x) \, d\mu(\omega) \, f(x) \, dx = \int \int \phi_\omega(x) \, f(x) \, dx \, d\mu(\omega),$$

so that "Fubini's theorem is true by definition."

Let us check that this definition jibes with the contexts where we have encountered integrals like (4.10) [$\equiv$ (4.5)]:

1. (4.5) is shorthand for

$$\langle 0 | \phi(f) \phi(g) | 0 \rangle = \text{const.} \int \frac{d\mu(j)}{2\omega_j} \left(\hat{\tilde{f}}^* \right)^* (j) \, \hat{\tilde{g}}(j), \qquad (4.12)$$

where $\hat{\tilde{g}}(j)$ (for example) is the restriction of the Fourier transform of the spectral transform of $g(s, y)$ to the "mass shell", $\omega = +\omega_j$. (The constant depends on the normalization of the Fourier transformation.)

Exercise 13: Derive (4.12) from the definition of $\phi(f)\phi(g)$.

2. The identification of twice the real part of (4.10) with the kernel of the operator $\cos\sqrt{K}(t-s)/\sqrt{K}$ comes via the above Fubini identity from the spectral calculation of

$$\frac{\cos\sqrt{K}(t-s)}{\sqrt{K}}f$$

for sufficiently well-behaved f (and similarly for the imaginary part).

The distributional integral defining G_+ can be worked into the form of *a differential operator acting on a continuous function*. The starting point is

$$\psi_j(y) = \frac{K^p\psi_j(y)}{\omega_j^{2p}}.$$

It follows that

$$G_+(\underline{x}; s, f) \equiv \int \frac{d\mu(j)}{2\omega_j} e^{-i\omega_j(t-s)}\psi_j(x)\int \psi_j(y)^* f(y)\,\rho(y)\,d^dy$$

$$= \int \frac{d\mu(j)}{2\omega_j^{2p+1}} e^{-i\omega_j(t-s)}\psi_j(x)\int \psi_j(y)^*\,(K^pf)\,(y)\,\rho\,dy$$

$$\equiv \int\left[\int \frac{d\mu(j)}{2\omega_j^{2p+1}} e^{-i\omega_j(t-s)}\psi_j(x)\psi_j(y)^*\right](K^pf)\,(y)\,\rho\,dy.$$

If p is sufficiently large, the inner integral converges classically to some function $G_p(\underline{x}, y)$ (and the equality leading to it is then an instance of the *genuine* Fubini theorem). [Here I have tacitly used the polynomial growth of the kernel of the spectral projection E_Λ associated with K:

$$\int_{\omega_j^2<\Lambda} d\mu(j)\,\psi_j(x)\psi_j(y)^* = O(\Lambda^{\frac{d}{2}}) \quad \text{as } \Lambda\to\infty. \tag{4.13}$$

See, e.g., Hörmander 1968, or the brief discussion in Sec. 2.2 of Fulling & Ruijsenaars 1987.] Thus we have

$$G_+(\underline{x}; s, f) = G_p(\underline{x}; s, K^pf) \equiv (K^p_{(y)}G_p)(\underline{x}; s, f).$$

(A consistent result will be obtained from this formula for all sufficiently large p.)

Typically, $K^p G_p$ will be a *function with singularities*. In the flat case, (4.11), G_p is a certain Bessel function. Define

$$2\sigma \equiv (t-s)^2 - (\mathbf{x} - \mathbf{y})^2. \tag{4.14}$$

(Note that σ, as a function of $\underline{x} - \underline{y}$, is positive inside the future and past light cones, is 0 *on* the cones, and is negative in the spacelike region.) In space-time dimension 4 one finds [see, e.g., Bjorken & Drell 1965, Appendix C, or Bogolubov & Shirkov 1959, Appendix I]

$$G^{(1)}(\underline{x},\underline{y}) = \frac{m}{4\pi\sqrt{2\sigma}} \theta(2\sigma) Y_1(m\sqrt{2\sigma}) - \frac{m}{2\pi^2\sqrt{-2\sigma}} \theta(-2\sigma) K_1(m\sqrt{-2\sigma})$$

$$\approx -\frac{1}{2\pi^2} \left[\frac{1}{2\sigma} - \frac{m^2}{2} \ln\left(\text{const.}\, m\sqrt{|2\sigma|}\right) + \frac{m^4}{4} + \cdots \right] \tag{4.15a}$$

and

$$G(\underline{x},\underline{y}) = \frac{m}{4\pi\sqrt{2\sigma}} \epsilon(t-s)\theta(2\sigma) J_1(m\sqrt{2\sigma}) - \frac{1}{2\pi} \epsilon(t-s)\delta(2\sigma), \tag{4.15b}$$

where

$$\epsilon(z) \equiv \operatorname{sgn} z \equiv \frac{z}{|z|}$$

and the precise meaning of the covariant Dirac delta function $\delta(2\sigma)$ will be explained later. We note that both functions have singularities on the light cone, $\sigma = 0$, and that G vanishes outside the cone, as expected.

MATHEMATICAL INTERPRETATION OF SINGULARITIES

What does it mean to say that a distribution *equals* a function, as in our claim that $G_+ = K^p G_p$?

1. If supp f lies inside the open region where $K^p G_p$ is nonsingular (more precisely, $K^p G_p \in \mathcal{L}^1_{\text{loc}}$), then

$$G_+(f) = \int \left(K^p G_p\right)(y)\, f(y)\, \rho\, dy,$$

as usual.

2. If supp f overlaps the singularities of $K^p G_p$, the "official" definition of $\left(K^p G_p\right)(f)$ is by integration by parts: $G_p(K^p f)$. (If K^p is not self-dual, it must be replaced by its transpose here.)

Let's look at some elementary, one-dimensional examples:

(1) For $x \neq 0$, one has $\frac{1}{x} = \frac{d}{dx} \ln|x|$. The logarithm is locally integrable. So one possible definition of $\phi(x) \equiv 1/x$ as a distribution on $C_0^\infty(\mathbf{R})$ is

$$\phi(f) = \int_{-\infty}^\infty \frac{d}{dx} \ln|x| f(x)\, dx$$

$$\equiv -\int_{-\infty}^\infty \ln|x| f'(x)\, dx$$

$$= -\lim_{\epsilon\downarrow 0} \left[\int_{-\infty}^{-\epsilon} \ln|x| f'(x)\, dx + \int_\epsilon^\infty \ln|x| f'(x)\, dx \right]$$

$$= +\lim_{\epsilon\downarrow 0} \left[\int_{-\infty}^{-\epsilon} \frac{1}{x} f(x)\, dx \right.$$

$$\left. + \int_\epsilon^\infty \frac{1}{x} f(x)\, dx + \ln\epsilon[f(-\epsilon) - f(\epsilon)] \right]$$

$$= \mathcal{P} \int_{-\infty}^\infty \frac{1}{x} f(x)\, dx,$$

since the factor multiplying $\ln\epsilon$ vanishes through order ϵ. ($\mathcal{P}$ indicates the *principal value* of the integral, which in the present case means that one integrates up to the singularity in such a way that the infinite contributions of its left and right sides cancel. In a moment we'll see a principal-value integral to which that interpretation does not apply.) On the other hand, we could equally well have written

$$\frac{1}{x} = \frac{d}{dx} \begin{cases} \ln x + 10 & \text{for } x > 0, \\ \ln|x| & \text{for } x < 0. \end{cases}$$

(The differentiated function here is not continuous, but it is in $\mathcal{L}^1_{\text{loc}}$. It is the derivative of a certain continuous function — call it $h(x)$ — so we can write $\frac{1}{x} = h''(x)$.) This construction would give

$$\phi(f) = -\int_{-\infty}^0 \ln|x| f'(x)\, dx - \int_0^\infty (\ln x + 10) f'(x)\, dx$$

$$= +\lim_{\epsilon\downarrow 0} \left[\int_{-\infty}^{-\epsilon} \frac{1}{x} f(x)\, dx + \int_\epsilon^\infty \frac{1}{x} f(x)\, dx \right] + 10\, f(0).$$

Thus $\phi(x) = \mathcal{P}\frac{1}{x} + 10\,\delta(x)$, whereas $\phi(x) = \mathcal{P}\frac{1}{x}$ according to the first definition. Which is right? Either. There is an inherent ambiguity in passing from the function $1/x$ to a distribution.

The point is that the continuous function G_p which (together with a differential operator K^p) defines a distribution G may contain information which is not present in the values of the function $K^p G_p$ in its nonsingular region. One must be alert to the possible presence of delta functions and derivatives thereof on the singular set (the light cone, in our application).

(2) Let's define $1/x^2$ as a distribution by

$$\frac{1}{x^2} \equiv -\frac{d}{dx} P \frac{1}{x} = -\frac{d^2}{dx^2} \ln |x|.$$

If $\psi(x) \equiv \dfrac{1}{x^2}$, then,

$$\psi(f) \equiv +P \int_{-\infty}^{\infty} \frac{1}{x} f'(x)\, dx$$

$$= \lim_{\epsilon \downarrow 0} \left[\int_{-\infty}^{-\epsilon} \frac{1}{x} f'(x)\, dx + \int_{\epsilon}^{\infty} \frac{1}{x} f'(x)\, dx \right]$$

$$= +\lim_{\epsilon \downarrow 0} \left[\int_{-\infty}^{-\epsilon} \frac{1}{x^2} f(x)\, dx \right.$$

$$\left. + \int_{\epsilon}^{\infty} \frac{1}{x^2} f(x)\, dx + \frac{1}{-\epsilon} f(-\epsilon) - \frac{1}{\epsilon} f(\epsilon) \right]$$

$$= \lim_{\epsilon \downarrow 0} \left[\int_{-\infty}^{-\epsilon} \frac{1}{x^2} f(x)\, dx + \int_{\epsilon}^{\infty} \frac{1}{x^2} f(x)\, dx - \frac{2}{\epsilon} f(0) + O(\epsilon) \right]$$

$$= \lim_{\epsilon \downarrow 0} \left[\int_{-\infty}^{-\epsilon} \frac{f(x) - f(0)}{x^2}\, dx + \int_{\epsilon}^{\infty} \frac{f(x) - f(0)}{x^2}\, dx \right].$$

This suggests the definition

$$P \int_{-\infty}^{\infty} \frac{f(x)}{x^2}\, dx \equiv \lim_{\epsilon \downarrow 0} \int_{|x| \geq \epsilon} \frac{f(x) - f(0)}{x^2}\, dx.$$

The generalization to higher powers is indicated in the next exercise. [*References:* Hilgevoord 1960, Sec. 3.3; Bogolubov et al. 1975, Sec. 2.3; Friedlander 1975, Chap. 2; Friedlander 1982, Chap. 2.]

Exercise 14: Define

$$P \int_{-\infty}^{\infty} \frac{f(x)}{x^n}\, dx$$

$$\equiv \lim_{\epsilon \downarrow 0} \int_{|x| \geq \epsilon} \frac{f(x) - f(0) - x f'(0) - \cdots - \frac{1}{(n-2)!} x^{n-2} f^{(n-2)}(0)}{x^n}\, dx.$$

Show that

$$\frac{d}{dx} \mathcal{P} \frac{1}{x^n} = -n \mathcal{P} \frac{1}{x^{n+1}}$$

(i.e., d/dx commutes with $\mathcal{P}$), and hence that, as distributions,

$$\mathcal{P} \frac{1}{x^n} = \frac{(-1)^{n-1}}{(n-1)!} \frac{d^n}{dx^n} \ln|x|.$$

3. In view of the foregoing, one adopts the following convention: If, as $x \to 0$ (with the obvious generalization to other points), the function ϕ has the asymptotic behavior

$$\phi(x) \sim \phi_0(x) + \sum_{n=1}^{N} \frac{a_n}{x^n},$$

where ϕ_0 is a locally integrable function, then

$$\phi(f) \equiv \int \phi_0(x) f(x) \, dx + \mathcal{P} \int \sum_{n=1}^{N} \frac{a_n}{x^n} f(x) \, dx.$$

(In particular, if $N = 1$ the second term is just

$$\lim_{\epsilon \downarrow 0} \int_{|x| \geq \epsilon} a_1 \frac{f(x)}{x} \, dx,$$

with no subtraction.)

Remark: An alternative approach to such distributions is based on an analytic continuation from σ to $\sigma \pm i\epsilon$, following the model of the well known identity

$$\lim_{\epsilon \downarrow 0} \frac{1}{x - i\epsilon} = \mathcal{P} \frac{1}{x} + i\pi \delta(x).$$

See the references cited in connection with Exercise 14.

4. We need to transplant these constructions to a multidimensional manifold, so as to define things such as $\delta(2\sigma)$ and $\mathcal{P} \frac{1}{2\sigma}$ (σ defined by (4.14), or its generalization defined by a curved pseudo-Riemannian metric (see Chap. 8)). This transplantation is discussed by Friedlander 1975 in Secs. 2.9 and 4.1–2. [See also Gel'fand & Shilov 1964, Chap. 3; Friedlander 1982, Chap. 7.] I outline the process briefly:

(A) Suppose ϕ is a distribution on $\mathbf{R}$ and S is a real-valued function on a manifold $\mathcal{M}$ with a volume element $d\mu$. (In our static field theory, $\mathcal{M} = \mathbf{R} \times \Omega$ and $d\mu = dt\, \rho\, d^d x$.) Suppose that $\nabla S \neq 0$ on the support of a given test function. Introduce a coordinate system with S as one coordinate; this induces a volume element (called a *Leray form*), $d\mu_c$, on each hypersurface $S = c = \text{const.}$:

$$d\mu = d\mu_c\, dS.$$

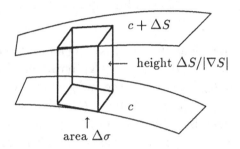

[More concretely, if $d\mu = \sqrt{g}\, dx^1 \wedge \cdots \wedge dx^n$ and $S \equiv x^1$, then $d\mu_c = \sqrt{g}\, dx^2 \wedge \cdots \wedge dx^n$, restricted to $S = c$. In Euclidean space,

$$d\mu_c = \frac{d\sigma}{|\nabla S|}$$

where $d\sigma$ (unrelated to (4.14) except by traditional choice of letter) is the natural volume element on the submanifold $\{S = c\}$.] Now define the distribution $\phi(S)(\,\cdot\,) \equiv \phi \circ S$ on $\mathcal{M}$ by

$$(\phi \circ S)(f) \equiv \int \phi(S(x))\, f(x)\, dx \equiv \phi\left[\int_{S=c} f\, d\mu_c\right].$$

Here ϕ acts on the variable c. One can check that (1) if ϕ is a function, this does reduce to the right thing:

$$\phi\left[\int_{S=c} f\, d\mu_c\right] = \int dS\, \phi(S) \int f\, d\mu_c = \int (\phi \circ S)\, f\, d\mu,$$

by definition of μ_c; (2) the chain rule is valid:

$$\nabla(\phi \circ S) = \phi'(S)\, \nabla S.$$

(B) Unfortunately, $S = 2\sigma$ violates the condition $\nabla_{(\underline{x})}S \neq 0$ at the point $\underline{x} = \underline{y}$ (but nowhere else), since $\nabla\sigma = (t - s, \mathbf{x} - \mathbf{y})$. So to define $\delta(2\sigma)$, for example, one must study a limit $\lim_{\epsilon \downarrow 0} \delta(2\sigma - \epsilon)$. (For $\mathcal{P}\frac{1}{2\sigma}$ we are already taking a limit of this nature, so we don't need to insert another.)

Finally, therefore, we are able to interpret (4.15). When we evaluate G and $G^{(1)}$ carefully, we get things of the form

$$K^p(\text{Bessel functions, etc.}).$$

Away from the light cone ($\sigma = 0$), the derivatives can be evaluated classically, yielding the J_1, Y_1, and K_1 terms in (4.15). In (4.15b), the J_1 term is in $\mathcal{L}^1_{\text{loc}}$; however, the integration by parts shows that $K^p(\ldots)$ also contains a $\delta(2\sigma)$ term. In (4.15a), the leading term of Y_1 or K_1 is $\frac{\text{const.}}{\sigma}$, which is not locally integrable and must be interpreted as a principal value; in this case there is no additional term supported on the cone. [This last can be verified by checking that $\mathcal{P}G^{(1)}$ satisfies $(\partial_t^2 + K)\mathcal{P}G^{(1)} = 0$ as a distribution. Since the equation is satisfied classically by the right-hand side of (4.15a) away from the cone, this means checking that the boundary terms in an integration by parts of $\int G^{(1)}(\underline{x}, \underline{y})(\partial_t^2 + K)f(\underline{y})\,d\underline{y}$ vanish in the limit as the boundary surface $|\sigma| = \epsilon$ approaches the cone.]

Fortunately, in renormalization theory one is mostly interested in $G^{(1)}(\underline{x}, \underline{y})$ for $\underline{x}$ *close to but not on* the light cone of $\underline{y}$. So the functional representation of $G^{(1)}$ away from the singularity is all one uses; whether there is another term supported *on* the singularity is irrelevant.

Remark: In this discussion we have tacitly observed and used the fact that $G_+(t, \mathbf{x}; s, \mathbf{y})$ (or any of its siblings) is a distribution in any one of the variables with the other three fixed. *A priori*, a quantum field theory involves an operator-valued distribution on space-time, and the definition of G_+ as a two-point function yields [only] a distribution in all the variables together (i.e., on a $(2d + 2)$-dimensional space). (This is an application of the *Schwartz kernel (nuclear) theorem* — see the books of Streater & Wightman and Bogolubov et al.) Our more explicit construction (which can be partially extended, with sufficient technical prowess, to more general space-times) has given us additional information.

RECONSTRUCTION OF THE FIELD THEORY

In a noninteracting field theory, the two-point functions determine the n-point functions:

Exercise 15: Show (by a Wick's-theorem calculation) that

$$\langle 0|\phi(\underline{x}_{2n})\phi(\underline{x}_{2n-1})\cdots\phi(\underline{x}_1)|0\rangle = \sum \prod_{i=1}^{n} G_+ \left(\underline{x}_{j_i}, \underline{x}_{k_i}\right),$$

where the sum is over all partitions of $\{1, 2, \ldots, 2n\}$ into an unordered set of ordered pairs (j_i, k_i), $j_i > k_i$. Show also that the vacuum expectation value $\langle 0|\phi(\underline{x}_{2n-1})\cdots|0\rangle$ of an *odd* number of noninteracting fields is 0.

This property hinges on the fact that $[\phi(\underline{x}), \phi(\underline{y})]$ equals a multiple of the identity operator, for a canonically quantized field satisfying a linear equation of motion.

In general, the n-point functions determine the quantum field theory. This is the *Wightman–GNS construction*. An abstract algebra of "operators" plus a vacuum expectation functional defined thereon implies the existence of a Hilbert space and a representation of the dynamical variables by genuine operators acting in it. In the literature [e.g., Streater & Wightman 1964, Bogolubov et al. 1975], this is shown for interacting theories in flat space, satisfying the standard Wightman axioms. It is clear that correspondingly weaker propositions hold in more general contexts such as curved space-time (irrelevant parts of the usual theorems, such as those concerning Poincaré invariance, being abandoned); but the rigorous literature is underdeveloped in this direction. (See, however, the work of Isham and Dimock cited in Chapter 6.)

Chapter 5
The Stress Tensor and the Casimir Effect

The introduction of "particles" in the previous chapter was rather artificial; it sprang from an analogy with the mathematics of the free field in flat space, not from physics. Later we shall see that under more general conditions (time-dependent dynamics or nonpositive K) a natural definition of particles does not exist (or is not unique, if one is more tolerant). It seems reasonable, therefore, to consider taking the field itself as the basic physical object.

The expectation value, $\langle \Psi | \phi(\underline{x}) | \Psi \rangle$, has an observable meaning for an electromagnetic field, at least. However, $\langle \Psi | \phi(\underline{x}) | \Psi \rangle$ is equal to 0 for any state of definite (even large) particle number. It appears that the expectation value of the field does not tell the whole story of the physics of any given quantum state.

How, then, is the presence of a large quantity of matter manifested in the field? Recall that in quantum theory the expectation value of the square of an observable is not the same thing as the square of the expectation value ($\langle A^2 \rangle \neq \langle A \rangle^2$). This comes about because measurements are probabilistic. $\langle A^2 \rangle$ depends on the standard deviation of A as well as on the mean of A, which is $\langle A \rangle$. Expectations of higher powers introduce still more independent objects. Therefore, such quantities as $\phi(\underline{x})^2$ are appealing candidates for nontrivial observables in eigenstates of particle number (including the vacuum). In fact, in interacting field theories such polynomial functions of fields are explicitly coupled to the other fields; this fact adds to the hope that they may have empirical significance. (Designing a "model detector" which unambiguously measures some particular function of a quantum field is no easy task, however.)

Here are three examples:

(1) Let ψ and ϕ be two scalar fields. If the Lagrangian or Hamiltonian density defining the dynamics of the theory contains a term proportional to $\psi \phi^2$ (evaluated at some point $\underline{x}$), then the classical

equation of motion of ψ contains a term proportional to ϕ^2. We defer for a moment the question of what happens to that equation of motion after ϕ is quantized.

(2) The electromagnetic field couples to the *charge-current density* of each electrically charged field in nature. For a charged scalar field the current-density vector is

$$J_\mu(\underline{x}) \equiv iq\,\phi(\underline{x})^* \overset{\leftrightarrow}{\partial}_\mu \phi(\underline{x}),$$

where $\overset{\leftrightarrow}{\partial}_\mu$ is an antisymmetrized derivative as in (3.12), and q is the charge carried by one of the particles described by the field ϕ. The *total charge* of the scalar field is (in flat space)

$$Q \equiv \int_{\mathbf{R}^d} J_0\, d^d x.$$

If ϕ is the only charged field in the theory under study, then Q is a conserved quantity related to gauge invariance (invariance under redefinition of the phase of ϕ), and correspondingly (by Gauss's theorem) J satisfies the local or differential conservation law,

$$\partial_\mu J^\mu = 0. \tag{5.1}$$

(When there are several charged fields, only the sum of the charges or currents of all the fields is necessarily conserved.)

(3) The gravitational field (the metric tensor) couples to the *energy-momentum tensor* or *stress tensor* of every other field in nature. This tensor, denoted $T_{\mu\nu}$, is already significant in flat-space field theories, where the metric does not enter as a dynamical variable. There T is introduced in connection with space-time translation symmetries and the related conservation laws. In a time-independent background metric,

$$H \equiv \int T_{00}\,\rho\, d^d x$$

is the conserved energy. (It happens to be equal to the Hamiltonian as previously defined.) In flat space there is also a conserved momentum,

$$P_j \equiv \int T_{0j}\, d^d x \qquad (j = 1, \dots, d).$$

In curved space-time these global conservation laws break down, but there remains a local conservation law analogous to (5.1):

$$\nabla_\mu T^\mu{}_\nu = 0, \tag{5.2}$$

where ∇ indicates a covariant derivative (see Chapter 8).

The specification of the stress tensor of a scalar field in flat space is complicated by the existence of two candidates. The "canonical" stress tensor is

$$T_{\mu\nu} \equiv \partial_\mu \phi \partial_\nu \phi - \tfrac{1}{2}\eta_{\mu\nu}\,\partial_\rho \phi \partial^\rho \phi + \tfrac{1}{2}m^2 \eta_{\mu\nu}\,\phi^2. \tag{5.3}$$

Here m is the mass of the field and $\eta_{\mu\nu}$ is the metric tensor of flat space:

$$g_{\mu\nu} = \eta_{\mu\nu} \equiv \begin{pmatrix} 1 & 0 & 0 & 0 \\ 0 & -1 & 0 & 0 \\ 0 & 0 & -1 & 0 \\ 0 & 0 & 0 & -1 \end{pmatrix} \tag{5.4}$$

in spatial dimension $d = 3$. (The rows and columns are numbered 0 (time), 1, 2, 3.) Thus, for instance, the energy density is

$$T_{00} = \frac{1}{2}\left[(\partial_0 \phi)^2 + \sum_{j=1}^{3}(\partial_j \phi)^2 + m^2 \phi^2 \right].$$

The "new improved" stress tensor [Callan et al. 1970] is

$$\overline{T}_{\mu\nu} \equiv \tfrac{2}{3}\partial_\mu \phi \partial_\nu \phi - \tfrac{1}{6}\eta_{\mu\nu}\,\partial_\rho \phi \partial^\rho \phi - \tfrac{1}{3}\phi\,\partial_\mu \partial_\nu \phi + \tfrac{1}{6}m^2 \eta_{\mu\nu}\,\phi^2. \tag{5.5}$$

(The numerical coefficients given here are specific to the case of 3 space dimensions. For a more general discussion see Chapter 6.) So

$$\overline{T}_{00} = \frac{1}{2}(\partial_0 \phi)^2 + \frac{1}{6}\sum_{j=1}^{3}(\partial_j \phi)^2 - \frac{1}{3}\phi\sum_{j=1}^{3}\partial_j^2 \phi + \frac{1}{2}m^2 \phi^2.$$

One notes that $T_{00} - \overline{T}_{00} = \frac{1}{3}\sum_j \partial_j(\phi \partial_j \phi) = \nabla \cdot (\ldots)$ is a divergence (in the sense of three-dimensional vector calculus), and hence both T's give the same $H = \int_{\mathbf{R}^3} T_{00}\, d^3x$ (under the standard assumption that the field configuration is sufficiently localized that no boundary terms arise in any integration by parts). This ambiguity in the energy *density* reflects the fact that — outside the context of general relativity — energy cannot be assigned to individual spatial points, but is a property of the state of a physical system as a whole. When gravity is introduced, however, energy must be localized unambiguously. Later we shall see that T and $\overline{T}$ represent two different (equally consistent, but inequivalent) ways of coupling ϕ to the gravitational field.

Remark: In relativistic physics there are two sign conventions for the metric tensor. I prefer (5.4), which is said to have *negative signature* (more specifically, signature -2) because the minus signs outnumber the plus signs. My feeling is that energies and energy densities are naturally positive and it is silly for them to change sign when an index is raised or lowered. However, a huge number of authors write the exact negative of (5.4) as $\eta_{\mu\nu}$, with a corresponding positive-signature convention for a general metric $g_{\mu\nu}$.

This whole idea of field observables runs into a huge problem when we pass from a classical to a quantum theory: Nonlinear functions of a quantum field don't make literal sense. Since $\phi(\underline{x})$ is a distribution, $\phi(\underline{x})^2$ isn't defined. (We may take ϕ^2 as a prototype of J_μ, $T_{\mu\nu}$, and any other quadratic expression in quantized fields, for which the same problem arises.) As we have seen, $\langle 0|\phi(\underline{x})\phi(\underline{y})|0\rangle \equiv G_+(\underline{x}, \underline{y})$ is singular as $\underline{y}$ approaches $\underline{x}$, so that $\langle 0|\phi(\underline{x})\phi(\underline{x})|0\rangle$ appears to be infinite! Of course,

$$\langle 0|\phi(f)^2|0\rangle = \int d\underline{x} \int d\underline{y} \, G_+(\underline{x}, \underline{y}) \, f(\underline{x}) f(\underline{y})$$

is perfectly OK for $f \in C_0^\infty$, but that quadratic functional of f isn't what we need. We want $\phi^2(\underline{x})$ to be a distribution, so that the *linear* functional $\langle 0|\phi^2(f)|0\rangle$, defined formally as $\int d\underline{x} \, G_+(\underline{x}, \underline{x}) \, f(\underline{x})$, makes sense somehow.

This may be seen as the MAIN PROBLEM of quantum field theory in curved space-time. (It may also be seen as the central problem of the rigorous construction of *interacting* quantum field theories in flat space-time corresponding to given classical Lagrangians [Wightman 1967]. Unfortunately, a solution to one problem does not entail a solution to the other, and the techniques which have been successfully applied in the two situations are rather different. Presumably we will some day have a constructive quantum field theory in curved space-time — or even a rigorous quantum gravity! — which will reunite the two programs.)

Our first strategy for defining a finite $\phi^2(\underline{x})$ will be this:

1. Calculate something that, from a naive point of view, ought to

have $\phi(\underline{x})^2$ as a limit. For example, we could look at $\phi(f)^2$ as $f \in C^\infty$ condenses toward $\delta(\underline{y} - \underline{x})$. One finds, however, that no flexibility is lost, and much technical simplification is gained, by dispensing with the test function and considering $\phi(\underline{x})\phi(\underline{y})$ as $\underline{y} \to \underline{x}$. (We must take this limit through vectors $\underline{y} - \underline{x}$ which are not null with respect to the indefinite space-time metric, since the singularity of G_+ arises along the entire light cone, not just at the origin.) Further elegance and simplification are attained by studying the symmetrized combination

$$\tfrac{1}{2}[\phi(\underline{x})\phi(\underline{y}) + \phi(\underline{y})\phi(\underline{x})].$$

2. Subtract the expectation value of this operator in the vacuum state.

3. Take the limit, hoping that the result will now be finite and well-defined. For instance, in the test-function approach, we would have

$$\phi^2(\underline{x}) \equiv \lim_{f \to \delta_{\underline{x}}} \left[\phi(f)^2 - \langle 0|\phi(f)^2|0\rangle\right].$$

Examples of this "point-splitting" approach will be presented in detail presently. In the jargon of quantum field theory, step 1 is a *regularization*; steps 2 and 3 are a *renormalization*.

Historically, the motivation of step 2 is that in nongravitational physics only *differences* of energies are significant; one is free to assign zero energy density to the vacuum as a convention. (In special-relativistic theories, Lorentz invariance provides a deeper reason for requiring the energy of the vacuum to be zero. But this is not relevant to our study of general time-independent field theories.) However, we may employ this strategy as a technical device without committing ourselves to the physical postulate that the vacuum energy is zero. Success in step 3 will reduce the problem of defining or calculating $\langle \Psi|\phi^2|\Psi\rangle$ or $\langle \Psi|T_{\mu\nu}|\Psi\rangle$ to the corresponding problem for $\langle 0|\phi^2|0\rangle$ or $\langle 0|T_{\mu\nu}|0\rangle$. We need more physics to determine whether the latter are zero, and if not, what they are. (In fact, we shall conclude that they are *not* generally zero, and the nonzero (but finite) values must be added to the finite differences of expectation values obtained in step 3.)

THE CASIMIR EFFECT

Let us consider the theory defined by the free-field operator, $K =$

$-\nabla^2 + m^2$, in the region

$$\Omega = \{\,\mathbf{x} \equiv (x,y,z) : 0 < z < L\,\} \subset \mathbf{R}^3, \qquad (5.6)$$

with the Dirichlet boundary condition ($\phi = 0$) at $z = 0$ and $z = L$. We shall treat this system in conjunction with the ordinary massive free field, defined by the same K but with $\Omega = \mathbf{R}^3$.

For each of these models, as a special case of the general construction in Chapter 3, we have field operators and a vacuum state. For the "slab" (5.6) the field expansion is

$$\phi(t,\mathbf{x}) = \frac{1}{2\pi} \int_{\mathbf{R}^2} dk_1\, dk_2 \sum_{n=1}^{\infty} \frac{1}{\sqrt{\omega_{\mathbf{k}} L}} \Big[\, a_{\mathbf{k}} e^{-i\omega_{\mathbf{k}} t} e^{i(k_1 x + k_2 y)} \sin k_3 z$$
$$+ a_{\mathbf{k}}^{\dagger} e^{i\omega_{\mathbf{k}} t} e^{-i(k_1 x + k_2 y)} \sin k_3 z \Big], \quad (5.7)$$

$$k_3 \equiv \frac{n\pi}{L}, \qquad \omega_{\mathbf{k}}^2 \equiv k_1^2 + k_2^2 + k_3^2 + m^2,$$

$$[a_{\mathbf{k}}, a_{\mathbf{k}'}^{\dagger}] = \delta(k_1 - k_1')\delta(k_2 - k_2')\delta_{nn'}.$$

In the natural notation $\mathbf{x}_\perp \equiv (x,y)$, etc., the connection with our general notation is

$$\psi_{\mathbf{k}}(\mathbf{x}) = \frac{1}{2\pi}\sqrt{\frac{2}{L}}\, e^{i\mathbf{k}_\perp \cdot \mathbf{x}_\perp} \sin k_3 z, \qquad \int \ldots d\mu(\mathbf{k}) = \sum_{n=1}^{\infty} \int_{\mathbf{R}^2} \ldots dk_\perp.$$

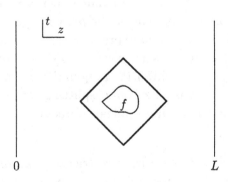

Let $|0_L\rangle$ be the vacuum of the slab system and $|0\rangle$ be the ordinary free-field vacuum. The slab is naturally embedded in $\mathbf{R}^3$, and thus for field operators with test functions whose support is away from the boundary hyperplanes, we may think of $|0_L\rangle$ and $|0\rangle$ as different states of the *same* physical system. (The diamond-shaped region in the diagram is intended to suggest a *causal domain*, bounded above and below in space-time by pieces of light cone or other null hypersurfaces. As explained in Chapter 6 — and in more detail in treatises on Lorentzian geometry and hyperbolic differential equations, such as Hawking & Ellis 1973 and Choquet-Bruhat 1968 — such a *globally hyperbolic* domain supports a self-contained field dynamics. The bare dynamics of a field theory inside such a domain is not influenced by which larger domain Ω contains it. On the other hand, which *states* of the quantum field theory are *physically most plausible* will depend on the global situation; the field inside the diamond has had an infinite amount of time to come into equilibrium with the walls, or the unlimited empty space, beyond it.) Therefore, it is possible to apply the previously announced strategy to $|0_L\rangle$, with $|0\rangle$ in the role of the "vacuum" to be subtracted: $\langle 0_L|T_{\mu\nu}(\underline{x})|0_L\rangle - \langle 0|T_{\mu\nu}(\underline{x})|0\rangle$ has a definite physical meaning for $\underline{x}$ inside the slab, Ω.

In calculating this quantity I shall make several simplifications for the sake of presentation: (1) I give the details only for $\langle 0_L|\phi^2(\underline{x})|0_L\rangle - \langle 0|\phi^2(\underline{x})|0\rangle$, not for $T_{\mu\nu}$. (2) I let the mass m equal 0. Although this is not allowed by our standing hypotheses on K, the construction of Fock space turns out to make perfect sense for the massless free field as long as the spatial dimension d is greater than 1. (This would not be true for the Neumann boundary condition ($\partial f/\partial z = 0$ at 0 and L), since in that case 0 is an eigenvalue of K. When 0 is merely in $\sigma_c(K)$, the negative powers of K become unbounded operators, but one sees from (5.7) that the transverse dimensions help the integral to converge at the lower limit so that the formal expression for the field operator makes sense. Such infrared niceties are discussed further in the appendix and in Fulling & Ruijsenaars 1987.) (3) I regularize by looking at $\phi(\underline{x})\phi(\underline{x}')$ with $\mathbf{x}_\perp = \mathbf{x}'_\perp$. That is, the "point-splitting" takes place entirely in the (t, z) plane, not in a fully arbitrary direction.

Exercise 16:

(a) Redo the calculation without restrictions (2) and (3). (This is a major undertaking, recommended only to lovers of Bessel functions.)

(b) Redo the calculation with periodic boundary conditions in z in place of Dirichlet. (This is easier than the Dirichlet calculation.)

When $m = 0$, (4.15a) reduces to its leading term:

$$\frac{1}{2} G^{(1)}(\underline{x}, \underline{x}') = \frac{-1}{8\pi^2 \sigma} \equiv \frac{-1}{4\pi^2[(t-t')^2 - (\mathbf{x}-\mathbf{x}')^2]} \equiv \operatorname{Re} G_+ .$$

Also, $\frac{1}{2} G \equiv \operatorname{Im} G_+ = 0$ off the light cone. (The massless field in even-dimensional space-time commutes at timelike as well as spacelike separations (*Huygens's principle*), leaving only terms proportional to $\delta(\sigma)$ on the two parts of the light cone.) Compare this with

$$G_+^L(\underline{x}, \underline{x}') \equiv \langle 0_L | \phi(\underline{x}) \phi(\underline{x}') | 0_L \rangle$$

$$= \frac{1}{(2\pi)^2 L} \int_{\mathbf{R}^2} d^2 k_\perp \sum_{n=1}^{\infty} \frac{1}{\omega} e^{-i\omega(t-t')} e^{i k_\perp \cdot (\mathbf{x}-\mathbf{x}')_\perp} \sin k_3 z \sin k_3 z'$$

$$= \frac{1}{4\pi L} \sum_{n=1}^{\infty} \int_{k_3}^{\infty} d\omega \, e^{-i\omega(t-t')} [\cos k_3(z - z') - \cos k_3(z + z')]. \quad (5.8)$$

Here we have appealed to (3) above and to the fact that, since $\omega^2 \equiv k_\perp^2 + k_3^2$,

$$\int_0^{\infty} \cdots \frac{|\mathbf{k}_\perp| \, d|\mathbf{k}_\perp|}{\omega} = \int_{k_3}^{\infty} \cdots d\omega.$$

Proposition 1: In the distribution sense, and modulo terms supported at the origin, one has

(a) $$\int_k^{\infty} e^{-i\omega\tau} \, d\omega = \frac{1}{i\tau} e^{-ik\tau} \qquad (k > 0);$$

(b) $$\sum_{n=1}^{\infty} e^{-in\lambda} = \frac{1}{e^{i\lambda} - 1}.$$

(There *are* delta-function terms at 0 which have been omitted here.)

Proof: First observe that if τ and λ had negative imaginary parts, the integral and sum would be classically convergent to the asserted values. [In (a), the integral is elementary, and the contribution from the upper limit vanishes since $\operatorname{Re}(-i\omega\tau) < 0$ approaches $-\infty$. For (b), use

$$\sum_{n=1}^{\infty} e^{-in\lambda} = \frac{e^{-i\lambda}}{1 - e^{-i\lambda}} = \frac{1}{e^{i\lambda} - 1},$$

the geometric series.] So we need only verify that, as distributions acting on smooth test functions $f(\tau)$ with support away from 0,

$$\lim_{\epsilon\downarrow 0}\int_k^\infty e^{-i\omega\tau}e^{-\epsilon\omega}\,d\omega = \int_k^\infty e^{-i\omega\tau}\,d\omega$$

and

$$\lim_{\epsilon\downarrow 0}\frac{1}{i\tau+\epsilon}e^{-ik\tau}e^{-\epsilon k} = \frac{1}{i\tau}e^{-ik\tau},$$

and similarly for the two sides of (b). But these limits simply follow from the dominated convergence theorem (or more elementary but more clumsy arguments for Riemann integrals). For example,

$$\left[\lim_{\epsilon\downarrow 0}\int_k^\infty e^{-i\omega\tau}e^{-\epsilon\omega}\,d\omega\right](f) \equiv \lim_{\epsilon\downarrow 0}\int_k^\infty e^{-\epsilon\omega}\left(\int_{-\infty}^\infty f(\tau)\,e^{-i\omega\tau}\,d\tau\right)d\omega$$

$$= \int_k^\infty \hat{f}(\omega)\,d\omega \equiv \left[\int_k^\infty e^{-i\omega\tau}\,d\omega\right](f).$$

Proposition 2: $\qquad F(i\lambda) \equiv \dfrac{1}{e^{i\lambda}-1} = \displaystyle\sum_{m=0}^\infty \dfrac{B_m}{m!}(i\lambda)^{m-1},$

where B_m are the *Bernoulli numbers*:

$$B_0 = 1, \qquad B_1 = -\tfrac{1}{2}, \qquad B_2 = \tfrac{1}{6},$$
$$B_3 = B_5 = B_7 = \ldots = 0, \qquad B_4 = -\tfrac{1}{30}, \qquad \ldots\,.$$

This Laurent series can be found in many handbooks.

Using the two parts of Proposition 1 in succession (recalling that $k_3 = n\pi/L$), we can evaluate (5.8) as

$$G_+^L = \frac{1}{4\pi L}\sum_{n=1}^\infty \frac{e^{-ik_3(t-t')}}{i(t-t')}$$

$$\times \frac{1}{2}\left[e^{ik_3(z-z')} + e^{-ik_3(z-z')} - e^{ik_3(z+z')} - e^{-ik_3(z+z')}\right]$$

$$= \frac{1}{8\pi L}\frac{1}{i(t-t')}\left\{F\left(\tfrac{i\pi}{L}[t-t'-z+z']\right) + F\left(\tfrac{i\pi}{L}[t-t'+z-z']\right)\right.$$

$$\left. - F\left(\tfrac{i\pi}{L}[t-t'-z-z']\right) - F\left(\tfrac{i\pi}{L}[t-t'+z+z']\right)\right\}. \qquad (5.9)$$

We are interested in the behavior of this function when $t \approx t'$ and $z \approx z'$.

According to Proposition 2, for $\lambda \approx 0$ we have

$$\frac{1}{i} F(i\lambda) = -\frac{1}{\lambda} + \frac{i}{2} + \frac{\lambda}{12} + \frac{\lambda^3}{720} + \cdots.$$

(Note that the second term is the only nonreal one.) This expansion may be used on the first two of the four terms inside the braces in (5.9); they are equal to

$$-\frac{1}{4\pi^2} \frac{1}{(t-t')^2 - (z-z')^2} + \frac{i}{8\pi L(t-t')} + \frac{1}{48L^2}$$
$$+ \text{ terms of order } (t-t')^2 \text{ or } (z-z')^2. \quad (5.10a)$$

Note that the first term here is precisely the function $\frac{1}{2}G^{(1)}$ for the massless free field — the term we intend to subtract.

To analyze the remaining two terms in the braces of (5.9), let

$$\delta \equiv \frac{\pi}{L}(t-t'), \qquad \zeta \equiv \frac{\pi}{L}(z+z'), \qquad C \equiv -\frac{1}{8L^2}.$$

For small δ the two terms are

$$\frac{C}{i\delta}[F(i\delta + i\zeta) + F(i\delta - i\zeta)]$$
$$= \frac{C}{i\delta}[F(i\zeta) + F(-i\zeta)] + C[F'(i\zeta) + F'(-i\zeta)] + O(\delta).$$

But note that

$$F(i\zeta) \equiv \frac{1}{e^{i\zeta} - 1} = F(-i\zeta)^*$$

for ζ real, and that $F'(i\zeta) = -e^{i\zeta}(e^{i\zeta} - 1)^{-2}$, so that

$$F'(-i\zeta) = \frac{-e^{-i\zeta}}{(e^{-i\zeta} - 1)^2} = F'(i\zeta)^* = \frac{-e^{i\zeta}}{(1 - e^{i\zeta})^2} = F'(i\zeta).$$

Thus $F'(i\zeta)$ is real, and the two terms in question reduce to

$$\frac{2C}{i\delta} \operatorname{Re} F(i\zeta) + 2CF'(i\zeta) + O(\delta). \quad (5.10b)$$

We can extract the real and imaginary parts of F by writing

$$F(i\zeta) = \frac{1}{\cos\zeta - 1 + i\sin\zeta} = -\frac{1}{2} - i\frac{\sin\zeta}{2(1 - \cos\zeta)}.$$

Also, for $z' \approx z$ we have $\zeta = 2\pi z/L + O(z - z')$, and hence

$$F'(i\zeta) = \frac{1}{4}\csc^2\left(\frac{\zeta}{2}\right) = \frac{1}{4}\csc^2\left(\frac{\pi z}{L}\right) + O(z - z').$$

Therefore, collecting the imaginary terms in the expressions (5.10), we see that, away from the light cone,

$$\mathrm{Im}[G_+^L - G_+] \equiv \tfrac{1}{2}G^L - \tfrac{1}{2}G$$

$$= \frac{1}{8\pi L(t - t')} + \frac{1}{4L^2} \times \frac{L}{\pi(t - t')} \times \left(-\frac{1}{2}\right) - 0$$

$$= 0 \quad (\text{exactly}).$$

This result was foreordained (in Chapter 4): The commutator function, G, is the same in all states. (It also could have been seen earlier from (5.9) and the fact that $\mathrm{Re}\, F(i\zeta) = -\frac{1}{2}$ for all real ζ.)

More interesting is

$$\mathrm{Re}[G_+^L - G_+] \equiv \tfrac{1}{2}G^{(1)L} - \tfrac{1}{2}G^{(1)}$$

$$\longrightarrow \frac{1}{48L^2} - \frac{1}{16L^2}\csc^2\left(\frac{\pi z}{L}\right) \quad \text{as } z' \to z,$$

where the first term comes from (5.10a) and the second from (5.10b). We make the usual, reasonable physical assumption that $\langle 0|\phi^2(\underline{x})|0\rangle = 0$ for all $\underline{x}$: after renormalization, the observable ϕ^2 vanishes identically in the "true" vacuum state. Therefore, the expectation value of ϕ^2 in the slab vacuum state is precisely the quantity we have just calculated:

$$\langle 0_L|\phi^2(\underline{x})|0_L\rangle = \frac{1}{48L^2}\left[1 - 3\csc^2\left(\frac{\pi z}{L}\right)\right]. \tag{5.11}$$

This function is the fruit of our labor. Let us examine it more closely:

(1) It is *negative*, although it purports to be the expectation value of the square of a real-valued field! This is typical of the loss of "formal properties" often suffered by field-theoretic quantities under renormalization.

(2) As $z \to 0$ or $z \to L$,

$$\langle \phi^2 \rangle \sim \frac{1}{48L^2}\left[1 - \frac{3L^2}{\pi^2 z^2} + O\left(\left(\frac{z}{L}\right)^0\right) + \cdots\right]$$

$$= -\frac{1}{16\pi^2 z^2} + O(z^0). \tag{5.12}$$

Note that the leading term is independent of L.

(3) The expansion (5.12) also applies as $L \to \infty$ with z fixed. Indeed, the whole calculation can be redone for the case of Ω equal to the half-space $\mathbf{R}^2 \times (0, \infty)$, with the result $-(16\pi^2 z^2)^{-1}$. This inverse-square term is thus the effect of a "Dirichlet wall" on the ground state of a quantized scalar field.

(4) On the other hand, at the midpoint, $z = L/2$, one has $\langle \phi^2 \rangle = -(24L^2)^{-1}$, whereas the effect of a single wall at that distance would be $-(4\pi^2 L^2)^{-1}$. Thus the ϕ^2 in the slab is not simply the naive sum of the contributions of the two walls, $-(2\pi^2 L^2)^{-1} \approx -(20L^2)^{-1}$.

(5) If you did Exercise 16(b), you got a nonzero constant answer proportional to L^{-2}, L being the circumference of the periodic universe. This result clearly can't be attributed to a wall, since no wall is present!

From these observations we may conclude that the nonzero vacuum expectation value of ϕ^2 (and of $T_{\mu\nu}$) is a combination of two quite separate physical effects: First, there is a kind of "fuzz" which hangs on the boundary (if any). The boundary condition tends to suppress the quantum fluctuations of the field in the immediate vicinity of the wall, so that the equilibrium condition of the field in that region is not the same as in open space. Second, there is an effect which is uniform throughout the space and seems to be associated with the *finite size* of the space with respect to one of its dimensions. Calculationally, this term is directly traceable to the *discreteness of the normal modes* of the field (i.e., the replacement of an integral by a sum), and its magnitude increases with the spacing of the eigenvalues (which varies inversely to the size parameter, L). It is, therefore, an effect of "quantization" in the original, literal sense of the word.

CALCULATIONS OF ENERGY DENSITY

The following discussion is based largely on DeWitt 1975, Sec. 2, which should be consulted for details and for additional physical discussion.

Since T is a quadratic functional of the field, it can be regarded as the restriction to the *diagonal set*, $\underline{x} = \underline{x}'$, of a bilinear functional $T(\underline{x}, \underline{x}') \equiv T(\phi(\underline{x}), \phi(\underline{x}'))$. Thus the ingredients of the renormalized $\langle T_{\mu\nu} \rangle$ are quantities such as

$$\lim_{\underline{x} \to \underline{x}'} \left[\langle 0_L | \partial_\mu \phi(\underline{x}) \, \partial_\nu \phi(\underline{x}') | 0_L \rangle - \langle 0 | \partial_\mu \phi(\underline{x}) \, \partial_\nu \phi(\underline{x}') | 0 \rangle \right].$$

One can move the differential operator $\partial_\mu \partial_{\nu'}$ outside the angular brackets to get formulas for these things in terms of G_+^L and G_+. *Warning:* In curved space the relation between $\langle T \rangle$ and derivatives of G_+ is more complicated: Differentiation does not exactly commute with a covariant "vacuum subtraction", and the result is the notorious *trace anomaly*. A careful prescription for renormalization of the energy-momentum tensor in curved space-time will be attained in Chapter 9.

In place of $\underline{x}$ and $\underline{x}'$ one can use the variables $\underline{x}$, σ, and t, where $t^\mu(\underline{x}, \underline{x}')$ is the unit vector pointing from $\underline{x}$ toward $\underline{x}'$. This is called the *unit tangent vector* determined by the two space-time points, because its generalization to curved space is a tangent vector to a geodesic joining the points. ("Unit" in the context of an indefinite metric means that $t^\mu t_\mu = \pm 1$. If $\underline{x}$ and $\underline{x}'$ have lightlike separation, then the unit tangent vector can't be defined.) In terms of the quantity

$$\sigma \equiv (\underline{x} - \underline{x}')^2 \equiv (t - t')^2 - (\mathbf{x} - \mathbf{x}')^2,$$

one has

$$t^\mu = - \frac{\eta^{\mu\nu} \partial_\nu \sigma}{\sqrt{2|\sigma|}}.$$

More transparently put, the gradient $\partial_\mu \sigma$ is, up to sign, a tangent vector from $\underline{x}$ to $\underline{x}'$ with "length" equal to the separation between those two points. This vector is well-defined (and smooth) even on the light cone, and it is quite convenient for calculations in curved space-time. Interpretation of a formula, however, is sometimes clearer if the coordinates σ and t^μ are used in place of $\partial_\mu \sigma$.

For the massless scalar field, a calculation yields

$$\langle 0|T_{\mu\nu}(\underline{x}, \underline{x}')|0\rangle = -\frac{1}{8\pi^2\sigma^2}(\eta_{\mu\nu} \mp 4t_\mu t_\nu), \qquad (5.13a)$$

where the sign is negative if t is timelike, positive if t is spacelike. The result for the "new improved" tensor, $\langle 0|\overline{T}_{\mu\nu}(\underline{x}, \underline{x}')|0\rangle$, is the same. In particular, if $t_\mu = (1, 0, 0, 0)$ ("point-splitting along the time axis"), then

$$\langle 0|T_{\mu\nu}|0\rangle = \frac{3}{2\pi^2}\frac{1}{(t-t')^4}\begin{pmatrix} 1 & 0 & 0 & 0 \\ 0 & \frac{1}{3} & 0 & 0 \\ 0 & 0 & \frac{1}{3} & 0 \\ 0 & 0 & 0 & \frac{1}{3} \end{pmatrix}. \qquad (5.13b)$$

It's amusing to note that this has the tensorial structure of the energy-momentum tensor of a gas in its rest frame. (The preferred frame arises because of the introduction of a particular vector t, which destroys Lorentz invariance.) Nevertheless, this term has no physical significance: Its form is an artifact of the regularization method, and it will disappear in the renormalization.

The point-split expectation values of the canonical and improved stress tensors in the Casimir slab vacuum state can be calculated in the same way as $\langle \phi^2 \rangle$. The leading term is always (5.13), which we subtract. For the improved tensor the result is

$$\langle 0_L|\overline{T}_{\mu\nu}|0_L\rangle - \langle 0|\overline{T}_{\mu\nu}|0\rangle = \frac{\pi^2}{1440L^4}\begin{pmatrix} -1 & 0 & 0 & 0 \\ 0 & 1 & 0 & 0 \\ 0 & 0 & 1 & 0 \\ 0 & 0 & 0 & -3 \end{pmatrix}$$
$$+ \text{ terms which vanish as } \underline{x}' \to \underline{x}. \quad (5.14)$$

For $\langle 0_L|T_{\mu\nu}|0_L\rangle - \langle 0|T_{\mu\nu}|0\rangle$ one gets (5.14) plus an extra term,

$$\frac{\pi^2}{48L^4}\frac{3 - 2\sin^2\left(\frac{\pi z}{L}\right)}{\sin^4\left(\frac{\pi z}{L}\right)}\begin{pmatrix} -1 & 0 & 0 & 0 \\ 0 & 1 & 0 & 0 \\ 0 & 0 & 1 & 0 \\ 0 & 0 & 0 & 0 \end{pmatrix}. \qquad (5.15)$$

Thus we have obtained renormalized values for $\langle 0_L|T_{\mu\nu}|0_L\rangle$ and $\langle 0_L|\overline{T}_{\mu\nu}|0_L\rangle$ which are independent of t^μ and are invariant under Lorentz transformations (and translations) in the $(t, \mathbf{x}_\perp)$ hyperplane. (In the following discussion, expectation values with respect to $|0_L\rangle$ are understood to be renormalized.)

Observe that $\langle 0_L|T|0_L\rangle$, like $\langle 0_L|\phi^2|0_L\rangle$, is dominated by a term, (5.15), concentrated near the walls, but that $\langle 0_L|\overline{T}|0_L\rangle$ is z-independent — it is purely a quantization effect. A number of related models have been calculated, and the variations on our theme which they provide are instructive:

(1) The energy density of an electromagnetic field is proportional to $\mathbf{E}^2 + \mathbf{B}^2$, the sum of electric and magnetic contributions. In the Casimir situation one finds that $\langle 0_L|\mathbf{E}^2|0_L\rangle$ and $\langle 0_L|\mathbf{B}^2|0_L\rangle$ separately are similar to the $\langle 0_L|T_{00}|0_L\rangle$ of the scalar field, but when added their boundary singularities cancel, leaving an electromagnetic $\langle 0_L|T_{00}|0_L\rangle$ which is constant in the z direction like $\langle 0_L|\overline{T}_{00}|0_L\rangle$.

(2) For the scalar field with a single boundary plane one gets the result of taking $L \to \infty$ in (5.14–15):

$$\langle 0_L|\overline{T}_{00}|0_L\rangle = 0, \qquad \langle 0_L|T_{00}|0_L\rangle = \frac{-1}{16\pi^2 z^4}.$$

(3) For periodic boundary conditions, either of the scalar field's stress tensors is, as one would expect, qualitatively similar to (5.14). However, for a given L (as circumference and as distance between the walls), the periodic $\langle 0_L|\overline{T}_{00}|0_L\rangle$ is 16 times as large as the Dirichlet $\langle 0_L|\overline{T}_{00}|0_L\rangle$. Calculationally, if not physically, we can quickly see why this should be: After separation of the relevant variable, the eigenfunctions of the Dirichlet problem are $\sin(n\pi z/L)$ with $n = 1, 2, 3, \ldots$ and $\omega_n{}^2 = n^2\pi^2/L^2$. Those of the periodic problem are $e^{2\pi i n z/L}$ with $n = 0, \pm 1, \pm 2, \ldots$; or, better for our present purpose, $e^{m\pi i z/L}$ with $m = 0, \pm 2, \pm 4, \ldots$ and $\omega_m{}^2 = m^2\pi^2/L^2$. Roughly speaking, there is conservation of the total number of modes: in the periodic case they are spaced twice as far apart in the variable ω, but the eigenvalues are doubly degenerate to compensate. [This equivalence is necessary for consistency with *Weyl's asymptotic formula*, which says in the one-dimensional case that $N(\omega^2)$, the number of independent modes

with eigenvalue less than ω^2, is asymptotic to $\frac{L}{\pi}\omega$ as $\omega \to \infty$. The latter follows from (4.13) (with $y = x$) in circumstances where that approximation is uniform in x.] This compensation breaks down, however, in the renormalization of expectation values. In that calculation we subtract an *integral* from a *series* involving essentially the same function. The Casimir effect is directly attributable, therefore, to the spacing between the terms in the series, which is proportional to L^{-1}. In our result

$$\langle 0_L|\overline{T}_{00}|0_L\rangle = \frac{-\pi^2}{1440L^4}$$

it is easy to see that one factor L^{-1} comes from the normalization of the eigenfunctions, leaving L^{-3} as the effect of the spacing between eigenvalues. If we double this spacing, we should therefore multiply the sum by $2^3 = 8$; another factor of 2 comes from the doubling of the modes in the periodic case. The two differences in spectral structure of the Dirichlet and periodic systems therefore work in the *same* direction here, rather than cancelling as in the Weyl formula.

(4) The argument just given may not be entirely convincing, since the two summands involved are not exactly the same. There are at least two contexts, however, where they *are* the same and the effect may be seen in its purity. One is the calculation of the *total energy* of the Casimir system, as presented, for instance, by Fierz 1960. The second is the Casimir stress tensor in *two-dimensional* space-time:

Exercise 17: In dimension 2 (that is, $d = 1$) the canonical and improved energy-momentum tensors are the same, and

$$T_{00} = \tfrac{1}{2}[(\partial_t\phi)^2 + (\partial_x\phi)^2 + m^2\phi^2].$$

Calculate the Dirichlet and periodic Casimir energy densities for a massless scalar field. (Ignore the discrete mode with $\omega = 0$ in the periodic case; see the appendix for treatment of such modes.) You should find that both densities are independent of the spatial coordinate, and that the periodic one is larger by a factor 4.

Let me take this opportunity to denounce the tendency in the physics literature to refer to Casimir-type effects as "topological". (Similar remarks have been made by Hu & O'Connor 1987.) After all,

$\mathbf{R}^2 \times (0, L)$ and $\mathbf{R}^3$ have the same topology, but in the role of Ω one of them has a Casimir vacuum energy and the other does not. On the other hand, $\mathbf{R}^2 \times (0, L)$ and $\mathbf{R}^2 \times S^1$ have different topology, but we have just seen that they have qualitatively similar vacuum energies. Numerically, the Casimir energy in a large periodic universe is closer to that of $\mathbf{R}^3$ than to that of a small universe with identical topology. As I remarked earlier, Casimir effects are really of two types: boundary effects and "quantization" or "discretization" effects. Neither of these has anything *directly* to do with topology.

Incidentally, each of these classes of effects has a counterpart of current interest in pure mathematics. The boundary effects are traceable to the high-frequency asymptotic behavior of the density of eigenvalues and of the eigenfunctions, as determined by the boundary geometry [Kac 1966; Clark 1967; Balian & Bloch 1970]. Integrated over the boundary of a compact region or manifold, the boundary corrections to the integral kernels of various operators yield contributions to the *index formulas* which relate the analysis, the geometry, and, yes, the topology of the space in question [Atiyah et al. 1973; see other references on the index theorem in Chap. 9]. The quantization effects stem ultimately from a tendency of the eigenvalue density to be *periodic* at high frequency. This tendency can be related to the existence of closed geodesics (or, more generally, closed "classical paths"); see Duistermaat & Guillemin 1975 and Balian & Bloch 1972, 1974. In a periodic universe we have closed paths of length L (the circumference), but in the Casimir slab the shortest path has length $2L$, since the "particle" must bounce off both walls before returning to its original position with its original momentum. This precisely agrees with our observation on the relative spacing of the eigenvalues in the two systems.

FORCE CALCULATIONS

The *total energy* per unit area between the walls, if we accept the new improved stress tensor for the scalar field, is

$$E(L) \equiv \int_0^L \langle 0_L | \overline{T}_{00} | 0_L \rangle \, dz = \frac{-\pi^2}{1440L^3} \, .$$

If the walls are themselves physical objects capable of motion, there should be an attractive force between them,

$$F = -\frac{\partial E}{\partial L} = \frac{-\pi^2}{480L^4} \, .$$

The result is similar for the more realistic case of an electromagnetic field, where the Dirichlet walls are replaced by perfectly conducting plates (see below).

If we attempt this calculation with the canonical stress tensor for the scalar field, (5.14) plus (5.15), we run into a z^{-4} singularity at each wall, which is nonintegrable. However, this term is independent of L, so it is fair play to ignore it in computing the force. Furthermore, the next-order terms, proportional to $(z^2 L^2)^{-1}$, miraculously cancel when the sine functions are expanded; so we do get a finite force.

SOME HISTORY

AND SOME CONTACT WITH THE REAL WORLD

A good review of the Casimir effect at the end of what might be called the subject's "classical" period is given by Boyer 1970.

In the beginning, Casimir 1948 was trying to calculate the Van der Waals force between two polarizable atoms. The charge fluctuations in one atom can create an electric field capable of polarizing the charge in the other atom, so that there is a net force between them:

To simplify the analysis, he considered the similar force between an atom and a conducting plate:

From there he was led to the problem of two parallel plates. It was found that to get agreement with experiment in any of these problems, it was necessary to take into account the finiteness of the speed of light, which implies that the influence of charge fluctuations in one part of the system reaches the other part only after a delay. This means that the energy stored in the electromagnetic field passing between the two bodies must be taken into account. In fact, it turned out that the *long-range limit* of the force can be associated *entirely* with the change in the energy of the field as the distance between the bodies varies; the charge fluctuations faded into the background of the analysis.

The field energy is, formally, a sum over the modes (eigenfunctions) of the field, of the sort we considered above. This sum is divergent. In calculating the energy *change* ΔE, therefore, Casimir (and other authors, such as Fierz and Boyer, who later simplified his treatment) inserted an *ad hoc* convergence factor, usually $e^{-\omega_n/\Lambda}$ with Λ a large constant, to make the two energies finite. Their difference, ΔE, then (fortunately) turned out to have a finite limit as $\Lambda \to \infty$. [From a mathematical point of view, this is a kind of *Abel summation* (see Hardy 1949). One finds, however, that replacing ω_n by n in the exponential would give a different answer! The choice of a particular summation prescription ultimately requires some physical justification.] This exponential cutoff is algebraically equivalent to an analytic continuation of our point separation along the time axis: take $\Lambda = i(t - t')$.

The physical interpretation offered for this procedure was this: No physical conductor is perfect at arbitrarily high frequencies. A very high-frequency wave should hardly notice the presence of the plates at all. Therefore, in a real experiment ΔE *must* be finite. An integral or sum defining it must have an effective cutoff depending on the detailed physics of the materials; $e^{-\omega_n/\Lambda}$ is a plausible model. The fact that the result is independent of Λ in the limit of large Λ suggests that this model is roughly correct and a more detailed model is unnecessary for a basic understanding.

The attraction between smooth, very close surfaces was eventually demonstrated experimentally. (Most instructive of the experimental papers is Tabor & Winterton 1969.) Actually, for technical reasons dielectric materials were used instead of conductors. The theory of the effect for dielectrics is more complicated than for conductors but leads to similar results.

The existence of literally pointlike charged particles has always been somewhat problematic within the framework of classical electrodynamics. In view of the negative vacuum energy in the configuration of conducting parallel plates, Casimir proposed a model of the electron as a charged sphere with properties like those of a macroscopic conductor. The hope was that an attractive force arising from the dependence of the vacuum energy on the sphere's radius would balance the electrostatic self-repulsion of the charge distribution, thereby holding the electron together stably. Boyer 1970 (and also Davies 1972) succeeded in calculating a conducting spherical shell's vacuum energy. (He actually put the shell inside a larger concentric sphere and found the total energy of that entire finite-volume configuration as a function of the radii.) Boyer's results were fatal to the Casimir electron model; the Casimir force turned out to be *repulsive* in this case:

$$F = -\frac{\partial E}{\partial R} > 0.$$

Furthermore, he found that the energy associated with the presence of a spherical conducting shell is infinite; that is, the energy difference between a configuration with an inner shell and the one without did not converge as the cutoff was removed. In the presence of a *curved* conducting boundary the electromagnetic field behaves as the scalar field with canonical stress tensor does at a flat boundary. (Boyer's calculation was not localized, so it was not then obvious that the divergent energy is concentrated at the sphere in analogy to (5.12) and (5.15). This does become clear from the work to be discussed next.)

The rise of interest in quantum field theory in curved space-time in the 1970s attracted renewed attention to the Casimir effect, as a more tractable model of field-theoretic effects associated with the geometry of space. It was in this new era that calculations of the energy-momentum tensor, not just the total energy, were made, and the question of the geometrical covariance of the cutoff procedure was raised.

Among those who approached the subject from a general-relativistic motivation were Deutsch & Candelas 1979; this work was completed in Candelas 1982. They consider boundary contributions (*not* quantization effects) for general curved surfaces, and for several types of quantum field. They avoid eigenfunction expansions by working directly with the Green functions of the elliptic operators involved.

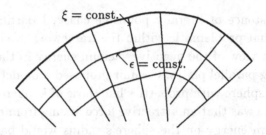

For $x \in \Omega$ and close to the boundary of Ω, let ξ be the point on the boundary closest to x and ϵ be the distance from x to ξ. Then (ξ, ϵ) provides a convenient coordinate system in the vicinity of the boundary (which we assume smooth, for the time being). Deutsch and Candelas *assume* that near the boundary the renormalized stress tensor has an expansion of the form

$$ T_{\mu\nu}(x) \sim \sum_{n=-4}^{\infty} A_{\mu\nu}^{(n)}(\xi)\, \epsilon^n . $$

Note that this is consistent with our findings for the flat plate, where $\epsilon = z$. They also *assume* that $A^{(n)}(\xi)$ depends only on the geometry of the boundary at ξ; that is, it can be expressed in terms of a function defining the surface and its derivatives, evaluated at ξ only. Geometrical covariance and dimensional analysis then imply that each $A^{(n)}$ is a linear combination of finitely many scalar quantities built out of the second fundamental form of the surface at ξ.

[For a quick definition of the *second fundamental form*, κ, let us assume without loss of generality that ξ is the origin and the inward normal is the positive z axis. Near the origin, the surface is given by some equation of the form $z = f(x, y)$, which has a Taylor expansion $z = Ax^2 + Bxy + Cy^2 +$ third-order terms. This can be diagonalized by a rotation:

$$ z = \kappa_1 x^2 + \kappa_2 y^2 + \cdots , $$

where the eigenvalues of the second fundamental form, κ_i, are also called the *principal curvatures* of the surface at ξ.]

I have little doubt that the assumptions of Deutsch and Candelas are correct, at least as regards negative values of n (the only values relevant to their arguments). They should be provable in analogy with similar facts about the kernel of the heat operator [McKean & Singer 1967, Greiner 1971, Gilkey 1979], but I do not believe that there is any rigorous treatment in the literature. The various terms in their expansion in fact correspond to terms in the extension of Weyl's asymptotic formula for the eigenvalue distribution $N(\omega^2)$; this extension is not a true asymptotic expansion but holds in various "averaged" senses [Brownell 1955; Hörmander 1968].

Deutsch and Candelas were then able to calculate the coefficients in their general series by matching it against various special cases for which the answers were known or could be easily found. (This part of the argument is quite similar to the way in which McKean & Singer and Gilkey investigate the integrated trace of the heat kernel on a manifold with boundary.) For $\Omega \subset \mathbf{R}^3$ they find a hierarchy of terms:

	Invariant		*Order in $T_{\mu\nu}(x)$*	*Order in $N(\omega^2)$*
1.	$\int 1$	(surface area)	ϵ^{-4}	ω^2
2.	$\int (\kappa_1 + \kappa_2)$	(trace)	ϵ^{-3}	ω
3a.	$\int \kappa_1 \kappa_2$	(determinant)	ϵ^{-2}	1
3b.	$\int (\kappa_1 - \kappa_2)^2$	[see below]	ϵ^{-2}	1
4.	cubic terms		ϵ^{-1}	ω^{-1}

Higher-order terms are of order ϵ^0, hence nonsingular. Note that $(\kappa_1 - \kappa_2)^2$ is a linear combination of the determinant and the square of the trace of the fundamental form. (Alternatively, it and the determinant can be expressed as linear combinations of $(\operatorname{tr} \kappa)^2$ and $\operatorname{tr}(\kappa^2)$.) Each of these terms is multiplied by a numerical coefficient which depends on the type of field under consideration.

Since the total vacuum energy is $E \equiv \frac{1}{2} \int \omega \, dN(\omega^2)$, the terms in $N(\omega^2)$ that go as ω^p with $p \geq -1$ cause divergences in E. Clearly, this is consistent with the result of integrating over ϵ: Negative powers of ϵ give infinities in E, though not in $T_{\mu\nu}(x)$ for $\epsilon \neq 0$.

It is quite illuminating to observe what happens to each of these terms in the special cases of Casimir's slab and Boyer's sphere.

For the slab, $\kappa_1 = 0 = \kappa_2$, so the only term in the asymptotic series is the first one. As already remarked, the coefficient of this term is zero for an electromagnetic field, and hence the Casimir energy (per unit area) is finite in Casimir's problem.

For the sphere:

1: Again, this term does not exist for the electromagnetic field (though it does for other fields of less importance to macroscopic physics).

2,4: Being odd functions of κ, these cancel between the inside and outside of any thin shell.

3a: This term is independent of the radius of the sphere, by the Gauss–Bonnet theorem [e.g., O'Neill 1966, Sec. 7.8]. Therefore, although it makes the energy infinite even after renormalization, it does not contribute to the force.

3b: This vanishes for a sphere ($\kappa_1 = \kappa_2$). In the general case, Candelas showed that the coefficient of this term is negative.

The conclusion is that the finiteness of the Casimir force for the slab and the sphere is an accident of their geometries. In general, the electromagnetic force on a perfect conductor will turn out infinite. Worse, the negativity of term 3b means that a spherical shell is (infinitely) unstable against wrinkling! The perfect-conductor boundary condition, therefore, must be judged to be a pathological idealization in this context. Good physics requires that the infinite terms be replaced by *cutoff-dependent* terms related to the detailed properties of realistic materials.

This conclusion is not universally accepted. An alternative viewpoint, presented by Kennedy et al. 1980, is that the infinite energy near the boundary is cancelled by a term of opposite sign concentrated *on* the boundary. This philosophy originated in a procedure of "zeta-function renormalization" which is guaranteed to make the total energies of reasonable, bounded systems always finite.

Another paper on Casimir-type calculations which deserves to be mentioned is that of Bender & Hays 1976. It presents another method of regularization with some computational advantages. The motivation of that paper was the "bag model" of elementary particles.

Lessons from the Casimir effect

First of all, it shows that renormalized vacuum energy is experimentally detectable! No worker in the field of overlap of quantum theory and general relativity can fail to point this fact out in tones of awe and reverence.

Second, it ratifies our working philosophy that field concepts have primacy over particle concepts. The "vacuum" of the slab system, Fock-quantized according to Chapter 3, is not really empty space. It can be thought of as a "sea" or "soup" of particles of the free theory in equilibrium with the walls. (Another instance of this picture of the vacuum occurs for the (scalar) meson field around two static point sources representing nucleons [Mandl 1959, Chap. 4]. The ground state of the theory with sources can be interpreted as containing a meson "cloud" around each nucleon. The variation of the total energy with the distance between the nucleons gives rise to the celebrated *Yukawa force*.)

In more general dynamical situations, however — in particular, in curved space — we have no well-defined "free" theory to compare against. Although we are not physically justified in regarding the vacuum state (or, perhaps better, "ground state") of such a model as *empty*, we also do not have a clear notion of *particles* to describe what the vacuum does contain. We therefore aim to develop a method for defining renormalized field observables, such as $T_{\mu\nu}$, which does not require subtracting the value of the observable in some fixed "true" vacuum state. Only then will we have escaped from the tyranny of the particle concept to a completely field-theoretic understanding of quantum field theory.

Chapter 6
Quantum Field Theory in General Space-Times

After these lengthy preliminaries, we are finally ready to discuss curved space-time.

CANONICAL FORMALISM

The standard free field in flat space-time is governed by the Klein–Gordon equation,

$$\frac{\partial^2 \phi}{\partial t^2} - \nabla^2 \phi + m^2 \phi = 0.$$

An action integral which generates this equation is

$$S = \int_{\mathbf{R}^N} d^N x \; \tfrac{1}{2} [\eta^{\mu\nu} \partial_\mu \phi(\underline{x}) \partial_\nu \phi(\underline{x}) - m^2 \phi(\underline{x})^2],$$

where η is the metric tensor of flat space-time, (5.4), and $N \equiv d + 1$ is the space-time dimension. (There was a time when to suggest that N could be something other than 4 was to mark oneself as a mathematician dabbling in physics. Nowadays, under the influence of Kaluza–Klein and superstring theories, no apology is necessary.)

The reader who is unfamiliar with the extension of the Lagrangian formalism of Chapter 1 to field theories would benefit from reading Hill 1951. Suffice it to say here that the integrand of S is called the *Lagrangian density*, $\mathcal{L}$, and that the space-time integration combines the summation over degrees of freedom, which creates a Lagrangian in the sense of Chapter 1, with the time integration which forms the action from the Lagrangian. Thenceforth, in much of the formalism the space and time variables are treated on an equal footing.

We shall generalize the Klein–Gordon Lagrangian [density] to curved space-time, and then quantize the resulting theory "canonically" in analogy with quantum mechanics.

A geometrically covariant generalization of $\mathcal{L}$ to a (pseudo-)Riemannian manifold is

$$L = \tfrac{1}{2} [g(\underline{x})^{\mu\nu} \nabla_\mu \phi(\underline{x}) \nabla_\nu \phi(\underline{x}) - m^2 \phi(\underline{x})^2 - \xi R(\underline{x}) \phi(\underline{x})^2], \qquad (6.1)$$

where ξ is a dimensionless constant (to be discussed later). $\nabla_\mu \phi$ is the covariant derivative of ϕ, which reduces to the literal partial derivative of ϕ since ϕ is a scalar function. (The specification of covariant derivatives by the Levi–Civita connection — alias the Christoffel symbols — becomes significant when second and higher derivatives appear.) R is the curvature scalar of the manifold (the trace of the Ricci curvature tensor). Since L is scalar, it must be multiplied by the square root of

$$g \equiv |\det(g_{\mu\nu})|$$

to produce a density suitable for integration over the manifold:

$$S = \int_{\mathcal{M}} L \sqrt{g}\, d^N x \equiv \int_{\mathcal{M}} \mathcal{L}\, d^N x.$$

Exercise 18: Show that the Euler–Lagrange equation associated with (6.1) is

$$\Box \phi + (m^2 + \xi R)\phi = 0, \tag{6.2}$$

where $\Box$ is the *covariant d'Alembertian operator* (the indefinite-metric analogue of the *Laplace–Beltrami operator*)

$$\Box \phi \equiv g^{\mu\nu} \nabla_\mu \nabla_\nu \phi \quad \left(= g^{\mu\nu}(\partial_\mu \nabla_\nu \phi - \Gamma^\rho_{\mu\nu} \nabla_\rho \phi)\right)$$

$$= \frac{1}{\sqrt{g}}\, \partial_\mu \left[g^{\mu\nu} \sqrt{g}\, \partial_\nu \phi\right]. \tag{6.3}$$

It will be noted that (6.1) is not the only covariant functional which reduces to the Klein–Gordon Lagrangian when the metric is flat (even when we restrict attention to actions which lead to *linear* field equations for ϕ). If we are to include a term proportional to $R\phi$ in (6.2), why not one proportional to $R^2 \phi$? Indeed, why not one proportional to $R^{\mu\nu} \partial_\mu \partial_\nu \phi$? ($R^{\mu\nu} \equiv R^\alpha{}_{\mu\alpha\nu}$ is the Ricci tensor (the trace of the Riemann tensor), in the notation usual in physics. Cf. Chapter 8.) This last proposal would be pathological in that it would move the characteristic curves of the hyperbolic field equation away from the light cones of the space-time, perhaps rendering the equation nonhyperbolic if the curvature became too strong. Generally speaking, one seeks for the *simplest* generalization of a special-relativistic theory to a curved-space setting, and considers any of the unlimited number of possible additional terms only upon encountering a good reason to do so (experimental, technical, or philosophical).

The proper question, therefore, is why one should bother to include the ξR term. There are at least two reasons. First, when $m = 0$ and ξ has a particular numerical value ($\xi_{\text{conf}} = \frac{1}{6}$ when $N = 4$; $\xi_{\text{conf}} = 0$ when $N = 2$), the action and equation of motion exhibit *conformal invariance*: A solution in space-time with the metric $g_{\mu\nu}(x)$ is also, after a simple rescaling, a solution in the metric $\Omega(x)g_{\mu\nu}(x)$, where Ω is any given positive function. (A precise theorem to this effect is stated at the end of this chapter.) Therefore, a case can be made that ξ_{conf} is a "simpler" or "better" choice of ξ than $\xi = 0$. Second, it is known [e.g., Bunch et al. 1980] that the renormalization of a theory of an *interacting* field (in particular, one with a term $\lambda\phi^4$ in the Lagrangian) in curved space-time will necessarily involve a *counterterm* proportional to $R\phi^2$. In that context it is mandatory to regard ξ as a free parameter of the theory, at least until after the renormalization (subtraction of counterterms from "bare" terms in the original action or field equation) has been carried out.

Definition: A scalar field satisfying (6.2) in N-dimensional space-time is said to be *minimally coupled* if $\xi = 0$, *conformally coupled* if

$$\xi = \xi_{\text{conf}} \equiv \frac{N-2}{4(N-1)}.$$

A Lagrangian, besides defining the dynamics of the field itself, tells how that field influences the behavior of the gravitational field (or metric tensor, or geometry). The latter is governed in classical general relativity by *Einstein's equation*,

$$G_{\mu\nu}(x) \equiv R_{\mu\nu} - \tfrac{1}{2}Rg_{\mu\nu} = GT_{\mu\nu}(x), \qquad (6.4)$$

where G is Newton's gravitational constant, suitably (but not quite traditionally) normalized. The tensor $T_{\mu\nu}$ represents the energy, momentum, and pressure of all forms of matter (all fields except the gravitational). This gravitational field equation arises as the Euler–Lagrange equation associated with variations with respect to $g_{\mu\nu}$ of the total action, $S = S_{\text{grav}} + S_{\text{matter}}$. The matter action consists of a term $\int \mathcal{L} \, d^N x$ for each dynamical matter field in the theory. Einstein's gravitational action is

$$S_{\text{grav}} = \frac{1}{G} \int R \sqrt{g} \, d^N x.$$

It follows that

$$T_{\mu\nu} = \frac{2}{\sqrt{g}} \frac{\delta}{\delta g^{\mu\nu}} S_{\text{matter}}.$$

[Later we'll see that quantum field theory strongly suggests that S_{grav} be modified by addition of terms quadratic in the curvature tensor, leading to new terms in the gravitational field equation that are of fourth order in the metric. The reason (as far as this book is concerned) is the same as that for the presence of the ξR term: Counterterms of these types arise in renormalizing the quantum theory of the matter fields. This issue concerns the precise form of the gravitational half (left-hand side) of the gravitational equation; it does not affect the formal expression for the energy-momentum tensor, which is our present concern.]

Exercise 19: Carry out the variation of S_{matter} for Lagrangian (6.1), and show that in the limit of flatness, with $N = 4$, T becomes the canonical energy-momentum tensor (5.3) if $\xi = 0$, but becomes the new improved tensor of Callan, Coleman, and Jackiw (5.5) when $\xi = \frac{1}{6}$.

To obtain canonical commutation relations we must break the relativistic union of space and time which the Lagrangian formalism has so obligingly assimilated. (This assertion will be qualified later in the chapter.) We need a time coordinate. Let us therefore suppose that a coordinate system $\{x^0, x^1, \ldots, x^d\}$ (possibly only local) is given in $\mathcal{M}$, and that the coordinate $x^0 \equiv t$ is *timelike*, in the sense that the surfaces $\{t = \text{constant}\}$ are spacelike. That is, the gradient $\nabla_{\tilde\mu} t$ is a timelike vector.

Remark: This condition is not the same thing — if g_{0j} is not 0 for all $j > 0$ — as timelikeness of the tangent vector to the curves of constant spatial coordinates, $d\tilde{x}^\mu/dt$. The former vector field is covariant, the latter contravariant; they happen to have the same coordinates, $(1,0)$, in the particular coordinate system concerned. See the appendix for further discussion. (Readers unaccustomed to the abstract index notation of classical differential geometry should be warned that the "μ" in this discussion does not refer to any particular coordinate, or even to any particular coordinate system. The expressions $\nabla_{\tilde\mu} t$ and $d\tilde{x}^\mu/dt$ indicate how these vector fields — well-defined abstract objects — can be computed in any given coordinate system. The tildes are intended to distinguish this general coordinate system from the one under discussion, which contains t.)

Now we can define the canonical momentum,

$$\pi(\underline{x}) \equiv \frac{\partial \mathcal{L}}{\partial(\partial_0 \phi)} = \sqrt{g}\, g^{0\nu} \partial_\nu \phi. \tag{6.5}$$

Note that this is essentially the normal derivative of ϕ off a hyper-surface $\{t = \text{constant}\}$. Relative to one of those submanifolds, π is a density (since it is the appropriate component of a $(N-1)$-form). That is, $\int_t \pi f\, dx^1 \cdots dx^d$ is a well-defined scalar integral, the Jacobian of a coordinate change in the hypersurface being absorbed in the $\sqrt{g}$. That factor also conspires with $g^{0\nu}$ to keep π unchanged under rescalings of t. On the other hand, changing the *direction* of time does reverse the sign of π.

It is therefore consistent to impose the canonical relation

$$[\phi(t,x), \pi(t,y)] = i\delta(x-y), \tag{6.6}$$

along with the usual conditions that $[\phi, \phi] = 0 = [\pi, \pi]$ at equal times. Here δ is an ordinary Dirac delta distribution with respect to the d-dimensional *coordinate* variables, x and y. As in (3.23), the canonical commutation relation can be "covariantized" by redefining δ and π by a factor which is a density on the submanifold.

Exercise 20: Calculate the Hamiltonian,

$$H \equiv \int d^d x\, (\pi \partial_0 \phi - \mathcal{L}),$$

and verify that the Hamilton–Heisenberg equations,

$$\frac{\partial \phi}{\partial t} = i[H, \phi], \qquad \frac{\partial \pi}{\partial t} = i[H, \pi],$$

are equivalent to (6.2) and (6.5), if (6.6) is applied formally to the classical expression H. (The integration is over a surface of constant t; ignore the "global" issues that this raises.)

GLOBAL HYPERBOLICITY

So far, all our discussion has been local. To grasp the general solution of (6.2), however, we need to have a well-posed Cauchy prob-lem. (This will result in a complete and nonredundant set of "normal modes" if we are so lucky as to be able to solve the equation explicitly by separation of variables.)

Definitions: A *causal* curve is one whose tangent vector is everywhere timelike or lightlike (null). A *Cauchy surface* for a pseudo-Riemannian manifold $\mathcal{M}$ is a spacelike hypersurface which is intersected exactly once by every inextendible causal curve in $\mathcal{M}$.

We shall assume that our space-time $\mathcal{M}$, as a whole, satisfies the conditions of the following theorem:

Theorem: The following are equivalent:

(A) $\mathcal{M}$ possesses a Cauchy surface.

(B) $\mathcal{M}$ is *globally hyperbolic*. (See the references for the definition of this technical concept. Roughly speaking, it says that the region enclosed between a forward light cone and a backward light cone is never a noncompact set, but it more precisely concerns the space of curves connecting the vertices, rather than the region of points between them.)

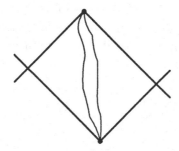

(C) $\mathcal{M}$ is of the form $\mathbf{R} \times M$ (at least homeomorphically, if not diffeomorphically). That is, $\mathcal{M}$ is *foliated* into $(N-1)$-dimensional submanifolds, all of which are topologically the same. Moreover, *each* leaf of this foliation is a Cauchy surface.

References: Geroch 1970; Choquet-Bruhat 1968; Hawking & Ellis 1973, Chap. 6; O'Neill 1983, Chap. 14.

We can assume that our Cauchy surfaces are represented, at least locally, by equations $t = $ constant.

To clarify the situation, here are some examples of spaces that are *not* globally hyperbolic. In each sketch, the surface S would like to be a Cauchy surface, but can't be.

1. Any space-time containing closed causal curves. In such a space solutions of the field equation will fail to exist for many perfectly normal Cauchy data sets, simply because the local solutions will not match up when they meet themselves "on the far side of the world". It is not clear that quantum field theory is possible in such a space. A pair of spacelike-separated points can also be timelike-separated; do the field operators at those two points commute?

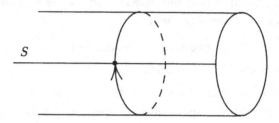

2. A fragment of $\mathbf{R}^N$ such as Casimir's slab. We already know that *boundary conditions* are necessary to get a well-defined dynamics here. If we were to attempt to solve the field equation on the basis of initial data only, it is clear from the embedding of the space inside $\mathbf{R}^N$ that waves, hence "news" unpredictable from the initial data, will run in and out of the region through the boundary.

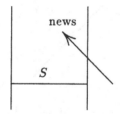

3. *Anti-deSitter space* [Fulling 1972, Sec. 3.6; Avis et al. 1978]: Two-dimensional anti-deSitter space is the covering space of a hyperboloid in pseudo-Riemannian $\mathbf{R}^3$, the time dimension being the erstwhile periodic one. (There are higher-dimensional analogues.) Unlike the previous example, this space-time is geodesically complete — there is no "edge" that a particle or wave will fall over after a finite distance of travel. Nevertheless, the light cones flare out in such a way that particles can exit from the space — and, more to the point, news can come into the space — within a finite time. The situation can be made clearer by a conformal mapping of the space onto a strip of finite width, on which 45-degree diagonal lines are light cones and horizontal lines are unsuccessful would-be Cauchy surfaces. This model is static, and one could therefore attempt to quantize fields in it along the lines of Chapter 4. However, the incompleteness of its dynamics will show up as the lack of self-adjointness of the associated elliptic operator K.

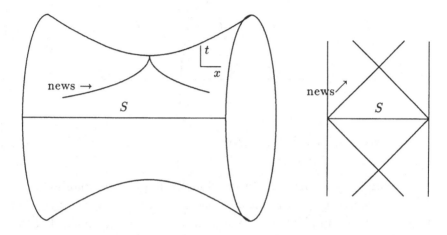

4. A *naked singularity* is the anticipated end of the evaporation [Hawking 1974, 1975] of a black hole, insofar as the quantization of gravity is neglected. The sketch (below) shows a collapsing cloud of matter, across which stretches what would be a Cauchy surface if the space-time were suitably truncated somewhere below the singular point. Points inside the future light cone of the singularity can be reached by news coming out of the singularity. We have no settled physical theory to tell us what comes out of this singularity (or any other naked singularity). We do not know how the commutator of a quantum field should behave there.

The situation can be summarized by a hierarchy of classes of space-time models.

First, we have the static space-times (excluding pathological cases such as anti-de Sitter space). In these, as I shall show in detail presently, (scalar) quantum field theory is a special case of the general construction in Chapters 4 and 5. The unsolved problems in that area are primarily technical ones, such as how to calculate $\langle T_{00} \rangle$ and other expectation values accurately and efficiently. [There is also the celebrated subject of how to interpret the situation when a space-time, or a part of it, is static with respect to two different time-translation symmetry

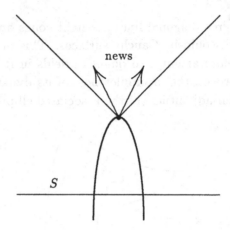

groups; see the long list of references on Rindler and Schwarzschild spaces later in this chapter.]

Second, there are nonstatic, globally hyperbolic manifolds, and also the nonstatic spaces that give rise to well-posed *mixed* hyperbolic problems when supplemented by proper boundary conditions. For these we still need to answer the questions, "What *states* are physical, and how are the renormalized physical observables *defined*?" (These issues will be investigated (and partially resolved) in Chapters 7 and 9.) Nevertheless, the field equation and the commutation relations are known. (In more technical terms, the *algebra* of the theory is understood — up to technical niceties such as the exact specification of the class of admissible test functions — and only the choice and physical interpretation of *representations* of the algebra remain to be decided.)

Finally, there are the space-times that are not globally hyperbolic. In this case, as remarked for several of the examples above, we do not know the algebra of the theory, because we don't know how to write down the commutators of the field.

Theorem: [Leray 1953; Choquet-Bruhat 1968; Dimock 1980] The Cauchy problem for the field equation (6.2) on $\mathcal{M}$ is well-posed for data (ϕ, π) on S if S is a Cauchy surface for $\mathcal{M}$. Indeed,

$$\phi(t, x) = -\int_S d^d y \left[G(\underline{x}, \underline{y})\, \pi(0, y) - \sqrt{g}\, g^{0\mu}(\underline{y})\, \frac{\partial}{\partial y^\mu} G(\underline{x}, \underline{y})\, \phi(0, y) \right],$$
$$(6.7)$$

where $G = G_{\text{adv}} - G_{\text{ret}}$, G_{adv} and G_{ret} being the advanced and retarded fundamental solutions, which satisfy

$$(\Box + m^2 + \xi R)G = \frac{\delta(\underline{x} - \underline{y})}{\sqrt{g}} .$$

Exercise 21: Show that the canonical commutation relations are equivalent to the *covariant commutation relation*

$$[\phi(\underline{x}), \phi(\underline{y})] = iG(\underline{x}, \underline{y}). \tag{6.8}$$

(Here x^0 and y^0 are not (necessarily) equal.) *Hint:* Since both sides of (6.8) satisfy the homogeneous field equation (6.2) in each variable, what needs to be proved is that the distributional initial data of (6.8) coincide with (6.6).

Peierls 1952 showed that (6.8) is related to the classical *Poisson brackets* of the field evaluated at various space-time points. DeWitt 1965, 1984 has stressed that this makes it possible in principle to formulate quantum field theory without making the relativistically repugnant splitting of $\mathcal{M}$ into time and space.

GENERAL CONSIDERATIONS

We have seen that a scalar field theory in a globally hyperbolic space-time model is defined by the field equation (6.2) and the commutation relations (6.6), or alternatively by (6.2) and the commutation relation (6.8). Both ingredients can be thought of as being specified by the Lagrangian, (6.1)

After questions of principle have been settled, the noninteracting scalar field is of little interest to physicists. The experimental world involves other kinds of fields, and those fields interact with themselves and each other, resulting in nonlinear and coupled equations of motion.

In the presence of nonscalar fields, the canonical quantization procedure requires some ad hoc modifications. In a consistent quantum theory of spinor fields (representing particles of spin $\frac{1}{2}$ or some other half-integer), the basic algebraic relations must involve *anti*commutators — more about this in the next chapter. Fields of zero mass and spin greater than or equal to 1 are gauge fields; extra, unphysical degrees of freedom must be suppressed. Furthermore, non-Abelian gauge fields and the gravitational field itself have inherently nonlinear self-interactions stemming from their geometrical interpretations, so that our preliminary game of developing "free" field theories seems inappropriate. There is a vast and growing literature on the quantization of higher-spin fields, and of interacting fields, in curved space. The details are beyond the scope of this book; I offer here only some philosophical remarks.

Most theoretical physicists, following Schwinger, regard the *action principle* as fundamental. Theories are defined by Lagrangians. DeWitt argues that the naive canonical approach should be generalized and superseded by a covariant treatment in the spirit of Peierls. Others, in the tradition of Dirac and Bergmann, treat gauge fields as infinite-dimensional analogues of mechanical systems with constraints.

There is, however, another point of view, more consistent with the spirit of axiomatic field theory. *Any* commutation or anticommutation relations consistent with the dynamics can define a possible model. (Recall from Chapter 1 the representations of the Lie algebra of the rotation group, which arise as observables whenever particles have spin. They are clearly an ingredient independent of the canonical apparatus of Poisson brackets and Heisenberg (position-momentum) commutators.) In this approach a formal theory consists of equations of motion plus commutation rules (or some more general algebraic relations). These need not determine each other, but it is a nontrivial requirement that they be mutually consistent. Together they determine an algebra of field operators (or more general observables), modulo technical questions about topologies, test functions, etc. The more mathematical aspect of the problem of theoretical physics is to make sense out of these formal objects (at some greater or lesser level of rigor). The more physical (i.e., experimentally oriented) aspect is to choose a model which is a correct description of the world.

This philosophy has the advantage and the disadvantage of any general framework. It admits any model that has a chance of being correct, but it provides no guidance on how to choose the correct model. The action principle is complementary to this: It suggests a narrowed range of theories to be investigated. It has often proved to be a trustworthy guide.

In the context of curved space-time, formulating even such a general framework involves nontrivial questions. A natural first step is to try to write down generalizations of the basic principles of special-relativistic quantum theory which are tenable in the presence of a background gravitational field. This was attempted in the present author's thesis (Fulling 1972, Chap. 4 and Sec. 7.6) and, more extensively, by Isham 1978 and Dimock 1980. Isham gave a curved-space version of the Wightman axioms for quantum field theory; Dimock gave a curved-space version of the Haag–Kastler axioms for general quantum theory (in terms of algebras of local observables).

Isham and Dimock proceeded to give a rigorous implementation of their axioms for the canonical linear scalar field. (The latter author also treats a spinor field, in Dimock 1982.) Their construction is similar to the treatment of static models in our Chapter 3: They start with an *arbitrary* representation of the canonical commutation relations on an arbitrary Cauchy surface, S. The time evolution dictated by the hyperbolic equation of motion (cf. (6.7)) is then applied to the test functions inside the arguments of the canonical field operators. By duality, this defines a time evolution of the operators, and one verifies that the result is an operator-valued solution of the equation of motion, satisfying the covariant commutation relation (6.8) — hence a Heisenberg-picture field. This completes a rigorous construction of the field *algebra*. But, of course, this construction is not unique; the resulting *representation* of the algebra is not guaranteed to have any physical relevance (in the nonstatic case). One is left with the problem of which representation to choose, whose resolution requires an investigation of observables and their expectation values (cf. later chapters).

This is a convenient place to remark that there are several ways (ultimately equivalent) to construct Fock-like representations of a field in curved space-time. Translations among various of these formalisms are worked out by Panangaden 1979. Within each such approach there are many inequivalent representations, which must be distinguished by physical considerations, not mathematical formalism.

1. The most straightforward generalization of the procedure employed in the free and static cases is to seek a splitting (direct-sum decomposition) of the space of complex-valued solutions of the field equation into two parts, generalizing the positive-frequency and negative-frequency solutions of the static case. Each half of the solution space will contain one solution for each admissible choice of initial data $\phi(t_0, x)$ on a Cauchy surface S. Usually this

splitting is combined with a choice of a basis for the space of initial data, resulting in a basis of "normal modes" for the classical solutions. The solutions of the two types are associated with annihilation and creation operators, respectively, in the operator solution of the field equation. Concrete examples of this construction will appear in the next chapter and the appendix. Literature on this approach includes DeWitt 1975, Hájíček 1977, and Birrell & Davies 1982.

2. These decompositions of the space of complex solutions are in one-to-one correspondence with *complex structures* on the space of real solutions. That formulation has been pursued by Segal 1974, Ashtekar & Magnon 1975a, and Kay 1978, among others.

3. Deutsch & Najmi 1983 have pointed out that the various choices of a "positive-frequency" subspace of the classical solutions can be parametrized by the operator relating the initial data for π on S to that for ϕ. This condition formally resembles the Robin boundary condition in potential theory, except that the normal derivative is in a timelike direction instead of in space.

4. As remarked in Chapter 4, a choice of representation is tantamount to a choice of symmetrized Green function, $G^{(1)}$. Attempts to take the Green function as the basic object go back to Lichnerowicz 1964.

In appraising the physical significance of these constructions (and especially of attempts to make them unique on a given Cauchy surface by adding conditions), it should be remembered that there are time-independent systems for which *no* Fock representation (or "particle definition") seems physically appropriate. See the appendix and Ashtekar & Magnon 1975b.

SPECIALIZATION OF THE GENERAL FORMALISM TO THE STATIC CASE

Definition: A metric is *static* if, in appropriate coordinates with $t \equiv x^0$ timelike,

(a) $g_{\mu\nu}(\underline{x})$ is independent of t, and

(b) $g_{0j}(\underline{x}) = 0$ for $j = 1, \ldots, d$.

In coordinate-free terms:

(a) There is a timelike Killing vector field. (A vector field on a Riemannian metric is a *Killing field* if the flow it generates consists of isometries.)

(b) There is a family of spacelike hypersurfaces *orthogonal* to the Killing vector everywhere. (We make these the $\{t = \text{constant}\}$ surfaces.) In a globally hyperbolic case these are Cauchy surfaces.

Definition: If (a) holds but not necessarily (b), then the metric or geometry is called *stationary*.

Field quantization in a stationary but nonstatic space-time seems to be most conveniently done in terms of a two-component, first-order formalism (in contrast to the one-component, second-order formulation we have been pursuing in this book). See Kay 1978, and also our appendix.

The foregoing terminology is quite standard. Less standard, but useful, are:

Definitions: A static metric is

i) *ultrastatic* if

   (c) $g_{00}(x) = 1$ for all x.

ii) *uniformly static* if $0 < c_1^2 \leq g_{00}(x) \leq c_2^2$ for some constants c_j.

Remark: $\left(g^{00}\right)^{-\frac{1}{2}}$ is called the *lapse function*, because $\left(g^{00}\right)^{-\frac{1}{2}} \Delta t$ measures to first order the distance between surfaces labeled by t and by $t + \Delta t$. [To relate this to earlier remarks (in this chapter and Chap. 4), note that $\left(g^{00}\right)^{+\frac{1}{2}}$ is the norm of the gradient vector ∇t. When $g_{0j} = 0$, $\left(g^{00}\right)^{-\frac{1}{2}} = \left(g_{00}\right)^{+\frac{1}{2}}$ is also the norm of the tangent vector $d\underline{x}/dt$. See also the discussion surrounding Fig. 10 in the appendix.] A stationary metric possesses not only a nontrivial lapse function, but also a nontrivial *shift vector field*, describing the slanting of the lines of constant spatial coordinates. The shift field is the same thing as $-g_{0j}(\underline{x})$, but the relation of the lapse and shift to g_{00} is more complicated [Kay 1978, Eq. (1.2)].

Following Kay 1978, let's introduce a convenient notation for the static metric determined by a given spatial metric and lapse function:

Definition: $\mathcal{M} = \text{Stat}(M, \gamma, g_{00}^{\frac{1}{2}}, 0)$ is the $(d + 1)$-dimensional manifold $\mathcal{M} = \mathbf{R} \times M$ equipped with the static Lorentzian metric, g, determined by the Riemannian metric γ on the d-dimensional manifold M together with the lapse function $g_{00}^{\frac{1}{2}}$ on M. In local coordinates adapted to the static structure, $\underline{g}$ is given by

$$ds^2 = g_{00}(x)\,dt^2 - \gamma_{jk}(x)\,dx^j\,dx^k,$$

or

$$g_{\mu\nu} = \begin{pmatrix} g_{00} & 0 \\ 0 & -\underline{\gamma} \end{pmatrix}.$$

Kay considers stationary geometries also; the last argument of Stat, which is 0 here, is the shift vector. Note that I shall often abuse notation by using the same letter to refer both to a bare manifold and to the entire system of manifold plus metric.

To avoid possible pathologies, let us assume that all the components of the metric tensor (in adapted coordinates) are smooth functions (at least C^2).

Definition: The *optical metric* on $\mathcal{M}$ is $g_{00}^{-1}\,\underline{g}$. In other words, $\mathcal{M}$ equipped with its optical metric is $\mathcal{M}' = \mathrm{Stat}(M, \underline{\gamma}' \equiv g_{00}^{-1}\underline{\gamma}, 1, 0)$.

$\mathcal{M}$ and $\mathcal{M}'$ are *conformally related*: the metrics differ only by a scalar factor. Hence they have the same causal structure (the same null geodesics, the same Cauchy surfaces). So $\mathcal{M}$ is globally hyperbolic if and only if $\mathcal{M}'$ is. The geodesics of $M' \equiv (M, \gamma')$ are the spatial projections of the null geodesics (light paths) of $\mathcal{M}$; this is the origin of the name "optical" [Gibbons & Perry 1978].

The following theorems are essentially due to Kay 1978:

Theorem 1: An ultrastatic $\mathcal{M} = \mathrm{Stat}(M, \underline{\gamma}, 1, 0)$ is globally hyperbolic if and only if (M, γ) is *complete* (geodesically or as a metric space).

Remark: Geodesic and metric completeness are equivalent for Riemannian manifolds, but *not* for pseudo-Riemannian ones (space-times) [Choquet-Bruhat et al. 1977, Sec. V.C.4].

Theorem 2: Consider two conformally related metrics on M: $\underline{\gamma} = \Omega(x)\underline{\gamma}'$. If $\Omega(x) \leq c_2^2$ and M is complete, then M' is complete. If $0 < c_1^2 \leq \Omega(x)$ and M' is complete, then M is complete.

Corollary 1: If $\mathcal{M}$ is uniformly static, then $\mathcal{M}$ is globally hyperbolic if and only if M is complete.

Corollary 2: If $\mathcal{M}$ is globally hyperbolic but M is not complete, then g_{00} doesn't have a positive lower bound (equivalently, g^{00} is unbounded).

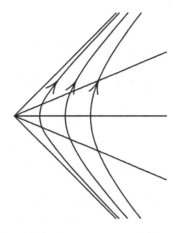

Important examples are *Rindler space* (a fragment of flat Minkowski space-time) and the exterior *Schwarzschild space* (the region surrounding a nonrotating black hole or large mass). After extra spatial
dimensions are suppressed, these spaces have the structure of a wedge
bounded by null geodesics (the *horizon*). The wedge is globally hyperbolic, but it and its Cauchy surfaces are not geodesically complete. In
keeping with the corollary, the lapse function approaches 0 at the horizon, as indicated by the merging of the Cauchy surfaces there. (The
curved arrows in the diagram indicate the Killing flow which makes
these spaces static.) The quantum field theory in these spaces, (and
in similar fragments of *deSitter space*) has had tremendous historical
importance in the development of our subject. A partial listing of the
relevant literature is: Fulling 1973, 1977; Boulware 1975a,b; Figari et
al. 1975; Hawking 1975; Unruh 1976; Hartle & Hawking 1976; Israel
1976; Candelas & Deutsch 1977, 1978; Christensen & Fulling 1977; Gibbons & Hawking 1977; Gibbons & Perry 1978; Dowker 1978; Bisognano
& Wichmann 1975; Sewell 1980, 1982; Bell & Leinaas 1983; York 1985;
Kay 1985; Fulling & Ruijsenaars 1987; Dimock & Kay 1986, 1987.

Corollary 3: If M is compact, then $\mathcal{M}$ is globally hyperbolic.

Now consider the field equation, (6.2), on a static $\mathcal{M}$. It takes the
form

$$g^{00}(x)\,\frac{\partial^2 \phi}{\partial t^2} = \text{terms not involving } \frac{\partial}{\partial t} \text{ or } t.$$

Division by the coefficient puts this into the familiar form

$$\frac{\partial^2 \phi}{\partial t^2} = K\phi,$$

with

$$K\phi \equiv g_{00} \left[\frac{1}{\sqrt{g}} \partial_j \left(\sqrt{g}\, g^{jk} \partial_k \phi \right) + (m^2 + \xi R)\phi \right]. \qquad (6.9)$$

K is formally self-adjoint with respect to the $\mathcal{L}_\rho^2$ inner product if

$$\rho = g^{00} \sqrt{g} = (g_{00})^{-\frac{1}{2}} \sqrt{\gamma}. \qquad (6.10)$$

Therefore, the theory of Chapters 3 and 4 applies.

If $\mathcal{M}$ is ultrastatic, ρ equals $\sqrt{\gamma}$ and

$$K = -\Delta_\gamma + (m^2 + \xi R), \qquad (6.11)$$

where the first term is the Laplace–Beltrami operator on M, and R is the curvature scalar on $\mathcal{M}$ (not M). (In this case, γ and γ' are the same thing.) In the general case, the field equation can be "cleaned up" by a change of dependent variable: $\tilde{\phi} \equiv g_{00}^{\frac{1}{2}} \phi$ in space-time dimension 4. (See Exercise 23 for general dimension.) This yields

$$-\frac{\partial^2 \tilde{\phi}}{\partial t^2} = \tilde{K}\tilde{\phi},$$

with

$$\tilde{K} \equiv -\Delta_{\gamma'} + g_{00}\,(m^2 + \xi R) + \tfrac{1}{2}\Delta_{\gamma'}(\ln g_{00}) + \tfrac{1}{4}\,\|\nabla(\ln g_{00})\|_{\gamma'}^2. \quad (6.12)$$

Exercise 22: Derive equations (6.9)–(6.12).

In any static case, therefore, the elliptic operator of the theory can be assumed to have the form

$$\tilde{K} = \Delta_{\gamma'} + V(x)$$

in $\mathcal{L}^2(M') \equiv \mathcal{L}^2_{\sqrt{\gamma'}}(M)$. The lore of Schrödinger operators in Chapter 2 has been in part extended to these operators, and in particular one has this basic theorem (see Chernoff 1973 and Kay 1978):

Theorem: If $\mathcal{M}$ is globally hyperbolic and $\tilde{K}$ is lower semi-bounded,

$$\langle \tilde{\phi}, \tilde{H}\tilde{\phi} \rangle_{\gamma'} \geq C\|\tilde{\phi}\|_{\gamma'}{}^2,$$

then $\tilde{K}$ is essentially self-adjoint on $C_0^\infty(M)$. If, moreover,

$$V(x) \geq \epsilon > 0 \qquad \text{and} \qquad g_{00}(x) \geq c_1{}^2 > 0,$$

then $\sigma(\tilde{K})$ has a positive lower bound, and hence the negative powers of $\tilde{K}$ exist and are bounded operators on $\mathcal{L}^2(M')$.

The origin of the transformation leading to (6.12) is

Theorem: Let g and g' be conformally related metrics:

$$g_{\mu\nu}(x) = \Omega(x)g'_{\mu\nu}(x), \qquad \Omega(x) > 0.$$

(The functions involved may be time-dependent.) Let $\phi(x)$ be a solution of $(\square + V(x))\phi = 0$. Then if $n = 4$, $\tilde{\phi}(x) \equiv \Omega(x)^{\frac{1}{2}}\phi(x)$ solves

$$(\square' + \Omega V + \tfrac{1}{6}R' - \tfrac{1}{6}\Omega R)\tilde{\phi} = 0. \tag{6.13}$$

(Here, of course, primed objects refer to the primed metric.) In particular, if $V = \tfrac{1}{6}R$, then the differential equation is conformally invariant. In dimension N, the key identity is

$$\square + \frac{N-2}{4(N-1)}R = \Omega^{-\frac{1}{4}(N-2)-1}\left[\square' + \frac{N-2}{4(N-1)}R'\right]\Omega^{\frac{1}{4}(N-2)}, \tag{6.14}$$

and hence the conformally invariant wave equation is the one with

$$V = \frac{N-2}{4(N-1)}R.$$

Exercise 23:

(a) Derive (6.14).

(b) Generalize (6.12) to arbitrary dimension.

Chapter 7

Quantum Field Theory
in an Expanding Universe

In this chapter we consider models where the classical field equation can be solved by separation of variables — i.e., by the ansatz $\phi(t, x) = \psi_j(x)\phi_j(t)$ — but $\phi_j(t)$ is not $e^{\pm i\omega_j t}$; instead, ϕ_j satisfies an ordinary differential equation with *time-dependent* coefficients. The physical consequence of this fact is that one is drawn to the conclusion that particles are created in pairs by the action of the time-dependent gravitational field. This project allows the interpretational problems of the general globally hyperbolic space-time background to be studied in the context of a solvable partial differential equation.

Although this subject goes back at least to Schrödinger 1939, the principal, thorough investigation was made by Parker 1966, 1968, 1969, 1971.

PRELIMINARIES

The metric of a *spatially flat Robertson–Walker space-time* is given by the line element

$$ds^2 = dt^2 - a(t)^2(dx^2 + dy^2 + dz^2). \tag{7.1a}$$

(To maintain contact with most of the literature, in this chapter we restrict attention to the physical dimension, $N = 4$.) Let

$$\eta \equiv \int a^{-1}\, dt, \qquad C(\eta) \equiv a(t)^2;$$

one obtains a frequently used alternative form,

$$ds^2 = C(\eta)(d\eta^2 - dx^2 - dy^2 - dz^2). \tag{7.1b}$$

This reveals the geometry to be *conformally static* (in fact, *conformally flat*) with a conformal factor depending only on time. When the field equation is conformally invariant (see the last theorem of Chapter 6), therefore, it is easy to solve in a Robertson–Walker background.

Remark: The range of t or η need not be $(-\infty, \infty)$. Models of cosmological interest typically develop singularities at a finite value of a time coordinate, usually taken for convenience as 0. The transformation to "conformal time" η may map a finite interval to an infinite one, or vice versa. The arbitrary constant in η is left open by our definition and is chosen for convenience in each separate model.

Remark: Usually one takes the range of the spatial variable **x** (i.e., the manifold M, in the notation of the previous chapter) to be $\mathbf{R}^3$. For technical reasons the case of periodic boundary conditions (the 3-torus) is often considered as well.

Two frequently studied generalizations of this class of models should be mentioned.

(1) *Bianchi Type I*, or *Kasner-like*, cosmologies:

$$ds^2 = dt^2 + \sum_{j=1}^{3} a_j(t) \left(dx^j \right)^2.$$

This is *not* conformally static if the functions a_j are independent. These models describe a universe whose cosmological expansion rate is not isotropic. (On the other hand, it should be noted that space at each time is a perfectly isotropic Euclidean $\mathbf{R}^3$. The numerical value of a_j at a fixed time has no significance, since it can be absorbed by rescaling x^j.) A *Kasner universe* in the strict sense is *a solution of the vacuum Einstein equations,* $R_{\mu\nu} - \frac{1}{2}Rg_{\mu\nu} = 0$, of the Bianchi I form, the a_j being powers of t. Quantum field theory in Bianchi Type I spaces has been studied by Zel'dovich & Starobinsky 1971, Fulling et al. 1974, Lukash et al. 1976, and others.

(2) A conformally static space-time may have any three-dimensional Riemannian manifold as its basic spatial section, (M, γ):

$$\begin{aligned} ds^2 &= dt^2 - a(t)^2 \gamma_{jk}(x)\, dx^j\, dx^k \\ &= C(\eta) \left(d\eta^2 - \gamma \right). \end{aligned}$$

By far the most important cases are the *Robertson–Walker* (or *Friedmann-like*) *universes of constant (spatial) curvature,* where γ

is the metric of a three-dimensional sphere or hyperboloid. (Again, a *Friedmann universe* is a solution of Einstein's equation, of either this form or form (7.1).)

The time and space variables can be separated in the field equation (6.2) whenever the metric is conformally static. In spatially flat Robertson–Walker and Bianchi I metrics, the spatial equation is trivial, and the eigenfunctions $\psi_j(x)$ are the familiar $e^{i\mathbf{k}\cdot\mathbf{x}}$. The ψ_j for the curved Robertson–Walker cases are the eigenfunctions of the Laplace–Beltrami operator on the sphere or hyperboloid, which have been extensively studied (see references listed in Appendix A of Parker & Fulling 1974).

For simplicity, consider henceforth only the flat Robertson–Walker case. To treat the temporal equation and its solutions $\phi_\mathbf{k}(t)$, we introduce some notation:

$$\Omega_\mathbf{k}{}^2 \equiv \mathbf{k}^2 + m^2 C;$$

$$D \equiv \partial_t a = \tfrac{1}{2}\partial_\eta(\ln C) \qquad \left(\partial_t \equiv \frac{\partial}{\partial t}, \quad \text{etc.}\right);$$

$$\chi_\mathbf{k}(\eta) \equiv a(t)\phi_\mathbf{k}(t).$$

Then the equation satisfied by $\chi_\mathbf{k}$ is

$$\frac{d^2\chi_\mathbf{k}}{d\eta^2} + \left[\Omega_\mathbf{k}(\eta)^2 + (6\xi - 1)(\partial_\eta D + D^2)\right]\chi_\mathbf{k} = 0. \qquad (7.2)$$

(The factor $\partial_\eta D + D^2$ is $\tfrac{1}{6}RC$.)

The general solution of the field equation is thus

$$\phi(\eta, \mathbf{x}) = (2\pi)^{-\frac{3}{2}} C(\eta)^{-\frac{1}{2}} \int d^3k \left[a_\mathbf{k}\chi_\mathbf{k}(\eta)e^{i\mathbf{k}\cdot\mathbf{x}} + a_\mathbf{k}^\dagger\chi_\mathbf{k}(\eta)^* e^{-i\mathbf{k}\cdot\mathbf{x}}\right].$$
$$(7.3)$$

As in Chapter 4, we shall at first think of $a_\mathbf{k}$ and $a_\mathbf{k}^\dagger$ as arbitrary complex coefficients, giving the general classical solution; later they will become annihilation and creation operators, giving the Heisenberg operator solution of the quantum field theory. Note that if χ is a solution of (7.2), then χ^* is also, and the two classes of solutions join together in (7.3) to make up a complete and nonredundant basis (in the continuum sense) of solutions of the partial differential equation.

Let's examine (7.2) under increasingly general conditions.

(1) If $m = 0$ and $\xi = \frac{1}{6}$ (the conformally invariant case), then the equation is $\partial_\eta^2 \chi_{\mathbf{k}} = \mathbf{k}^2 \chi_{\mathbf{k}}$, whose solutions are elementary. With the choice

$$\chi_{\mathbf{k}}(\eta) = \frac{1}{\sqrt{2k}} e^{-ik\eta} \qquad (k \equiv |\mathbf{k}|),$$

(7.3) is the image, under the conformal transformation, of the standard solution of the free massless field theory. This result can be interpreted as saying that no particles are created by the gravitational field in this model. However, the physics of this theory is not just a trivial transplantation of the free field onto the expanding universe: later we shall see that the "vacuum" energy density does not vanish, but is a nontrivial functional of $C(\eta)$.

(2) If $m \neq 0$ or $\xi \neq \frac{1}{6}$, the equation has the form

$$\frac{d^2 \chi_{\mathbf{k}}}{d\eta^2} + [\mathbf{k}^2 + (\text{function of } \eta)]\chi_{\mathbf{k}} = 0.$$

An obvious consequence of the nonconstant coefficient is that the solutions can't, in general, be written down in terms of elementary functions. A less obvious but more important consequence is that there is no natural choice of basis for the space of solutions, analogous to the positive- and negative-frequency exponentials in the conformally trivial case. This makes it impossible to rely on a formal analogy with the free field in arriving at an intepretation of this field theory in terms of particles.

(3) In the anisotropic Bianchi Type I case, the generalization of (7.2) is

$$\frac{d^2 \chi_{\mathbf{k}}}{d\eta^2} + \left[a^2 \sum_{j=1}^{3} \left(\frac{k_j}{a_j} \right)^2 + (6\xi - 1)(\text{fn. of } \eta) + \xi(\text{fn. of } \eta) \right] \chi_{\mathbf{k}} = 0.$$

Here we note, first, that the value $\xi = \frac{1}{6}$ is no longer special. Furthermore, the coefficient involving $\mathbf{k}$ is also η-dependent! This commonplace observation turns out to have devastating effects on naive and at-first-sight attractive attempts to define particles within time-dependent quantum field theories [Zel'dovich & Starobinsky 1971; Fulling 1979].

In cases 2 and 3, $\chi_{\mathbf{k}}$ (or $\phi_{\mathbf{k}}$) can be normalized so that the canonical commutation relations for ϕ and π translate into the standard annihilation-creation commutation relations

$$[a_{\mathbf{k}}, a_{\mathbf{k}'}^\dagger] = \delta(\mathbf{k} - \mathbf{k}'), \qquad [a_{\mathbf{k}}, a_{\mathbf{k}'}] = 0.$$

However, this condition does not make χ_k — hence a_k — unique. It gives us a whole class of natural, appealing Fock representations. (Cf. the papers of Isham, Dimock, and Deutsch & Najmi cited in Chapter 6.)

Exercise 24: Find these normalization conditions on the χ and the ϕ functions. (They involve the Wronskian of the function with its complex conjugate.)

ASYMPTOTICALLY STATIC MODELS;
BOGOLUBOV TRANSFORMATIONS AND PARTICLE CREATION

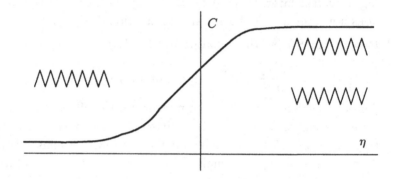

Suppose that

$$C(\eta) = \begin{cases} \text{constant} & \text{for } \eta < -T, \\ \text{constant} & \text{for } \eta > +T. \end{cases}$$

(The two constants do not have to be the same, but they may be, so long as the function does something nontrivial in the interval $(-T, T)$.) Define $\chi_k^{\text{in}}(\eta)$ to be the (unique) solution of (7.2) with the asymptotic behavior

$$\chi_k^{\text{in}} \equiv \sqrt{C} \, \phi_k^{\text{in}}(\eta) = \begin{cases} \frac{1}{\sqrt{2\Omega}} e^{-i\Omega\eta} & \text{for } \eta < -T, \\ \frac{1}{\sqrt{2\Omega}} \left[\alpha_k e^{-i\Omega\eta} + \beta_k e^{i\Omega\eta} \right] & \text{for } \eta > +T. \end{cases} \qquad (7.4)$$

Here Ω is $\Omega_k(\eta)$ evaluated in the appropriate region. The constancy of the Wronskian (cf. Exercise 24) implies that

$$|\alpha_k|^2 - |\beta_k|^2 = 1. \qquad (7.5)$$

(Since all the functions involved depend only on the magnitude of $\mathbf{k}$, I shall no longer stress its vectorial nature in the notation.)

These functions are very similar to scattering wave functions in quantum mechanics (see Chap. 2), except that the variable is the time, not the space coordinate. Also, the natural normalization is different: In scattering theory one of the terms in the asymptotic region with two terms is normalized to unity, because it represents an incoming beam. Here, in contrast, the part of the solution in the region with only one term is normalized. As in scattering theory, there is an alternative basis of solutions, χ_k^{out}, each of which is proportional to the appropriate function $e^{-i\Omega\eta}$ in the region $\eta > T$.

When $\eta < -T$, our space-time is flat. This region contains entire Cauchy surfaces. The dynamics of the field there is that of the free field, and the physics of any experimental operation is surely the same as in the globally flat space of nongravitational physics. Therefore, $\phi_k^{\text{in}} e^{i\mathbf{k}\cdot\mathbf{x}}$ is surely the wave function of a real particle, as far as that epoch of time is concerned. But identical reasoning shows that $\phi_k^{\text{out}} e^{i\mathbf{k}\cdot\mathbf{x}}$ must be the wave function of a real particle in the region $\eta > T$. Since $\phi^{\text{in}} \neq \phi^{\text{out}}$, a one-particle state in the far past is not the same thing as a one-particle state in the far future.

(This argument is most convincing in the present case, where γ is the Euclidean metric. In more general cases, the geometry is static but not flat in the asymptotic regions. The argument then applies to the extent that one accepts the operational validity of the "particle" concept in static space-times, as developed in Chapters 4 and 6.)

So, we have *two* natural quantizations of the field, associated with two Fock spaces, which we can call $\mathcal{F}^{\text{in}}$ and $\mathcal{F}^{\text{out}}$. Each theory continues to make sense as a definition of $\phi(t, \mathbf{x})$ for t outside its respective static region. (Equation (7.3) expresses the Heisenberg-picture field in terms of the respective annihilation and creation operators; the construction is made rigorous by the arguments of Isham 1978 and Dimock 1980.) Nevertheless, the operators in each case are not directly related to particle observables outside their respective epochs.

Now we need to investigate quantitatively the relation between these two notions of "particles".

Exercise 25: Show from (7.4) that

$$\phi_k^{\text{in}} = \alpha_k \phi_k^{\text{out}} + \beta_k \phi_k^{\text{out}*},$$

and hence

$$a_{\mathbf{k}}^{\text{out}} = \alpha_k a_{\mathbf{k}}^{\text{in}} + \beta_k^* a_{-\mathbf{k}}^{\text{in}\dagger},$$
$$a_{\mathbf{k}}^{\text{in}} = \alpha_k^* a_{\mathbf{k}}^{\text{out}} - \beta_k^* a_{-\mathbf{k}}^{\text{out}\dagger}. \tag{7.6}$$

This kind of redefinition of annihilation and creation operators is called a *Bogolubov transformation*.

We consider the representation of both kinds of operators within the Fock space $\mathcal{F}^{\text{in}}$. To postpone a subtle side issue, we take M to be the torus, so that the modes are discrete and labeled by $\mathbf{k} = 2\pi\mathbf{n}/L$, $\mathbf{n} \in \mathbf{Z}^3$. Now

$$N_{\mathbf{k}}^{\text{out}} \equiv a_{\mathbf{k}}^{\text{out}\dagger} a_{\mathbf{k}}^{\text{out}}$$

is the operator representing the observable "number of particles in mode $\mathbf{k}$ when $\eta > T$". Let us calculate this number when the initial state is the vacuum. (Hereby we do not intend to make the physical hypothesis that the Universe was created in an empty condition. We simply need to choose a definite state for our first calculation.) We have

$$\langle 0^{\text{in}} | N_{\mathbf{k}}^{\text{out}} | 0^{\text{in}} \rangle = \langle 0^{\text{in}} | (\alpha_k^* a_{\mathbf{k}}^{\text{in}\dagger} + \beta_k^* a_{-\mathbf{k}}^{\text{in}})(\alpha_k a_{\mathbf{k}}^{\text{in}} + \beta_k a_{-\mathbf{k}}^{\text{in}\dagger}) | 0^{\text{in}} \rangle$$
$$= |\beta_k|^2,$$

the last step following by a simple calculation using $[a_{-\mathbf{k}}, a_{-\mathbf{k}'}^{\dagger}] = \delta_{\mathbf{k}\mathbf{k}'}$. Therefore,

$$\langle 0^{\text{in}} | N_{\text{total}}^{\text{out}} | 0^{\text{in}} \rangle = \sum_{\mathbf{k}} |\beta_k|^2.$$

By our previous argument, this is the expectation value of the number of particles present in the final state.

A natural next question is whether this total number of created particles is finite. It is a well known theorem of asymptotic analysis [see, e.g., Olver 1961 or Littlewood 1963] that if $C'(\eta) \in C_0^\infty$, then β_k approaches 0 faster than any power k^{-N} as $k \to \infty$. Thus the sum does converge in such a case. In fact, solvable models typically yield $|\beta_k| \sim e^{-(\text{const})k}$, which is suggestive of a thermal spectrum [Parker 1976]. On the other hand, if $C(\eta)$ is not smooth, generally one will find that $|\beta_k| \sim k^{-N}$ with N determined by the order of singularity of C. A sudden expansion of the universe would create an infinite density of particles.

But what if $M = \mathbf{R}^3$? Then we have $[a_{-\mathbf{k}}, a_{\mathbf{k}'}^\dagger] = \delta(\mathbf{k} - \mathbf{k}')$, so $|\beta_k|^2$ is replaced by $\delta(0)|\beta_k|^2$ in the foregoing calculation. That is, $\langle N_{\mathbf{k}}^{\mathrm{out}} \rangle = \infty$, even for the smoothest and gentlest expansion of the universe. The physical meaning of this divergence is quite clear and nonfrightening, however. A particle-creation process is taking place throughout a homogeneous, infinite space. The total number of particles produced, if nonzero, has to be infinite.

Exercise 26: Returning to the torus, divide the total number by the volume of the space and take the limit of infinite volume. Conclude that the total *density* of particles created in the infinite universe is

$$(2\pi)^{-1} C(T)^{-\frac{3}{2}} \int d^3k \, |\beta_k|^2.$$

Thus the particle-number density will be finite for well-behaved metrics. Similar remarks apply to energy density, etc. The energy density in the far future can be computed by normal-ordering the formal expression for the energy-density operator with respect to the *out* particle operators and calculating the expectation value in the *in* vacuum:

$$\langle 0^{\mathrm{in}}|{:}T_{00}(x){:}|0^{\mathrm{in}}\rangle \equiv \langle 0^{\mathrm{in}}|T_{00}|0^{\mathrm{in}}\rangle - \langle 0^{\mathrm{out}}|T_{00}|0^{\mathrm{out}}\rangle$$

(cf. Chapter 5).

There is an alternative approach to the particle-creation calculation, which is instructive in a complementary way. This time one works in $\mathcal{F}^{\mathrm{out}}$, looking for a vector that can be identified with $|0^{\mathrm{in}}\rangle$ so that its out-particle content can be read off. Such a vector must satisfy

$$a_{\mathbf{k}}^{\mathrm{in}}|0^{\mathrm{in}}\rangle = 0 \quad \text{for all } \mathbf{k}. \tag{7.7}$$

Suppose for the moment that it exists and is unique. Acting on it repeatedly with $a_{\mathbf{k}}^{\mathrm{in}\,\dagger}$, one can build up a representation of the field algebra which is clearly unitarily equivalent to (or "a copy of") the representation in $\mathcal{F}^{\mathrm{in}}$. That is, each n-particle basis state in $\mathcal{F}^{\mathrm{in}}$ is identified with some definite vector in $\mathcal{F}^{\mathrm{out}}$ (albeit not a vector of definite out-particle number).

If $|0^{\mathrm{in}}\rangle$ exists in $\mathcal{F}^{\mathrm{out}}$, it must be of the form

$$\{\phi_0, \phi_1(\mathbf{k}_1), \ldots, \phi_n(\mathbf{k}_1, \ldots, \mathbf{k}_n), \ldots\}$$

for some functions ϕ_n. From (7.7) and (7.6) one sees that

$$\alpha_k{}^* a_\mathbf{k}^{\text{out}} |0^{\text{in}}\rangle = \beta_k{}^* a_{-\mathbf{k}}^{\text{out}\dagger} |0^{\text{in}}\rangle.$$

Thus

$$a_\mathbf{k}^{\text{out}} |0^{\text{in}}\rangle = \gamma_k a_{-\mathbf{k}}^{\text{out}\dagger} |0^{\text{in}}\rangle, \tag{7.8}$$

where

$$\gamma_k \equiv \left(\frac{\beta_k}{\alpha_k}\right)^*.$$

Now use formulas (3.13), with $\tilde{u}(\mathbf{k}') = \delta_{\mathbf{kk}'}$: The right-hand side of (7.8) is

$$\gamma_k\{0, \delta_{-\mathbf{kk}_1}\phi_0, \ldots, \sqrt{n}\,\text{sym}[\delta_{-\mathbf{kk}_1}\phi_{n-1}(\mathbf{k}_2,\ldots,\mathbf{k}_n)],\ldots\},$$

while the left-hand side is

$$\left\{\sum_{\mathbf{k}'}\delta_{\mathbf{kk}'}\phi_1(\mathbf{k}'),\ldots,\sqrt{n+1}\sum_{\mathbf{k}'}\delta_{\mathbf{kk}'}\phi_{n+1}(\mathbf{k}',\mathbf{k}_1,\ldots,\mathbf{k}_n),\ldots\right\}$$

$$= \{\phi_1(\mathbf{k}),\ldots,\sqrt{n+1}\,\phi_{n+1}(\mathbf{k},\mathbf{k}_1,\ldots,\mathbf{k}_n),\ldots\}.$$

Thus we have a sequence of equations

$$\phi_1(\mathbf{k}) = 0,$$
$$\sqrt{2}\,\phi_2(\mathbf{k},\mathbf{k}_1) = \gamma_k \delta_{-\mathbf{kk}_1}\phi_0,$$
$$\vdots$$
$$\sqrt{n+1}\,\phi_{n+1}(\mathbf{k},\mathbf{k}_1,\ldots,\mathbf{k}_n) = \sqrt{n}\,\text{sym}[\delta_{-\mathbf{kk}_1}\phi_{n-1}(\mathbf{k}_2,\ldots,\mathbf{k}_n)]$$
$$\vdots$$

Consequently,

$$\phi_n = 0 \quad \text{for all odd } n;$$

ϕ_0 is an arbitrary constant (which should be chosen eventually to make $\langle 0^{\text{in}}|0^{\text{in}}\rangle = 1$); and the higher even coefficients are determined by recursion relations

$$\phi_2(\mathbf{k}_1,\mathbf{k}_2) = \frac{1}{\sqrt{2}}\gamma_{k_1}\delta_{-\mathbf{k}_1\mathbf{k}_2},$$
$$\vdots$$
$$\phi_{n+2}(\mathbf{k}_1,\ldots,\mathbf{k}_{n+2}) = \frac{\sqrt{n+1}}{\sqrt{n+2}}\gamma_{k_1}\text{sym}[\delta_{-\mathbf{k}_1\mathbf{k}_2}\phi_n(\mathbf{k}_3,\ldots,\mathbf{k}_{n+2})]$$
$$\vdots$$

The recursion can be solved: For n even,

$$\phi_n(\mathbf{k}_1, \ldots, \mathbf{k}_n)$$
$$= \frac{1}{(n/2)!} \sqrt{\frac{n!}{2^n}} \, \mathrm{sym}[\gamma_{k_1} \delta_{-\mathbf{k}_1 \mathbf{k}_2} \gamma_{k_3} \delta_{-\mathbf{k}_3 \mathbf{k}_4} \cdots \gamma_{k_{n-1}} \delta_{-\mathbf{k}_{n-1} \mathbf{k}_n}].$$

The result is more commonly written (creation operators understood to be in the *out* representation)

$$(\phi_0)^{-1}|0^{\mathrm{in}}\rangle = \sum_n \frac{1}{\sqrt{n!}} \sum_{\mathbf{k}_1, \ldots, \mathbf{k}_n} \phi_n(\mathbf{k}_1, \ldots, \mathbf{k}_n) a^\dagger_{\mathbf{k}_1} \cdots a^\dagger_{\mathbf{k}_n} |0^{\mathrm{out}}\rangle$$

$$= |0^{\mathrm{out}}\rangle + \frac{1}{2} \sum_k \gamma_k a^\dagger_{-\mathbf{k}} a^\dagger_{\mathbf{k}} |0^{\mathrm{out}}\rangle$$

$$+ \frac{1}{8} \left(\sum_k \gamma_k a^\dagger_{-\mathbf{k}} a^\dagger_{\mathbf{k}} \right)^2 |0^{\mathrm{in}}\rangle + |6\text{-particle state}\rangle + \cdots$$

$$+ \left[\left(\frac{n}{2}\right)! \, 2^{\frac{n}{2}} \right]^{-1} \left(\sum_k \gamma_k a^\dagger_{-\mathbf{k}} a^\dagger_{\mathbf{k}} \right)^{\frac{n}{2}} |0^{\mathrm{out}}\rangle + \cdots$$

$$= \exp\left(\frac{1}{2} \sum_k \gamma_k a^\dagger_{-\mathbf{k}} a^\dagger_{\mathbf{k}} \right) |0^{\mathrm{out}}\rangle. \tag{7.9}$$

It can be checked that this vector has finite norm if and only if $\sum_\mathbf{k} |\beta_k|^2 < \infty$. (Recall that $\mathbf{k}$ is *discrete* in the present discussion.)

GENERAL BOGOLUBOV TRANSFORMATIONS

The Bogolubov transformation in (7.6) is called *diagonal*, because it operates on each normal mode independently, mixing its creation operator with its annihilation operator. (As written, the transformation mixes mode $\mathbf{k}$ with mode $-\mathbf{k}$, but this pair of modes can trivially be decoupled by passing from complex exponential to real, trigonometric basis eigenfunctions.) This is a convenient place to digress on Bogolubov transformations which are not thus diagonalized. The construction of a ground state, (7.9), of one representation within the Fock space of the other has a generalization, as does the existence condition that $|\beta_k|^2$ be summable.

This exposition is taken in part from Fulling 1972, Appendix F. Valuable original references include Kristensen et al. 1967 and Wald 1975.

A Bogolubov transformation on a and $a^\dagger$ is equivalent to a *symplectic transformation* on the canonical variables, x and p or ϕ and π. I shall not pursue the formal details here.

Consider an arbitrary asymptotically static, globally hyperbolic $\mathcal{M}$ (not necessarily conformally static). Again there will be *in* and *out* representations, and some linear relationship between the annihilation operators of one representation and the annihilation and creation operators of the other. Even more generally, whenever one considers two different expressions for the general solution of the (linear) field equation — schematically,

$$\phi(\underline{x}) = \sum_j \left(a_j \psi_j(\underline{x}) + a_j^\dagger \psi_j(\underline{x})^*\right) = \sum_k \left(b_k \phi_k(\underline{x}) + b_j^\dagger \phi_k(\underline{x})^*\right) \quad (7.10)$$

— there will be such a linear relationship, which is essentially the transpose of the linear mapping expressing one of the complete sets of normal modes in terms of the other (cf. Exercise 25). [Normally such a set of normal modes will be constructed in terms of a basis for some $\mathcal{L}^2_\rho(S)$, where S is a Cauchy hypersurface and the basis is given by the spectral decomposition of some self-adjoint operator there, related somehow to the dynamics. To get a complete set of solutions for the second-order equation, one needs two solutions for each basis element, and this leads to the problem of "positive and negative frequencies" described in the previous chapter. A complicating factor is that often in practice (in particular, in the "wedge" scenarios mentioned in Chap. 6) the domains of validity of the two expansions in (7.10) are different, and on the intersection of those domains the normal modes of each individual expansion are not independent. This creates technical problems which will be ignored in the present discussion.]

Let us make (7.10) somewhat less schematic by replacing the sums by integrations over appropriate measure spaces, $\int \ldots d\nu(j)$ and $\int \ldots d\mu(k)$. (Often these measure spaces are the same, but not always. Usually they are determined by spectral decompositions — see Chap. 2.) Then we will have some relationship of the form

$$a_j = \int d\mu(k)\, U_{jk}\, b_k + \int d\mu(k)\, V_{jk}\, b_k^\dagger, \quad (7.11a)$$

where U and V are integral kernels, possibly distributional. (Of course, the variables could really be discrete, as the conventional matrix notation appears to imply.) If we smear with a test function, we get

$$a(f) = b(U^t f) + b^\dagger(V^t f). \quad (7.11b)$$

Here "t" indicates transposition of an operator (taking the adjoint with respect to a real inner product), which arises naturally from the pairing of test function with distribution. From a mathematical point of view, it is preferable to start from (7.11b), with given operators U^t and V^t; the awkward transpose notation is necessary to make (7.11a) look like the equations one sees in the physics literature. Recall from Chapter 4 that the space of test functions for a set of annihilation and creation operators can be taken to be a Hilbert space; thus $f \in \mathcal{L}_\nu^2$ in (7.11b). For technical convenience, in theorem-proving one usually assumes that U^t and V^t are bounded operators from $\mathcal{L}_\nu^2$ to $\mathcal{L}_\mu^2$. (Unfortunately, unbounded operators are sometimes encountered in practice.)

Proposition: The a and b operators both satisfy the canonical commutation relations with their respective adjoints, if and only if

$$UU^\dagger = 1 + VV^\dagger \qquad \text{and} \qquad UV^t = VU^t. \qquad (7.12)$$

(This generalizes (7.5). "$\dagger$" indicates the ordinary (complex) adjoint.) Furthermore, the inverse Bogolubov transformation is

$$b(g) = a(\overline{U}g) - a^\dagger(Vg)$$

($g \in \mathcal{L}_\mu^2$; "$\overline{}$" indicates matricial complex conjugation — i.e., the transpose of the adjoint.)

The argument leading from (7.7) to (7.9) generalizes. Let $|0^a\rangle$ be the vacuum vector of the Fock space $\mathcal{F}^a$ associated with the a operators, etc.

Theorem: The following are equivalent:

(A) V is a Hilbert-Schmidt operator:

$$\|V\|_{\text{HS}}^2 = \int d\nu(j) \int d\mu(k)\, |V_{jk}|^2 < \infty.$$

(B) The [expectation value of the] total number of created particles is finite. (It equals $\|V\|_{\text{HS}}^2$.) Indeed, we have

$$\langle 0^b|N_{\text{total}}^a|0^b\rangle \equiv \langle 0^b| \int d\nu(j)\, a_j^\dagger a_j |0^b\rangle = \int d\nu(j) \int d\mu(k)\, |V_{jk}|^2.$$

(C) $\mathcal{F}^a$ is naturally isomorphic to $\mathcal{F}^b$, in the sense that $\mathcal{F}^a$ contains a vector identifiable with $|0^b\rangle$, and hence one identifiable with each element of any bases for the N-particle subspaces of $\mathcal{F}^b$. Namely, $|0^b\rangle$ corresponds to the Fock-space vector $\{\phi_n\}$ which has $\phi_n = 0$ for all odd n and

$$\phi_n(j_1, \ldots, j_n)$$
$$= \sqrt{n!}\, 2^{-\frac{n}{2}} \left[\left(\frac{n}{2} \right)! \right]^{-1} \text{sym}[\gamma(j_1, j_2) \cdots \gamma(j_{n-1}, j_n)] \phi_0$$

for even n, where $\gamma(j, j')$ is the integral kernel of $\gamma \equiv -U^{-1}V$.

Exercise 27: Assume that the situation described in the theorem obtains. Show from (7.12) that γ is a symmetric ($\gamma^t = \gamma$) Hilbert-Schmidt operator with *operator* norm $\|\gamma\|_2 < 1$. (The symmetry is necessary for the formula $\phi_2(j_1, j_2) = \frac{1}{\sqrt{2}} \phi_0 \gamma(j_1, j_2)$ to make sense in a boson Fock space. The norm condition is precisely what guarantees that

$$\sum_{n=0}^{\infty} \|\phi_n\|^2 < \infty$$

for finite ϕ_0, so that we really have a vector in $\mathcal{F}^a$.)

Remarks:

(1) If (C) of the theorem holds for $a = $ out and $b = $ in, one says that "the *in* and *out* representations are unitarily equivalent," or that "the S-matrix exists."

(2) A nontrivial (i.e, $V \neq 0$) *diagonal* Bogolubov transformation on a *continuous* spectrum of modes is *never* Hilbert-Schmidt: Its kernel is of the form

$$V_{jj'} = \beta_j\, \delta(j - j'),$$

which is not even a function. We saw this phenomenon earlier: A spatially infinite Robertson–Walker universe (if nonstatic) implies an infinite number of created particles and a non-Fock final state. (Nevertheless, the particle *density* and related local observables can remain finite. Thus such situations need not always be rejected as *physically pathological*.)

(3) A positive result on existence of the S-matrix was obtained by Wald 1979b and independently, by different methods, by Dimock 1979: If $\mathcal{M} = \mathbf{R}^4$ and the metric, g, is the flat metric, η, plus a *smooth perturbation of compact support*, then the conclusions of the theorem

hold. Wald's proof extends to a globally hyperbolic, C^∞ space-time with *compact spatial section* M [Fulling et al. 1981].

(4) When the S-matrix exists, so that $\mathcal{F}^{\text{in}} = \mathcal{F}^{\text{out}}$, one can define a certain Green function for the field equation by the formula

$$G_F(\underline{x}, \underline{y}) = \frac{\langle 0^{\text{out}} | T[\phi(\underline{x})\phi(\underline{y})] | 0^{\text{in}} \rangle}{\langle 0^{\text{out}} | 0^{\text{in}} \rangle}.$$

It has many of the formal properties of the Feynman propagator of a static theory [see (4.8) and (4.3)], and it figures strongly in the work of Schwinger, DeWitt, and others. In the infinite-volume limit of a Robertson–Walker model (in which limit the S-matrix ceases to exist — see remark (2)), the numerator and denominator both approach 0. Sense can still be made of the quotient; as remarked in Chapter 4, Rumpf has offered a formulation of quantum field theory in curved space-time based on Green functions which reduce to G_F when the S-matrix exists.

Unfortunately, nondiagonal Bogolubov transformations are rather difficult to work with in most concrete cases. The "matrices" U and V are usually distributions, and cavalier formal calculations with them are rather unconvincing. In my experience, the study of renormalized local observables (such as the stress tensor) within a fixed representation is more trustworthy and more productive.

STIMULATED EMISSION

As remarked earlier, there is no reason beyond arbitrary convenience for assuming that the initial state in a dynamical calculation is the vacuum. Fortunately, once the vacuum problem has been solved, the general case follows, in principle, easily. (The same Bogolubov transformation is a central ingredient in the analysis for other initial states. Later we shall see that once the stress tensor's vacuum expectation value has been renormalized, no additional renormalization is needed to get the expectation value with respect to some other state.)

If N particles are already present in a given mode of a boson field, then the creation of particles in that particular mode is amplified by a factor $2N + 1$ relative to the vacuum case. The extra radiation is called *stimulated emission*, as opposed to the *spontaneous emission* from the vacuum. The relation between stimulated and spontaneous emission has been known since Einstein in 1916 (see Pais 1982, Sec. 21b). In the context of gravitational particle creation (especially by black holes) it was treated by Wald 1976 and Bekenstein & Meisels 1977.

Fermions obey the exclusion principle (no more than one particle per mode). Not surprisingly, one finds that the presence of a fermionic particle *suppresses* creation in that mode, rather than stimulating it.

Particle *destruction* clearly is possible. (The theory is, after all, invariant under time reversal.) If our (Heisenberg-picture) state is $|0^{out}\rangle$, then the scenario is that an initial state with a complicated probability of having various numbers of pairs of particles in various modes [see (7.9)] evolves into the empty state. However, particle destruction requires a fine-tuning of the probability amplitudes in the initial state which presumably could not be set up by a plausible physical process. (In this connection, note that the creation of particles has no special connection with *expansion* of the universe as opposed to *contraction*. The common reference to "expansion" traces purely to the observed fact that the Universe we live in is expanding.)

Spin and statistics

A field in physics is characterized by its *spin*, an integer or half-integer which indexes its transformation properties under the rotation and Lorentz groups. The founders of quantum field theory quickly discovered that fields of half-integer spin, such as the Dirac electron field, could not be quantized canonically; inconsistencies appeared unless *anticommutators* ($\{A, B\} \equiv [A, B]_+ \equiv AB + BA$) were used in place of commutators. Conversely, fields of integer spin demanded commutators, not anticommutators. Later, deep arguments involving analytic continuation of the n-point functions showed that *all* special-relativistic field theories in flat space-time (including interacting theories) must satisfy this connection between spin and statistics (see Streater & Wightman 1964 or Bogolubov et al. 1975). (Commutator quantization leads to Fock spaces based on symmetrized wave functions, hence a certain statistical mechanics; anticommutators lead to antisymmetrized wave functions, hence a different statistics. This is the distinction between *bosons* and *fermions*.)

For the free scalar field in flat space-time, the necessity for a bosonic quantization follows easily from the relativistic requirement that field operators at spacelike-separated points commute or anticommute with one another [e.g., Bjorken & Drell 1965, Sec. 16.12]. The vacuum expectation value of the commutator, $\langle 0|[\phi(\underline{x}), \phi(\underline{y})]|0\rangle$, must be a relativistically invariant solution of the wave equation and must be antisymmetric in its two arguments. The vacuum expectation value of the anticommutator must be a relativistically invariant solution which is *symmetric* in its arguments. But we saw in Chapter 4 that the antisymmetric solution G vanishes at spacelike separations, while the symmetric solution $G^{(1)}$ does not. Therefore, it must be the commutator that vanishes.

On the other hand, the relativistically invariant solutions of the Dirac equation are obtained from those of the Klein–Gordon equation by applying a first-order differential operator (see the previously cited appendices of Bjorken & Drell 1965 and Bogolubov & Shirkov 1959). This turns symmetric functions into antisymmetric ones. Therefore, the function which vanishes at spacelike separation must be the anticommutator in this case.

Within the formalism describing particle creation in time-dependent geometries, Parker 1969, 1971 and Wald 1975 found more-elementary reasons why the noninteracting scalar and Dirac fields must satisfy the standard connection between spin and statistics.

We have noted that the Bogolubov coefficients of the scalar field in a conformally static cosmology satisfy $|\alpha_k|^2 - |\beta_k|^2 = 1$. This is simply a mathematical consequence of the differential equation satisfied by the normal modes, independent of quantum field theory; nevertheless, it is *necessary* for the canonical commutation relations to be preserved under the time evolution, and it is *inconsistent* with anticommutation relations. Parker showed that the Bogolubov coefficients of the Dirac field satisfy, instead, $|\alpha_k|^2 + |\beta_k|^2 = 1$ (suitably generalized to include spin indices). This is precisely the condition needed for dynamical consistency of the *anticommutation* relations expected to be satisfied by the Dirac field. Note that the boson and fermion Bogolubov transformations are analogous to Lorentz transformations and rotations, respectively; the fermion transformations are unitary, while the boson transformations preserve a *symplectic* quadratic form, which is not positive definite.

Presumably a similar remark concerning (7.12) holds in the general case. Wald argues the point, however, from the symmetry of the scalar two-particle wave function (Exercise 27). Conversely, a fermionic field must have $\phi_2(j_1, j_2)$ antisymmetric.

PARTICLE OBSERVABLES AT FINITE TIMES

An asymptotically static geometry is cosmologically implausible. Even if we did live in such a universe, we would expect to have a language to describe the physics taking place during epochs when the background geometry is not static. It is natural, therefore, to expect a description of a configuration of particles at each instant of time to be hidden in the mathematical apparatus of quantum field theory. Here we follow the fate of this idea.

Let $\mathcal{M} = \mathbf{R} \times M$. To each $S \equiv \{\, t_0 = \text{constant}\,\}$, diffeomorphic to M, is attached a canonical field algebra (a representation of the relations $[\phi, \pi] = \delta$, etc.). It is tempting to interpret each such algebra as a Fock algebra describing the number and state of "particles" existing at time t_0. (It happens that this notion is doomed to failure, but a study of it is essential to understand what has replaced it. The following remarks are based in part on Fulling 1979.) How, then, should $\phi(t_0, \cdot)$ and $\pi(t_0, \cdot)$ be related to operators $a_j(t_0)$, $a_j^{\dagger}(t_0)$ describing particles? We have here no elementary positive-frequency functions $e^{-i\omega_j t}$.

One's first thought is to distinguish a subspace of solutions of the field equation as having *instantaneous positive frequency*, by examining their behavior as functions of time near the hypersurface S in question. The fact that a positive-frequency solution at S no longer has purely positive frequency at some other hypersurface is not, in itself, a problem; rather, it is evidence that the theory predicts creation of particles by the time-dependent gravitational field. (The earlier distinction between *in* and *out* solutions would then become a special case of this.)

Consider the flat Robertson–Walker model, (7.1), and its anisotropic generalization, the Bianchi Type I. Recall that we made the change of variables

$$\eta = \int a(t)^{-1}\, dt, \qquad \chi_{\mathbf{k}}(\eta) = a(t)\phi_{\mathbf{k}}(t),$$

to put the equation for the time dependence of the normal modes into a form

$$\frac{d^2 \chi_{\mathbf{k}}}{d\eta^2} + W_{\mathbf{k}}(\eta)^2 \chi_{\mathbf{k}} = 0.$$

For Robertson–Walker the coefficient is

$$W_{\mathbf{k}}(\eta)^2 = W_k(\eta)^2 = \mathbf{k}^2 + m^2 a^2 + (\xi + \tfrac{1}{6})Ra^2. \qquad (7.13a)$$

In the general Bianchi I case, where $a \equiv (a_1 a_2 a_3)^{\frac{1}{3}}$, one has

$$W_{\mathbf{k}}(\eta)^2 = a^2 \sum_{j=1}^{3} \left(\frac{k_j}{a_j}\right)^2 + m^2 a^2$$

$$+ \left(\xi - \frac{1}{6}\right)Ra^2 + \frac{1}{18}\sum_{i<j}\left(\frac{da_i}{d\eta} - \frac{da_j}{d\eta}\right)^2. \qquad (7.13b)$$

If a were a constant, we would take as basic solution the one with the data

$$\chi_{\mathbf{k}}(\eta_0) = \frac{1}{\sqrt{2W_{\mathbf{k}}(\eta_0)}}, \qquad \frac{d\chi_{\mathbf{k}}}{d\eta}\bigg|_{\eta_0} = -i\sqrt{\frac{W_{\mathbf{k}}(\eta_0)}{2}}. \qquad (7.14)$$

This yields the standard positive-frequency solution,

$$\chi_{\mathbf{k}}(\eta) = \chi_{\mathbf{k}}(\eta_0)e^{-i\Omega_{\mathbf{k}}\eta}.$$

Therefore, imposing (7.14) at a different value of η_0 would yield the same result up to a phase. (The phase factor cancels out of the number operator, $N_{\mathbf{k}} = a_{\mathbf{k}}^\dagger a_{\mathbf{k}}$.) Let us try to use (7.14) even in the case of a variable a: For *each* η_0 we will get a function $\chi_{\mathbf{k}}^{(\eta_0)}(\eta)$ (or, equivalently, $\phi_{\mathbf{k}}^{(t_0)}(t)$). To keep the notation uncluttered, let's drop the superscript (η_0) on the mode function but attach it to the respective creation and annihilation operators. Then the field expansion is

$$\phi(t,x) = (2\pi)^{-\frac{3}{2}}\int d^3k \,[a_{\mathbf{k}}(t_0)\phi_{\mathbf{k}}(t)e^{i\mathbf{k}\cdot\mathbf{x}} + a_{\mathbf{k}}(t_0)^\dagger\phi_{\mathbf{k}}(t)^*e^{-i\mathbf{k}\cdot\mathbf{x}}].$$

From (7.14) and Exercise 24 we have $[a_{\mathbf{k}}, a_{\mathbf{k}'}^\dagger] = \delta(\mathbf{k} - \mathbf{k}')$, etc. (which is the motivation for the normalization of (7.14)).

Between any two times there will be a Bogolubov transformation connecting the sets of particle operators thus defined:

$$a_{\mathbf{k}}(t) = \alpha_{\mathbf{k}}(t,t_0)a_{\mathbf{k}}(t_0) + \beta_{\mathbf{k}}(t,t_0)^*a_{-\mathbf{k}}(t_0)^\dagger.$$

A great deal now hinges on the behavior of β as a function of $\mathbf{k}$. Earlier, we appealed to the theorem that in a smooth, asymptotically static model $\beta_{\mathbf{k}}(-T,T)$ will fall off faster than any negative power of k ($\equiv |\mathbf{k}|$) when T is sufficiently large, and we concluded that $\int d^3k\,|\beta_{\mathbf{k}}|^2 < \infty$. However, it is crucial that $\pm T$ be in the asymptotic regions. For general times, and even for smooth $a(t)$, one finds that

$$\beta_{\mathbf{k}}(t_1,t_2) \sim \text{const.} \times k^{-3} \quad \text{for } (7.13a),$$

where the $\mathbf{k}$-dependent term is independent of t, and

$$\beta_{\mathbf{k}}(t_1,t_2) \sim \text{const.} \times k^{-1} \quad \text{for } (7.13b),$$

where the $\mathbf{k}$-dependent term is also t-dependent. With a three-dimensional $\mathbf{k}$-space, therefore, the crucial integral $\int d^3k\,|\beta_{\mathbf{k}}|^2$ converges for (7.13a) but diverges for (7.13b).

The anisotropic cosmology is therefore afflicted by infinite particle creation [Zel'dovich & Starobinsky 1971]. More precisely, we can say the following:

(1) Even if $\mathbf{k}$ is a discrete variable, the Fock representations for different finite times can be unitarily inequivalent. This is so even though the asymptotic representations are equivalent: $\mathcal{F}^{\text{in}} = \mathcal{F}^{\text{out}}$. This contrast suggests strongly that there is something wrong about our *definition* of particles at the intermediate times.

(2) *Local* observables (densities) are badly behaved. For example, suppose we define an energy density at t_2 by normal-ordering the formal expression T_{00} with respect to the t_2-particle operators, so that

$$\langle 0^{t_2}|{:}T_{00}(t_2,x){:}|0^{t_2}\rangle = 0.$$

Then we find that

$$\langle 0^{t_1}|{:}T_{00}(t_2,x){:}|0^{t_1}\rangle = \infty.$$

That is, the vacuum states for two different times differ at any given time by an infinite quantity of matter per unit volume. Physically this outcome is even more unacceptable than (1). (We were prepared already to lose unitary implementability of the dynamics as soon as the volume of space became infinite.) Again, the problem is surely an artifact of the finite-time particle definition; it is not present in *in–out* calculations (if the metric is smooth).

Historical remark: Traditionally the variables η and χ were used only in the study of the conformal coupling, $\xi = \frac{1}{6}$. For $\xi = 0$ the usual procedure was to set

$$\tau \equiv \int a(t)^{-3}\, dt,$$

getting in Robertson–Walker the differential equation

$$\frac{d^2 \phi_{\mathbf{k}}(\tau)}{d\tau^2} + (k^2 a^6 + m^2 a^4 + \xi R a^4)\phi_{\mathbf{k}} = 0.$$

Consequently, an infinite density of particle creation was found even for the isotropic cosmology. In hindsight, the salient lesson is that the instantaneous particle concept is *ambiguous*, as well as implausible in its predictions. This point is developed further in Fulling 1979.

Various refinements of the positive-frequency gambit have been attempted, usually expressed in terms of *diagonalizing an instantaneous Hamiltonian operator*, $H(t_0)$. These superficially appear less *ad hoc*, and they are extendible to models that are not solvable by separation of variables (so that the Bogolubov transformations between various times are nondiagonal). However, they are subject to the same physical criticisms as the naive positive-frequency construction [see Fulling 1979].

THE ADIABATIC APPROACH

What I believe to be the correct resolution of the problem of choosing the field representation was presented already in Parker's original work (1966, 1969). It has been developed further by Parker & Fulling 1974; Fulling et al. 1974; Hu 1978, 1979; Bunch et al. 1978; Birrell 1978; Fulling 1979; Suen 1987.

The root of the problem with the ansatz of "instantaneous positive frequency" is that the *effective frequency* of oscillation of the solutions of an equation of the form

$$\frac{d^2\phi}{dt^2} + \omega(t)^2 \phi = 0 \tag{7.15}$$

is *not* ω, if ω depends on t. Of course, one would expect that use of a wrong frequency in (7.14), etc., would lead to an unphysical quantization.

This key point is made clearer by the analogy of a *damped harmonic oscillator*. The equation

$$\frac{d^2\phi}{dt^2} + 2\gamma\frac{d\phi}{dt} + \omega^2\phi = 0$$

has as a basis of solutions

$$\phi^{\mp}(t) \equiv e^{-\gamma t}\exp\left(\pm i\sqrt{\omega^2 - \gamma^2}\,t\right).$$

Note the shift of the frequency relative to the solutions of the undamped problem, $e^{\pm\omega t}$. It causes the solutions to get rapidly out of phase with the undamped functions. A solution matching $e^{-i\omega t}$ at $t = 0$ (i.e., one with the data $\phi(0) = 1$, $\phi'(0) = -i\omega$) will at $t = t_0 > 0$ match a solution $\alpha e^{-i\omega t} + \beta e^{+i\omega t}$ with some nontrivial β. In a quantum field theory based on the ansatz (7.14), this mathematical behavior would be gratuitously interpreted as particle creation. At the other extreme, if $\phi^+(t)$ itself were accepted as the fundamental mode function at all time, there would be no particle creation at all!

There is no point in debating which of these quantizations is correct, since a quantum field theory based on the damped oscillator equation is probably unacceptable for other reasons. We return to the undamped, time-dependent equation (7.15) and demonstrate that a frequency shift takes place there, too. For a general function ω, (7.15) does not have an explicit exact solution, but approximations which are asymptotically correct in the limit of high frequency are available. (That is the appropriate regime to examine, since our difficulties are coming from the behavior of quantities at large k, which implies large ω, as we see from (7.13).) The effect we are seeking arises in the second- and higher-order terms of asymptotic expansions which generalize the approximation labeled "WKB" (Wentzel–Kramers–Brillouin) by physicists, but more often known to mathematicians as the phase-integral or Liouville–Green approximation.

Theorem: If $\omega \in C^\infty$, the solutions of (7.15) are linear combinations of basis solutions which can be approximated as follows:

$$\phi^{\mp}(t) = \frac{1}{\sqrt{2W(t)}}\exp\left(\pm i\int^t W(t')\,dt'\right) + O\left(\omega^{-N}\right),$$

$$W(t)^2 = \omega(t)^2[1 + \delta_2(t)\omega^{-2} + \delta_4(t)\omega^{-4} + \cdots];$$
(7.16)

$\delta_n(t)$ is a function of $\omega(t)$ and is derivatives at t up through $\omega^{(n)}(t)$, and $\delta_n(t)$ is bounded as $\omega \to \infty$. (N is determined by the order at which the series is truncated.) If $\omega \notin C^\infty$, the process fails beyond a certain N determined by the order of singularity of ω.

This expansion can be regarded as a power-series expansion (of the effective frequency W) in a single parameter, T^{-2}, as $T \to \infty$. The expansion parameter can be introduced into (7.15) in either of two ways: replace $\omega(t)$ by $T\omega(t)$, or by $\omega(t/T)$. These are equivalent under an obvious redefinition of t (rescaling). Thus one says sometimes that the approximation refers to the limit of *large* ω, and sometimes that it is valid when ω is *slowly varying*. The invariant expression of the condition for validity is that the oscillations characteristic of the solutions are rapid compared to the time variation of ω.

Exercise 28: Show that

$$\delta_2 = \frac{3}{4}\left(\frac{\omega'}{\omega}\right)^2 - \frac{1}{2}\frac{\omega''}{\omega}.$$

Hint: Postulate a series for W and derive recursion relations.

The particular version (7.16) of the higher-order WKB approximation was developed by Fröman 1966; see also Campbell 1972 and Fulling 1982. Most of the physics literature relating to our problem (such as Parker & Fulling 1974) cites the paper of Chakraborty 1973, which presents an alternative (to the hint in Exercise 28) way of deriving the expansion, based on successive transformations of the Liouville type on the equation (7.15). More-general references on higher-order WKB approximations include Olver 1961 and many textbooks on asymptotic analysis.

Observations upon (7.16):

(1) The amplitude function $[2W(t)]^{-\frac{1}{2}}$ does not involve any integrals; it depends *locally* on $\omega(t)$. This will prove to be important in the renormalization theory (Chap. 9).

(2) The error term is uniform in t on finite intervals.

(3) Dimensional analysis (or the equivalence of the two methods of introducing T) implies that each time derivative in the numerator of $\delta_n \omega^{-n}$ must be compensated by a factor of ω in the denominator. This is borne out by the recursion relation derived in Exercise 28.

In our application [see (7.13)], ω goes to infinity asymptotically linearly with either k or m. The expansion is really in terms of one of the dimensionless parameters $(Tk)^{-1}$ or $(Tm)^{-1}$, depending on context.

(4) Only a certain finite, easily generated list of polynomial combinations of derivatives of ω may appear in the numerator of a given δ_n. In our application this translates into a list of combinations of derivatives of the metric function, a. For example, δ_2 (see Exercise 28) contains a'' and $(a')^2$, while δ_4 contains $a^{(4)}$, $a^{(3)}a'$, $(a'')^2$, $a''(a')^2$, and $(a')^4$. (For details see the last equation of Parker & Fulling 1974; the very last coefficient is a misprint — it should be $\frac{1}{8}$, not 18.)

We can now state the ansatz of the "adiabatic vacuum": In the field expansion (7.3), choose $\phi_{\mathbf{k}}(t)$ $\left(\equiv C(\eta)^{-\frac{1}{2}}\chi_{\mathbf{k}}(\eta)\right)$ to be any exact solution [of the ordinary differential equation (7.2), or (7.15) with ω equal to $W_{\mathbf{k}}$ from (7.13)] whose WKB approximation is of the form ϕ^+ in (7.16). Then define a vacuum state in the traditional way: $a_{\mathbf{k}}|0\rangle = 0$ for all $\mathbf{k}$.

There are two variants of this procedure, to be used in different contexts:

(A) Fix m. (In particular, m may be 0.) Implement the ansatz for all sufficiently large $\mathbf{k}$. (Note that if $\mathbf{k}$ is too small, (7.16) may be meaningless because the radicand is negative or zero. This is not important. For small $\mathbf{k}$, choose $\phi_{\mathbf{k}}$ in any nonsingular way.) This specification of the asymptotic behavior of the mode functions gives one control over the convergence of infinite sums over $\mathbf{k}$. As a consequence, the renormalization prescription of Chapter 9 can be guaranteed to give finite answers for this state and for a dense set of other states in the Fock space built on it.

(B) Consider m so large that the ansatz can be implemented for all $\mathbf{k}$. This stronger adiabatic construction will yield series expansions in m^{-1} of such quantities as $\langle 0|T_{00}|0\rangle$.

See Bunch et al. 1978 for further discussion.

Let us summarize the consequences of this construction:

(1) The procedure determines a Hilbert space for the theory (that is, a representation of the field algebra), modulo the complications introduced by infinite-volume divergences. The reason for this is

that any two solutions agreeing with the expansion (7.16), with $W_{\mathbf{k}}$ given up through the term δ_4, will differ by a Bogolubov transformation with $\beta_{\mathbf{k}} = O(k^{-3})$. This β is square-summable. A similar remark applies to ambiguities in the definition of ϕ^+ itself: Instead of expanding $W_{\mathbf{k}}{}^2$ in a power series, we could have expanded $W_{\mathbf{k}}{}^p$ for any $p \in \mathbf{R}$. The results would depend on p, but asymptotically they would be identical, the deviation decreasing as N increases.

(2) If $a(t)$ is asymptotically static, then $\phi_{\mathbf{k}}^{\text{in}}$ and $\phi_{\mathbf{k}}^{\text{out}}$ are in the class of solutions approximated by ϕ^+, for any N. Therefore, the adiabatic approach recovers the "right" Hilbert space for such systems, whereas Hamiltonian diagonalization gave the *wrong* representation, in general.

(3) On the other hand, the adiabatic construction applies to an arbitrary $a(t)$, whether or not $|0^{\text{in}}\rangle$ and $|0^{\text{out}}\rangle$ are definable! This is of the utmost importance, since the real universe was almost surely not static in its initial moments.

(4) After a covariant renormalization procedure to be described in Chapter 9, $\langle 0|T_{\mu\nu}(\underline{x})|0\rangle$ (or any other similar observable, such as electric charge density) comes out finite at all times. The same will be true of all other "nice" states in the Fock space built on $|0\rangle$.

It is important to understand what this theory does *not* do:

(a) It does not give a precise definition of "vacuum" and "particles" at each time. Many solution families (parametrized by $\mathbf{k}$) have the same asymptotic expansion in k, even to infinite order; so there is always an inherent "fuzziness" in a definition of particles based on (7.16).

(b) It does not give a framework for describing a dynamical evolution from time t_1 to time t_2, simply because (7.16) makes no reference to any particular moment of time. A solution consistent with (7.16) may be specified by data $\phi^+(t_1)$ and $\phi^{+\prime}(t_1)$, but imposing similar data at any other time yields a function in the *same* asymptotic equivalence class of solutions. Any particle-creation effect is therefore lost in the noise of the inherent nonuniqueness of the adiabatic definition of particles. [In an asymptotically static (and smooth) situation, recall that the particle creation falls off to infinite order in k. Any asymptotic expression (7.16) is written down to some finite order only, so the intrinsic error swamps the effect of particle creation.]

So, this is not a theory of particles. What it *does* do is to fix the correct Fock space, once and for all (modulo the infinite-volume problem, which does not affect the finiteness of local observables). The physical implications of the dynamics must be sought in terms of *field* observables. This forces us again to the problem of defining a meaningful energy-momentum tensor in curved space-time; it will occupy our attention for most of the rest of the book.

Of course, there is a major piece of unfinished business in the adiabatic approach: It has not yet been extended to models which can't be solved by separation of variables. In a nonseparable case, the ordinary differential equation with respect to time does not arise; the adiabatic analysis must somehow be applied to the entire partial differential equation, which, with luck, we can cast into the form

$$\frac{\partial^2 \phi}{\partial t^2} = K(t)\phi$$

for some time-dependent self-adjoint, elliptic operator, K. Although there is a book [Krein 1971] devoted to such mathematical problems (under possibly awkward technical conditions on $K(t)$, such as that its domain be independent of t), the difficulties of rigorous analysis in this area have deterred serious investigation so far. Presumably the lowest-order term in the adiabatic approximation should be based on the spectral decomposition of $K(t)$. One can foresee that serious problems will occur when the spectrum changes (e.g., when eigenvalues cross).

Chapter 8
Some Geometrical Apparatus

Up to now our treatment of differential geometry has been rather informal. The reader is assumed to have the modicum of background knowledge of manifolds and tensors needed to follow the discussion at the modest level of rigor appropriate to the occasion. In the study of renormalization in the next chapter, however, covariant derivatives and curvature tensors will enter in an essential and technical way. It seems prudent, therefore, to pause for a comparatively thorough and precise discussion of those concepts, which are central to contemporary physics — not only in general relativity, but also in the gauge theories of elementary particles.

We are concerned here only with purely "local" matters. Global topology, which leads in physics to monopoles, the Aharonov–Bohm effect, instantons, and so on, has been the subject of a vast amount of recent expository writing and is beyond the scope of this book.

COVARIANT DERIVATIVES

Let's start with a conceptual introduction to, or review of, the two most familiar and elementary instances of covariant differentiation in physics. They correspond to the fundamental coupling of matter to the electromagnetic and gravitational fields, respectively.

In nonrelativistic quantum mechanics the basic dynamical object is the wave function, $\psi(\underline{x}) = \psi(t, \mathbf{x})$. Insofar as the operational significance of ψ lies in the probability density, $|\psi(\underline{x})|^2$, the *phase* of ψ is irrelevant. At least, the overall phase is. At first sight it might appear that even the *relative* phase of the wave function at different points is unobservable:

$$\tilde{\psi}(\underline{x}) \equiv e^{i\theta(\underline{x})} \, \psi(\underline{x}) \equiv [U\psi](\underline{x})$$

contains the same information as ψ. This is not quite correct, however, as soon as one considers the momentum observable (or anything else which does not commute with position). A component of the momentum operator is

$$p^j = -i\partial_j \equiv -i\frac{\partial}{\partial x^j} \qquad (j = 1, \ldots, d),$$

so its expectation value in the state ψ is

$$\langle p^j \rangle = -i \int \psi^* \partial_j \psi \, d^d x.$$

But

$$-i\partial_j \psi = e^{-i\theta}[-i\partial_j \tilde{\psi} - (\partial_j \theta)\tilde{\psi}],$$

and because of the extra term,

$$\int \psi^* \partial_j \psi \, d^d x \neq \int \tilde{\psi}^* \partial_j \tilde{\psi} \, d^d x.$$

Thus ψ and $\tilde{\psi}$ are not physically equivalent. A Schrödinger equation of the elementary form is not invariant under U:

$$i\partial_0 \psi = -\frac{1}{2m} \sum_{j=1}^{d} \partial_j^2 \psi + V(\mathbf{x})\psi$$

transforms to

$$i\partial_0 \tilde{\psi} = -\frac{1}{2m} \sum_{j=1}^{d} [\partial_j - i(\partial_j \theta)]^2 \, \tilde{\psi} + [V + (\partial_0 \theta)] \, \tilde{\psi}.$$

Suppose, however, that each differentiation operator in the Schrödinger equation was already accompanied by a function:

$$i\partial_0 \psi = -\frac{1}{2m} \sum_{j=1}^{d} [\partial_j + iA_j(\mathbf{x})]^2 \psi + V(\mathbf{x})\psi. \qquad (8.1)$$

Then the equation satisfied by $\tilde{\psi}$ will be of the same form (8.1), but with different functions $\mathbf{A}$ and V, because

$$U[\partial_\mu + iA_\mu]U^{-1} = \partial_\mu + i(A_\mu - \partial_\mu \theta) \equiv \partial_\mu + i\tilde{A}_\mu.$$

(If we allow θ to depend on t, then $\mathbf{A}$ and V must also.) Of course, $\mathbf{A}$ really does exist; it is the electromagnetic vector potential. In this way, a theoretical physicist who had never heard of magnetism might be led to predict its existence, on the basis of the purely aesthetic requirement that quantum mechanics be invariant under the local phase transformations, $U = U\{\theta(\underline{x})\}$.

This observation makes possible the following new point of view. Refusing to commit ourselves to any local phase convention, we regard the wave function in an abstract way: ψ maps each point $\underline{x}$ onto some point in a certain space — call it $F_{\underline{x}}$ — about which we can say only that it has the same linear and metric structure as the complex plane. We specifically deny that this space can be identified with $\mathbf{C}$ in a fixed way; any such identification would be tantamount to an arbitrary phase convention. On these abstract wave functions there is defined an operation of *covariant differentiation*,

$$\psi \mapsto \nabla_\mu \psi.$$

Given any particular local phase convention, this operation is expressible by a concrete formula among complex-valued functions:

$$\nabla_\mu \psi(\underline{x}) = [\partial_\mu + iA_\mu(\underline{x})]\psi(\underline{x}). \tag{8.2}$$

If the phase convention is changed, $\tilde{\psi} \equiv e^{i\theta(\underline{x})}\psi$, then the *connection coefficients*, A_μ, change according to

$$\tilde{A}_\mu = A_\mu - \partial_\mu \theta. \tag{8.3}$$

A particular local phase convention is called "a gauge" (or "a choice of gauge"). The expectation value

$$\langle p^j - A^j \rangle = \int \psi^*(-i\nabla_j \psi)\, d^d x$$

is gauge-invariant. (In the Lagrangian formalism of particle mechanics, $\mathbf{p} - \mathbf{A}$ equals $m\mathbf{v}$, which is indeed the quantity with direct physical significance.) Similarly, the gauge-covariant Schrödinger equation is

$$i\nabla_0 \psi = -\frac{1}{2m} \sum_{j=1}^{d} \nabla_j^2 \psi, \tag{8.4}$$

which is (8.1). The Klein–Gordon and Dirac equations can be treated in the same way. ("*Covariant*" means that the *form* remains unchanged, while the particular coefficients change according to (8.3).)

Notational remarks:

(1) Our sign conventions are those of Messiah 1961, Chap. 20. In particular, $A_0 = +V$ and the components of $\mathbf{A}$ are $A^j = -A_j$. The quantum correspondence rule is $p_\mu \mapsto +i\partial_\mu$ for $\mu = 0, 1, \ldots, d$. Because of our metric-signature convention, the components of both spatial vectors, $\mathbf{p}$ and $\mathbf{A}$, change sign when indices are raised and lowered. People who use the metric signature $(-+++)$ also usually reverse the relative sign of ∂_μ and iA_μ in (8.2).

(2) Units are chosen so that c and $\hbar$ equal 1. Moreover, the unit electrical charge, e, is absorbed into A_μ. (It will come back eventually in the denominator of the kinetic Lagrangian of the electromagnetic field, $e^{-2}F_{\mu\nu}F^{\mu\nu}$.)

The second example is the differentiation of vector fields on a manifold. Any definition of such a derivative which can be covariantly expressed in all coordinate systems and satisfies some basic formal properties (to be reviewed briefly below) leads in each particular coordinate system to a formula of the type

$$\nabla_\mu v^\nu \equiv v^\nu{}_{;\mu} = \partial_\mu v^\nu + \Gamma^\nu_{\rho\mu} v^\rho. \tag{8.5}$$

where the connection coefficients or *Christoffel symbols* $\Gamma^\nu_{\rho\mu}$ are functions of $\underline{x}$. When the manifold is Riemannian or pseudo-Riemannian, there is a preferred connection whose coefficients are constructed from first-order derivatives of the metric tensor, but our preliminary remarks here are valid more generally.

To see the near inevitability of (8.5) and to give the Christoffel symbols a more intrinsic meaning, suppose that a basis has been chosen for the space of (contravariant) vectors at each point in the manifold: $\{\mathbf{e}_\alpha(\underline{x})\}_{\alpha=1}^n$. (The α labels different vectors in the basis, not the components of a single vector $\mathbf{e}$.) Suppose also that $\mathbf{e}_\alpha$ varies smoothly with $\underline{x}$, so that it makes sense to differentiate it. Each vector field $\mathbf{v} \equiv \{v^\mu(\underline{x})\}$ can be written as a linear combination, $\mathbf{v} = v^\alpha \mathbf{e}_\alpha$. We assume that covariant differentiation of the product of a scalar function and a vector field satisfies the Leibnitz rule. Then $\nabla \mathbf{v}$, the tensor whose components are standardly written $\nabla_\mu v^\nu$ or $v^\nu{}_{;\mu}$, must be calculable by

$$\nabla_\mu(v^\alpha \mathbf{e}_\alpha) = \left(\nabla_\mu(v^\alpha)\right)\mathbf{e}_\alpha + v^\alpha \nabla_\mu \mathbf{e}_\alpha,$$

where $\nabla_\mu(v^\alpha)$ is the covariant derivative of the *scalar function* v^α, assumed to reduce to the ordinary derivative, $\partial_\mu v^\alpha \equiv v^\alpha{}_{,\mu}$. *Define the*

Γ's by

$$\nabla_\mu \mathbf{e}_\alpha \equiv \Gamma^\beta_{\alpha\mu} \mathbf{e}_\beta .\qquad(8.6)$$

Then our calculation shows that

$$\partial_\mu \mathbf{v} = (v^\alpha{}_{,\mu} + \Gamma^\alpha_{\beta\mu} v^\beta) \mathbf{e}_\alpha ,$$

which is a restatement of (8.5).

Now let's pass from these examples to a general framework. Let M be a manifold of dimension n. Usually for us it will be equipped with a metric tensor, $g_{\mu\nu}(\underline{x})$. We want to consider derivatives of objects $\phi(\underline{x})$, which are examples of some type of (classical) "field" on M; technically, they are sections of a vector bundle over M. Each ϕ is a kind of function, whose value at a point $\underline{x}$ belongs to some vector space $F_{\underline{x}}$ (the *fiber* at $\underline{x}$). Physically, the elements of this space may be numbers or vectors, as in the two examples, or tensors, spinors, isovectors (of a non-Abelian gauge theory), etc. $F_{\underline{x}}$ is *isomorphic* to any other fiber, $F_{\underline{y}}$, but not *canonically identified* with it. Let the common dimension of the fiber spaces be r. They may be real or complex vector spaces, depending on application. Let F be a "fiducial" vector space isomorphic to the fibers. (Thus $F = \mathbf{C}$ in our first example, $\mathbf{R}^N$ (N = dimension of manifold) in the second.)

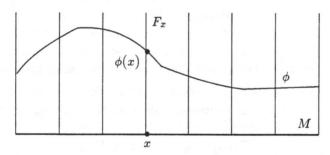

Ordinarily I use an index-free vector and matrix notation in F, but a classical index notation for tensors over M. For the latter, we will always be using a *coordinate basis*, rather than a more general "moving frame" (see, e.g., Schutz 1985, Chap. 5).

To do calculations with ϕ it helps to represent $\phi(\underline{x})$ by numbers. Therefore, we introduce a basis in each $F_{\underline{x}}$, so that

$$\phi(\underline{x}) = \phi^j \mathbf{e}_j \equiv \sum_{j=1}^r \phi^j(\underline{x}) \mathbf{e}_j(\underline{x}).$$

Relative to the basis $\{\mathbf{e}_j\}$, each field ϕ is identified with a sequence $\{\phi^j(\underline{x})\}_{j=1}^r$ of ordinary real- or complex-valued functions. A change of basis, or *gauge transformation*, is specified by an equation of the form

$$\tilde{\mathbf{e}}_j(\underline{x}) \equiv \sum_{k=1}^r \mathbf{e}_k(\underline{x})[U(\underline{x})^{-1}]^k{}_j \tag{8.7a}$$

or, equivalently,

$$\tilde{\phi}^j(\underline{x}) = \sum_{k=1}^r U(\underline{x})^j{}_k \phi^k(\underline{x}) \equiv (U\phi)^j. \tag{8.7b}$$

We assume that all "admissible" bases belong to an equivalence class such that the matrices $U^j{}_k$ relating them are smooth functions.

Now to the main business at hand, defining the derivative of ϕ. The literal partial derivatives $\partial\phi_j/\partial x^\mu$ do not fit together as the components of an intrinsically meaningful object, because they do not include information on the $\underline{x}$-dependence of the basis vectors. (In general there are no "constant" $\mathbf{e}$'s against which other things may be compared.) Consequently, $\partial_\mu\phi^j$ has a complicated, inhomogeneous transformation law involving derivatives of U. As the derivative of a scalar function is a covariant vector field, one would prefer the derivative of a section ϕ to be a covector-valued section (or $F_{\underline{x}}$-valued covector), $\nabla_\mu\phi$, which continues to behave like (8.7b) under gauge transformations (and also like an ordinary covector field under coordinate transformations in M).

Therefore, we define *a covariant differentiation* to be any mapping of ordinary sections into covector-valued sections which satisfies

$$\nabla_\mu(\phi_1 + \phi_2) = \nabla_\mu\phi_1 + \nabla_\mu\phi_2$$

and

$$\nabla_\mu(f\phi) = (\partial_\mu f)\phi + f\nabla_\mu\phi$$

(where f is any ordinary, scalar-valued function). Applying these axioms to the expansion $\phi = \phi^j\mathbf{e}_j$, we get

$$\begin{aligned}\nabla_\mu\phi &= (\partial_\mu\phi^j)\mathbf{e}_j + \phi^j\nabla_\mu\mathbf{e}_j \\ &\equiv (\partial_\mu\phi^j)\mathbf{e}_j + \phi^j(\nabla_\mu\mathbf{e}_j)^k\mathbf{e}_k \\ &= [\partial_\mu\phi^j + \phi^k(\nabla_\mu\mathbf{e}_k)^j]\mathbf{e}_j,\end{aligned}$$

where we had to do some index relabeling in the last step. Therefore, if $w_\mu(x)$ (which is, for each x and each μ, a matrix) is defined by

$$\nabla_\mu(\mathbf{e}_k) \equiv [w_\mu]^j{}_k \mathbf{e}_j, \tag{8.8}$$

there follows

$$(\nabla_\mu \phi)^j(x) = \left(\partial_\mu \phi^j\right)(x) + [w_\mu(x)]^j{}_k \phi^k(x),$$

usually abbreviated to

$$\nabla_\mu \phi = \partial_\mu \phi + w_\mu \phi. \tag{8.9}$$

Conversely, any derivative defined by such a formula satisfies the linearity and Leibnitz conditions from which we started.

Note that the covariant derivative (or the associated *connection form*, w_μ), is *extra structure* — it is not uniquely determined by the manifold and vector bundle.

Exercise 29: What happens to w under a gauge transformation,

$$\tilde{\mathbf{e}} = \mathbf{e}U^{-1}, \qquad \tilde{\phi} = U\phi?$$

(a) Show that

$$\tilde{w}_\mu = U[w_\mu - U^{-1}\partial_\mu U]U^{-1} = Uw_\mu U^{-1} - (\partial_\mu U)U^{-1}$$

and hence

$$\tilde{\nabla}_\mu = U\nabla_\mu U^{-1}.$$

(I.e., $\tilde{\nabla}$ and ∇ represent the same geometrical operation with respect to different bases.)

(b) Since we use coordinate bases, a coordinate transformation in M determines a gauge transformation, in the present extended sense, on the vector fields on M. Show that the application of (a) to contravariant vector fields yields the transformation law of Christoffel symbols found in classical texts on differential geometry,

$$\tilde{\Gamma}^\rho{}_{\mu\nu} = \frac{\partial \tilde{x}^\rho}{\partial x^\gamma}\frac{\partial x^\alpha}{\partial \tilde{x}^\mu}\frac{\partial x^\beta}{\partial \tilde{x}^\nu}\Gamma^\gamma{}_{\alpha\beta} - \frac{\partial^2 \tilde{x}^\rho}{\partial x^\alpha \partial x^\beta}\frac{\partial x^\alpha}{\partial \tilde{x}^\mu}\frac{\partial x^\beta}{\partial \tilde{x}^\nu}.$$

The transformation of the index μ under the coordinate change must also be taken into account here.

Gauge theories in physics involve two additional elements which are not emphasized here. First, the connection form itself becomes a dynamical field, satisfying its own field equations (Maxwell, Einstein, or Yang–Mills equations). This gives the theory a true gauge *in*variance, not just a covariance. Second, in the gauge theories of particle physics, not all smooth choices of local basis are admissible; or, as it is more often expressed, the group of allowed gauge transformations at a point is not the entire general linear group of dimension r. Rather, one considers an equivalence class of bases related among themselves by gauge transformations whose local values belong to a subgroup that (typically) preserves some quadratic form. The group, such as U(1), SU(3), SO(10), etc., is prescribed before the vector bundles supporting it as a gauge group are constructed, and, regardless of the bundle, the connection matrices w_μ always belong to some representation of the corresponding Lie algebra. We have already seen this situation in the electromagnetic case, where only multiplication by functions of modulus 1 (the group U(1)) was considered, and the connection cofficients consequently were purely imaginary. Since the choice of group determines the possible multiplets of elementary particles, particle physicists place great emphasis on the group in expounding gauge theories. For our purposes, however, the group is of secondary importance.

CURVATURE

The presence of a nonvanishing electromagnetic field means that there is *no* basis with respect to which $A_\mu = 0$. Similarly, the presence of a nontrivial gravitational field means that there is no coordinate system in which the Christoffel symbols are identically zero. (Either of these conditions could be enforced at a single point, but not throughout an open region.) Both of these fields are manifestations of *curvature* of vector bundles (more precisely, curvature of particular connections on bundles).

There are a few more facts about connections that need to be placed on record before we proceed.

First, if F^* is the dual space of F, then the covariant derivative of F^*-valued fields is defined by postulating that a Leibnitz rule holds for differentiation of the pairing of an F^*-valued field with an F-valued field (that is, a scalar function $\Psi(\phi) \equiv \Psi_j \phi^j$, where $\Psi(\underline{x})(\,\cdot\,) \in F_{\underline{x}}^*$ and $\phi(\underline{x}) \in F_{\underline{x}}$). This amounts to saying that the connection matrices for F^*-valued fields are the negatives of the transposes of those for F-valued fields. For example, the formula for differentiation of a covector field is

$$\nabla_\mu v_\nu \equiv v_{\nu;\mu} = \partial_\mu v_\nu - \Gamma^\rho_{\nu\mu} v_\rho \, ;$$

in the comparison with (8.5), the transposition is on the indices (ν, ρ).

Exercise 30: Prove the assertion ("This amounts to saying ... ").

Second, similar reasoning establishes that the connection form for the derivative of any kind of tensor should be the sum of the connection forms for the various factors of the tensor-product space in question. That is, the formula for the covariant derivative of a tensor contains, besides the literal derivative, one term for each index of the tensor, and the form of that term is the same as it would be if that were the only index the tensor possessed. This refers both to contravariant and covariant space-time indices (Greek superscripts and subscripts) and to primordial and dual bundle indices (referring to components of vectors in F and F^*). Example:

$$\nabla_\mu B^j_{\nu k} \equiv \partial_\mu B^j_{\nu k} - \Gamma^\rho_{\nu\mu} B^j_{\rho k} + [w_\mu]^j_{\ l} B^l_{\nu k} - [w_\mu]^l_{\ k} B^j_{\nu l} \, .$$

Charged scalar fields (for which $w_\mu = iA_\mu$) are included in this formalism, although the "index" involved has a range of only one value and would not normally be written as such.

Exercise 31: Let $M(\underline{x})$ be a "matrix-valued" field — i.e., its value at each point $\underline{x}$ is a linear operator mapping $F_{\underline{x}}$ into itself. Show that the covariant derivative of such an object is

$$M_{;\mu} = M_{,\mu} + [w_\mu, M].$$

Everything that has been said about differentiation on bundles applies, of course, to these tensor bundles. That is, the index j in the formalism may (and often does) stand for a whole string of indices, some of which are Greek and others are the bundle indices of one or more elementary field types such as spinors, charged scalars, isovectors, etc. (It should also be kept in mind that the elementary field type could simply be the ordinary vector field; inclusion of a connection in the formalism does not necessarily imply presence of a gauge field in the narrowest physical sense.) The derivative of a tensor is a tensor with one additional covariant space-time index.

Knowing how to differentiate a general tensor, we are now able to calculate higher-order derivatives of the original sections. The second derivative is

$$\phi_{;\mu\nu} \equiv \nabla_\nu \nabla_\mu \phi \qquad [\text{sic!}]$$
$$= (\phi_{;\mu})_{,\nu} + w_\nu \phi_{;\mu} - \Gamma^\alpha_{\mu\nu} \phi_{;\alpha}$$
$$= (\phi_{,\mu\nu} + w_{\mu,\nu}\phi + w_\mu\phi_{,\nu}) + (w_\nu\phi_{,\mu} + w_\nu w_\mu\phi)$$
$$- \Gamma^\alpha_{\mu\nu}\phi_{,\alpha} - \Gamma^\alpha_{\mu\nu} w_\alpha \phi.$$

(Some mathematicians consider the classical index notation particularly offensive or misleading in this context; see Remark below.) We note that this expression is not symmetric in μ and ν; covariant differentiations in different directions need not commute. To be precise,

$$\phi_{;\nu\mu} - \phi_{;\mu\nu} = (w_{\nu,\mu} - w_{\mu,\nu} + [w_\mu, w_\nu])\phi + (\Gamma^\alpha_{\mu\nu} - \Gamma^\alpha_{\nu\mu})\phi_{;\alpha}.$$

Since the values of ϕ, $\phi_{;\alpha}$, and the left side of the equation at any particular $\underline{x}$ are all tensors, and since the values of ϕ and $\phi_{;\alpha}$ could be anything in the respective fiber spaces, the objects

$$Y_{\mu\nu}(\underline{x}) \equiv w_{\nu,\mu} - w_{\mu,\nu} + [w_\mu, w_\nu] \qquad (8.10)$$

and

$$T^\rho_{\mu\nu}(\underline{x}) \equiv \Gamma^\alpha_{\mu\nu} - \Gamma^\alpha_{\nu\mu} \qquad (8.11)$$

must be tensors themselves, being linear maps from one tensor space into another. Y is called the *curvature* tensor of the bundle to which ϕ belongs. Note that for *each* μ and ν, $Y_{\mu\nu}(\underline{x})$ is a linear map from $F_{\underline{x}}$ into itself, hence, in component language, an $r \times r$ matrix, or a tensor with one contravariant bundle index and one covariant bundle index (both suppressed in our notation). T is the *torsion* tensor; it has nothing specifically to do with the bundle with fiber F, but rather is completely determined by the connection defining covariant derivatives of vector fields on M. If the Christoffel symbols of that connection are symmetric in their two subscripts, as is usually assumed, then the torsion is zero. To get the curvature of this manifold connection, let the sections ϕ in our general formalism be the contravariant vector fields; then the w's in (8.10) are the Christoffel symbols, and one has

$$[Y_{\mu\nu}]^\alpha_\beta = R^\alpha_{\beta\mu\nu} \equiv \Gamma^\alpha_{\beta\nu,\mu} - \Gamma^\alpha_{\beta\mu,\nu} + \Gamma^\alpha_{\gamma\mu}\Gamma^\gamma_{\beta\nu} - \Gamma^\alpha_{\gamma\nu}\Gamma^\gamma_{\beta\mu}. \qquad (8.12)$$

This is the *Riemann curvature tensor*. (See also (8.14).) The curvature tensor for covectors is the same with a minus sign (and a transposition, implemented automatically by the rule that a subscript can be contracted (summed over) only with a superscript). Finally, note that in the electromagnetic case, where $w_\mu = iA_\mu$,

$$Y_{\mu\nu} = i(A_{\nu,\mu} - A_{\mu,\nu}) \equiv iF_{\mu\nu}$$

is essentially the electromagnetic field-strength tensor.

Exercise 32: Verify directly from (8.10) that $Y_{\mu\nu}$ transforms under gauge transformations as a tensor with one covariant and one contravariant bundle index:

$$\tilde{Y}_{\mu\nu}(x) = U(x)Y_{\mu\nu}(x)U(x)^{-1}.$$

The equation that resulted from our second-derivative calculation can be rewritten

$$\phi_{;\nu\mu} = \phi_{;\mu\nu} + Y_{\mu\nu}\phi + T^\rho_{\mu\nu}\phi_{;\rho}. \tag{8.13}$$

This *Ricci identity* tells one how to unscramble repeated covariant derivatives into any desired order. Once again, tensor fields of any type are covered by this formula if $Y\phi$ is suitably interpreted; in general it will be a sum of terms, some involving Riemann tensors and some involving the Y of some elementary type of field (the *gauge curvature* or *gauge field strength* of a physical gauge theory). For example, let G_μ be the object $M_{;\mu}$ of Exercise 31; then

$$G_{\mu;\rho\nu} = G_{\mu;\nu\rho} - R^\alpha{}_{\mu\nu\rho}G_\alpha + [Y_{\nu\rho}, G_\mu],$$

if there is no torsion and if Y now denotes the elementary gauge field strength.

Remark: A tensor identity such as

$$\nabla_\mu\nabla_\nu v^\rho = \nabla_\nu\nabla_\mu v^\rho + R^\rho{}_{\sigma\mu\nu}v^\sigma \tag{8.14}$$

can be interpreted either as an equation between abstract tensors, or as an equation between the concrete components of the tensors with respect to some basis. In the latter case, the correct interpretation of the left-hand side, for instance, is

$$(\mathbf{e}_\mu)^\alpha (\mathbf{e}_\nu)^\beta \nabla_\alpha \nabla_\beta \mathbf{v}$$

and *not*

$$(\mathbf{e}_\mu)^\alpha \nabla_\alpha \left((\mathbf{e}_\nu)^\beta \nabla_\beta \mathbf{v} \right).$$

There is thus a dangerous ambiguity in the notation if one insists on regarding "∇_μ" as an abbreviation for a directional derivative along a particular concrete basis-vector field, since one will then be led to the second interpretation, which differs from the first by a term involving the derivative of $\mathbf{e}_\nu$. Many modern texts on differential geometry give the Ricci identity in a directional-derivative form, in which an extra term must be subtracted off to account for the derivatives of the vector fields along which one is taking the directional derivatives. With the notation $\mathbf{u} \cdot \nabla$ for $u^\mu \nabla_\mu$, etc., that equation is (in the case of zero torsion)

$$R(\mathbf{v}, \mathbf{u}, \mathbf{w}) \equiv \{ R^\alpha{}_{\beta\mu\nu} v^\beta u^\mu w^\nu \}$$
$$= [\mathbf{u} \cdot \nabla, \mathbf{w} \cdot \nabla]\mathbf{v} - ([\mathbf{u}, \mathbf{w}] \cdot \nabla)\mathbf{v},$$

where $[\mathbf{u}, \mathbf{w}]$ is the vector field with components $u^\mu w^\alpha{}_{,\mu} - w^\mu u^\alpha{}_{,\mu}$ (which is, incidentally, the commutator of $\mathbf{u} \cdot \nabla$ and $\mathbf{w} \cdot \nabla$ when acting on *scalars*). For a *coordinate* basis, $[\mathbf{e}_\mu, \mathbf{e}_\nu]$ will always be zero, and so the form of the Ricci identity is unchanged after all; for more general basis-vector fields, the commutator term is nontrivial. In any case, however, the classical Ricci identity (8.14) (or (8.13)) is a valid equation when properly understood as referring to components of tensors with respect to an arbitrary basis *at a point*, not to basis-vector fields; and that is how we shall always understand it.

The curvature and torsion tensors have certain symmetries, which must be taken into account when attempting to simplify expressions involving them into a unique and compact form. From the definition, it's obvious that

$$Y_{\nu\mu} = -Y_{\mu\nu} \tag{8.15a}$$

and hence

$$R^\alpha{}_{\beta\nu\mu} = -R^\alpha{}_{\beta\mu\nu}. \tag{8.15b}$$

The torsion, also, is antisymmetric in its two subscripts. The *Bianchi identity* is

$$Y_{\alpha\beta;\gamma} + Y_{\beta\gamma;\alpha} + Y_{\gamma\alpha;\beta} = -T^\rho_{\alpha\beta}Y_{\gamma\rho} - T^\rho_{\beta\gamma}Y_{\alpha\rho} - T^\rho_{\gamma\alpha}Y_{\beta\rho}; \qquad (8.16\mathrm{a})$$

as a special case,

$$R^\mu{}_{\nu\alpha\beta;\gamma} + R^\mu{}_{\nu\beta\gamma;\alpha} + R^\mu{}_{\nu\gamma\alpha;\beta} = \text{torsional terms}. \qquad (8.16\mathrm{b})$$

There is also the so-called *cyclic identity*,

$$\begin{aligned}
R^\mu{}_{\alpha\beta\gamma} &+ R^\mu{}_{\beta\gamma\alpha} + R^\mu{}_{\gamma\alpha\beta} \\
&= -T^\mu_{\alpha\beta;\gamma} - T^\mu_{\beta\gamma;\alpha} - T^\mu_{\gamma\alpha;\beta} + T^\rho_{\alpha\beta}T^\mu_{\rho\gamma} + T^\rho_{\beta\gamma}T^\mu_{\rho\alpha} + T^\rho_{\gamma\alpha}T^\mu_{\rho\beta}. \qquad (8.17)
\end{aligned}$$

This may be regarded as a Bianchi identity for the torsion; but it is more familiar in the context where the torsion is zero, in which case (8.17) with (8.15b) simply says that R is annihilated by complete antisymmetrization on its last three indices. Many authors call (8.17) the *first Bianchi identity*; (8.16b) is then the *second Bianchi identity*. Finally, the Ricci identity (8.13), applied to any of these tensors or any covariant derivative (to any order) of one of them, belongs in the list of symmetries.

   Proof of the Bianchi identities: Covariant derivatives satisfy the *Jacobi identity*

$$[\nabla_\mu, [\nabla_\nu, \nabla_\rho]]\phi + [\nabla_\nu, [\nabla_\rho, \nabla_\mu]]\phi + [\nabla_\rho, [\nabla_\mu, \nabla_\nu]]\phi = 0.$$

(This is a purely formal, or metamathematical, consequence of the antisymmetry of the definition of a commutator. The commutator here is just an abbreviation for the difference of two mixed second-order covariant derivatives in opposite order. It is not necessary to interpret the ∇ symbols as directional-derivative operations, so as to apply the Jacobi identity for *operators*. Therefore, it is correct to carry out this proof in the abstract index formalism.) Writing out all the derivatives and performing a few index manipulations, one arrives at

$$\begin{aligned}
0 = Y_{\nu\rho;\mu}\phi &+ T^\lambda_{\nu\rho;\mu}\phi_{;\lambda} + T^\lambda_{\nu\rho}Y_{\mu\lambda}\phi + T^\sigma_{\nu\rho}T^\lambda_{\mu\sigma}u_{;\lambda} - R^\lambda_{\mu\nu\rho}u_{;\lambda} \\
&+ \text{cyclic permutations on the free indices.}
\end{aligned}$$

Since ϕ and $\phi_{;\lambda}$ are arbitrary (and independent at any one $\underline{x}$), their respective coefficients may be set equal to 0. The results are versions of (8.16a) and (8.17).

In Riemannian geometry and gravitational theory one almost always deals only with connections which are *metric-compatible*:

$$\nabla_\rho g_{\mu\nu} = 0. \qquad (8.18)$$

This condition has the happy consequence that differentiation commutes with raising and lowering of indices. It also implies a large number of additional symmetries for the Riemann tensor. Differentiating (8.18) once more and antisymmetrizing leads to the conclusion that R is antisymmetric in its *first* two indices:

$$R_{\beta\alpha\mu\nu} = -R_{\alpha\beta\mu\nu}. \qquad (8.19)$$

(The first index here is the erstwhile superscript, lowered by the metric tensor.) This, together with the cyclic identity, implies the *pair symmetry*:

$$R_{\gamma\delta\alpha\beta} = R_{\alpha\beta\gamma\delta} + \text{torsion terms.} \qquad (8.20)$$

In addition, one usually takes the torsion to be zero. This uniquely determines the metric-compatible connection to be

$$\Gamma^\rho_{\mu\nu} = \tfrac{1}{2} g^{\rho\tau}(g_{\nu\tau,\mu} + g_{\mu\tau,\nu} - g_{\mu\nu,\tau}); \qquad (8.21)$$

the resulting Riemann tensor (8.12) is a sum of terms linear in second (literal) derivatives of g and terms quadratic in its first derivatives. I have included the torsional terms in most of the foregoing formulas simply to make them more useful for reference. However, it turns out that the techniques to be described in the remainder of this chapter also apply to theories with torsion, provided that the torsion tensor is *totally antisymmetric*. By convention, the superscript of T is regarded as the last index, so that

$$T_{\alpha\beta\gamma} \equiv T^\rho_{\alpha\beta} g_{\rho\gamma}.$$

The condition in question is then that this totally covariant tensor be antisymmetric under all permutations of the six indices.

Exercise 33: Consider a generic derivative of the Riemann tensor,

$$\nabla^p R = \{R_{\mu_1\mu_2\mu_3\mu_4;\mu_5\cdots\mu_{4+p}}\}.$$

(Assume that the connection is metric-compatible.) Show that after all the symmetries are taken into account, there are $\frac{1}{2}(p+1)(p+4)$ independent permutations of the indices, modulo torsion and lower-order derivatives of R. Namely, let $(\alpha, \beta, \gamma, \delta, \epsilon, \dots)$ be a canonical ordering of the formal index list; then a nonredundant list of independent components of the tensor $\nabla^p R$ is

(a) $R_{\alpha\beta\gamma\delta;\epsilon\dots}$

(b) $R_{\alpha\beta\gamma\mu;\delta\dots}$ for p choices of μ

(c) $R_{\alpha\gamma\beta\delta;\epsilon\dots}$

(d) $R_{\alpha\gamma\beta\mu;\epsilon\dots}$ for p choices of μ

(e) $R_{\alpha\mu\beta\nu;\gamma\dots}$ for $\frac{1}{2}p(p+1)$ choices of pairs $(\mu\nu)$ of canonically ordered distinct indices chosen from $(\delta, \epsilon, \dots)$.

PARALLEL TRANSPORT AND GEODESIC DISTANCE;
COVARIANT POWER SERIES

Let $x(\tau)$ be a curve in M, and let $\phi(\underline{x})$ be a section, defined at least on the [image of the] curve. One says that the values of ϕ are *parallel along* $x(\tau)$ if (with respect to any gauge)

$$\frac{d}{d\tau}\phi\big(x(\tau)\big) + w_\mu\big(x(\tau)\big)\phi\big(x(\tau)\big)\frac{dx^\mu}{d\tau} = 0. \qquad (8.22)$$

The left-hand side is called the *absolute derivative* of ϕ along the curve. If ϕ is defined in an open neighborhood of the curve (at least), then its covariant derivative is defined and the absolute derivative can be expressed in terms of it as

$$\dot{x}(\tau)\cdot\nabla\phi \equiv \frac{dx^\mu}{d\tau}\nabla_\mu\phi\big(x(\tau)\big).$$

Suppose that $x(0) \equiv \underline{x}'$, $x(1) \equiv \underline{x}$, and we are given $\phi_0 \in F_{\underline{x}'}$. Then the result of *parallel transport* of ϕ_0 to $\underline{x}$ (along this particular curve) is defined to be $\phi(1)$, where $\phi(\tau)$ is the solution of the differential equation (8.22) with initial value $\phi(0) = \phi_0$. ($\phi(\tau)$ is in $F_{x(\tau)}$ for each τ.) Following DeWitt 1965, we write

$$\phi(1) \equiv I_x(\underline{x}, \underline{x}')\phi_0. \qquad (8.23)$$

I_x is a one-to-one, linear mapping from $F_{\underline{x}'}$ onto $F_{\underline{x}}$; in concrete terms, it is a matrix-valued function of the two points, whose left index pertains to the fiber at $\underline{x}$ while its right index lives at $\underline{x}'$. (Consequently, when I is covariantly differentiated with respect to $\underline{x}$, for example, a connection form $w_\mu(\underline{x})$ attaches to the left index, but no connection form is needed for the right index.)

What is parallel transport in our two standard, elementary examples? In the electromagnetic case, (8.22) is

$$\frac{d\phi}{d\tau} + i\dot{x}^\mu A_\mu \phi = 0,$$

which is solvable in closed form:

$$\phi_1 = \exp\left(-i \int_{\underline{x}'}^{\underline{x}} A_\mu(\underline{y})\, dy^\mu\right) \phi_0,$$

where the line integral is along the curve $y = x(\tau)$. Thus

$$I_x(\underline{x}, \underline{x}') = \exp\left(-i \int_{\underline{x}'}^{\underline{x}} A_\mu(\underline{y})\, dy^\mu\right).$$

(In non-Abelian gauge theories A_μ is a matrix, (8.22) is a first-order differential *system*, and hence it generally can't be solved explicitly. Physicists sometimes denote the parallel-transport matrix in that case by

$$I_x(\underline{x}, \underline{x}') = \mathcal{P} \exp\left(-i \int_{\underline{x}'}^{\underline{x}} A_\mu(\underline{y})\, dy^\mu\right),$$

where $\mathcal{P}$ indicates a *path-ordered exponential* summarizing the solution of the equation by perturbation theory (i.e., as a power series in the magnitude of A).) In the gravitational case, the equation of parallel transport for a contravariant vector field is, from (8.5),

$$\frac{dv^\nu}{d\tau} = -\Gamma^\nu_{\rho\mu} v^\rho \dot{x}^\mu. \tag{8.24}$$

DeWitt 1965 and Christensen 1976 denote the resulting parallel-transport matrix by $g^\mu{}_{\nu'}$, where primed indices refer to $\underline{x}'$ and unprimed ones to $\underline{x}$.

Intuitively, one might prefer to take parallel transport as the fundamental notion and define covariant differentiation from it: In the context of a vector bundle, the problem with defining a derivative by a difference quotient,

$$\left.\frac{d\phi}{d\tau}\right|_{\tau=0} \equiv \lim_{\Delta\tau\to 0} \frac{\phi(x(\Delta\tau)) - \phi(x(0))}{\Delta\tau},$$

is that we don't know how to subtract values of ϕ at two different $\underline{x}$'s. But I_x gives us an identification of F_y with $F_{\underline{x}'}$, for each y on the curve x. (This identification depends on the curve!) Therefore, it is possible to modify the formula so that it makes sense: the absolute derivative at 0 is

$$\lim_{\Delta\tau\to 0} \frac{\phi(x(\Delta\tau)) - I_x(x(\Delta\tau), x(0))\phi(x(0))}{\Delta\tau}.$$

Exercise 34: Verify this formula for the absolute derivative within the framework of the definitions we've adopted. Would it matter if we put the inverse of I on the first term instead of I on the second?

The curve $x(\tau)$ is a *geodesic* if its own tangent vector is parallel along it. From (8.24), this condition is

$$\frac{d^2 x^\nu}{d\tau^2} = -\Gamma^\nu_{\rho\mu} \frac{dx^\rho}{d\tau} \frac{dx^\mu}{d\tau}. \tag{8.25}$$

This is a nonlinear second-order differential equation, which may be treated like the equation of motion of a classical mechanical system.

Remark: The antisymmetric part of Γ, which is the torsion tensor, makes no contribution to (8.25). Therefore, it is often said that the geodesics of a manifold are independent of the torsion. There is a complication, however: If a connection is metric-compatible and has a torsion tensor which is *not* totally antisymmetric, then the symmetric part of that connection is *not* the metric-compatible connection (8.21). In such a situation there are two sets of geodesics, those defined by the metric through (8.21), and those defined by the torsional, metric-compatible connection. This is the reason why our discussions below of Synge–DeWitt tensors and asymptotic expansions of Green functions are limited to totally antisymmetric torsions.

The *arc length* along $x(\tau)$ from $\underline{x}'$ to $\underline{x}$ is

$$s \equiv \int_0^1 \sqrt{g_{\mu\nu}\dot{x}^\mu\dot{x}^\nu}\, d\tau.$$

(To avoid complications, assume momentarily that the metric is positive definite.) We are interested in this quantity almost exclusively in the case that $x(\tau)$ is a geodesic. In practice we shall almost never use the integral expression just given; certain formal properties of the geodesic arc length suffice in all the many calculations involving it.

We wish to regard the geodesic arc length (or *geodesic distance*) as a function of two arguments, $\underline{x}$ and $\underline{x}'$. Some questions of principle need to be disposed of first. Two points in M may be joined by more than one geodesic. (For example, if M is a 2-sphere with the usual metric, then antipodal points are joined by infinitely many geodesics of equal length, and a generic pair of points is joined by two arcs of a great circle, one relatively short and one long.) Moreover, if the manifold is incomplete, two points may not be joined by any geodesic at all. (Think of two points on opposite sides of a "hole" cut out of the plane.) In fact, the same is true for some *complete* pseudo-Riemannian manifolds; the anti-DeSitter space described in Chapter 6 is an example. But it is intuitively plausible that all $\underline{x}$ *sufficiently close* to a given $\underline{x}'$ can be joined to $\underline{x}'$ by a geodesic, which can be uniquely and continuously defined by characterizing it as the *shortest* geodesic between those two points. In fact, more is true [Whitehead 1932; Friedlander 1975, Sec. 1.2]:

Theorem: Every point of M has a neighborhood D which is *geodesically convex* (or *normal*); that is, any two points in D are joined by precisely one geodesic segment which lies entirely in D.

(Note that a comma after "segment" would turn this proposition into a falsehood!)

Henceforth we tacitly restrict our discussion to a normal neighborhood. Define

$$\sigma(\underline{x},\underline{x}') \equiv \frac{1}{2}\,s^2.$$

This quantity was dubbed the *world function* by Synge 1960. Hadamard 1952, Synge 1960, and DeWitt 1965 have made marvelously productive use of it. See also the expositions of Garabedian 1964, esp. Sec. 2.6 and Chap. 5, and Friedlander 1975, esp. Sec. 1.2 (both of whom write 2Γ rather than σ). The world function is a C^∞ function (given a smooth metric to start from), unlike s, which has a conical singularity at 0. Furthermore, σ has a natural extension to manifolds of indefinite metric, in which case its values are (in my sign convention) positive for timelike separation, negative for spacelike separation, and zero on the light cone. In flat space, σ reduces to $\frac{1}{2}\|\underline{x}-\underline{x}'\|^2$.

Differentiating this last expression, we see that

$$\eta^{\mu\nu}\partial_\nu\sigma = (\underline{x}-\underline{x}')^\mu$$

in flat space. The analogous statement in curved space is that $-g^{\mu\nu}\nabla_\nu\sigma$ *is the tangent vector* (to the geodesic) *at $\underline{x}$ pointing toward $\underline{x}'$ with length equal to s*. Since the vector $\nabla_\nu\sigma$ is even more important than the scalar σ, it is customary to omit the semicolon in writing covariant derivatives of σ:

$$\sigma_\mu \equiv \sigma_{;\mu}, \quad \text{etc.}$$

Another significant object is

$$\hat\sigma^\mu \equiv -\sigma^{\mu'},$$

which is *the tangent vector at $\underline{x}'$ pointing toward $\underline{x}$ with length s*. (The primed index denotes a derivative with respect to x'.)

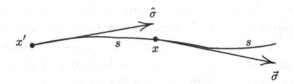

Thus the geodesics set up a one-to-one correspondence between a neighborhood of $\underline{x}'$ in M and a neighborhood of 0 in the tangent space $T_{\underline{x}'}(M)$:

$$\underline{x} \longleftrightarrow \hat\sigma(\underline{x}, \underline{x}').$$

The components of $\hat\sigma$ form a special coordinate system, the *Riemann normal coordinates at $\underline{x}'$*. (Riemann normal coordinates can be constructed without reference to the world function; see, e.g., Parker 1979 or many textbooks on general relativity.) Alternatively, one may stick to general coordinates, or to coordinate-free methods, and think of $\hat\sigma$ as just a certain vector field. This latter approach combines the best of both worlds: the advantages of a special coordinate system in a manifestly covariant, or geometrically intrinsic, formalism.

From the geometric interpretation of the quantities involved, it is obvious that

$$\sigma = \frac{1}{2}\sigma_\mu\sigma^\mu. \tag{8.26}$$

This first-order, nonlinear partial differential equation is also the Hamilton–Jacobi equation associated with the mechanical system whose "Newtonian" equation of motion is the geodesic equation, (8.25).

Restricting attention to normal neighborhoods and to geodesics within them renders the parallel-transport operator unique, so that we may write $I(\underline{x},\underline{x}')$ without ambiguity. Recall that I is, in effect, defined by the differential equation $\dot{x}\cdot\nabla I = 0$ and the initial condition $I(\underline{x}',\underline{x}') = 1$. In the present context, $\dot{x}$ is proportional to the σ vector, and so the equation can be written

$$\sigma^{\mu}I_{;\mu} = 0. \tag{8.27}$$

In dealing with functions of two variables, $f(\underline{x},\underline{x}')$, a convenient shorthand is

$$[f] \equiv f(\underline{x},\underline{x}).$$

Such a quantity is often called a *coincidence limit* in the physics literature, although the word "limit" is misleading — the function is simply to be *evaluated* at $\underline{x} = \underline{x}'$, with no commitment as to its continuity there. (The preferred jargon in the mathematical literature is "value on the diagonal".) When f is a covariant derivative of something, primes and the lack thereof on the indices are used to indicate which argument was the subject of the differentiation before the arguments were set equal:

$$[B_{;\mu\nu'}] \equiv \nabla_{\mu}\nabla_{\nu'}B(\underline{x},\underline{x}')\big|_{\underline{x}=\underline{x}'}.$$

Note that $[B]_{;\mu}$ is not the same thing as $[B_{;\mu}]$, since the former implicitly involves differentiation with respect to $\underline{x}'$ as well as $\underline{x}$. From the multivariable chain rule (traditionally called **Synge's theorem** in this context), we have

$$[B]_{;\mu} = [B_{;\mu}] + [B_{;\mu'}]. \tag{8.28}$$

This principle can be used to obtain primed (and mixed) derivatives easily from unprimed ones, and vice versa. Note that primed and unprimed derivatives involve independent connections, so their order with respect to each other is arbitrary. Therefore, the derivative to be eliminated can always be assumed to act last. In particular,

$$[\hat{\sigma}^{\mu}{}_{\nu\rho\tau...}] \equiv -[\sigma^{\mu'}{}_{\nu\rho\tau...}] = -[\sigma_{\nu\rho\tau...}{}^{\mu'}] = \nabla^{\mu}[\sigma_{\nu\rho\tau...}] - [\sigma_{\nu\rho\tau...}{}^{\mu}]. \tag{8.29}$$

Consider now a scalar function, $f(\underline{x},\underline{x}')$ (where "scalar" means not only that the dependent variable has only one — real or complex — component, but also that the bundle connection is trivial). By the multivariable version of Taylor's theorem, f (to whatever order it is smooth) possesses a power-series expansion in the Riemann normal co-ordinates of $\underline{x}$ about $\underline{x}'$. The coefficients in the series are the coincidence limits of the *literal* partial derivatives of f with respect to $\underline{x}$. I claim, however, that they may be replaced by the covariant derivatives:

$$f(\underline{x},\underline{x}') \sim [f] + [f_{;\mu}]\hat{\sigma}^{\mu} + \frac{1}{2!}[f_{;\mu\nu}]\hat{\sigma}^{\mu}\hat{\sigma}^{\nu} + \cdots . \qquad (8.30)$$

Equivalently, the totally symmetric part of the covariant derivative is equal to the literal derivative in normal coordinates. (In (8.30) and the following discussion, the coincidence limits are evaluated at $\underline{x}'$. Equivalently, one can reverse the roles of the two points, putting primes on all the derivatives.)

To see this, regard $f(\underline{x},\underline{x}')$ as a function of the arc-length param-eter, s, of the geodesic from $\underline{x}'$ to $\underline{x}$. Apply the single-variable Taylor theorem and the chain rule:

$$\begin{aligned}
f(\underline{x},\underline{x}') &= f(x(s),\underline{x}') \\
&\sim \cdots + \frac{1}{2}\frac{d^2}{d\tilde{s}^2}\left(f(x(\tilde{s}),\underline{x}')\right)\big|_{\tilde{s}=0}\, s^2 + \cdots \\
&= \cdots + \frac{1}{2}\frac{d}{d\tilde{s}}\left[\nabla_{\mu}f(x(\tilde{s}),\underline{x}')\frac{dx^{\mu}}{d\tilde{s}}\right]_{\tilde{s}=0} s^2 + \cdots .
\end{aligned}$$

Absolute differentiation along the curve obeys the Leibnitz rule; we use that fact to calculate the needed second derivative. The absolute derivative of $dx^{\mu}/d\tilde{s}$ vanishes because the curve is a geodesic [(8.25)]. Thus the term is

$$\frac{1}{2}\nabla_{\nu}\nabla_{\mu}f(x(\tilde{s}),\underline{x}')\frac{dx^{\nu}}{d\tilde{s}}\frac{dx^{\mu}}{d\tilde{s}}\,s^2 \qquad (\tilde{s}=0).$$

But $dx^{\mu}/d\tilde{s}$ is the *unit* tangent vector to the geodesic, which becomes $\hat{\sigma}$ upon absorbing a factor s. So we have reproduced the second-order term in (8.30). The argument continues by induction to higher orders.

If f is replaced by a section of a vector bundle, (8.30) as it stands does not make sense, because its two sides belong to different fibers. However, inserting a parallel-transport operator makes it correct:

$$\phi(\underline{x},\underline{x}') \sim \sum_{p=0}^{\infty}\frac{1}{p!}I(\underline{x},\underline{x}')[\phi_{;\mu_1\ldots\mu_p}]\hat{\sigma}^{\mu_1}\cdots\hat{\sigma}^{\mu_p} . \qquad (8.31)$$

The proof is the same as for (8.30), except that it must be applied to $I(\underline{x}, \underline{x}')^{-1}\phi(\underline{x}, \underline{x}')$ and that the absolute derivative makes its appearance already in the first order. The absolute derivative of I vanishes [(8.27)].

Now take a covariant derivative of (8.30) with respect to $\underline{x}$, recalling that the coincidence limits there are independent of $\underline{x}$. Take the coincidence limit of the result, noting that $[\hat{\sigma}^\mu] = 0$:

$$[f_{;\nu}] = [f_{;\mu}][\hat{\sigma}^\mu{}_\nu].$$

Since ∇f is arbitrary, it must be true that $[\hat{\sigma}^\mu{}_\nu] = \delta^\mu_\nu$, or, equivalently, $[\sigma_{\mu\nu}] = g_{\mu\nu}$. Considering higher derivatives of (8.30) in the same way, one concludes that *the totally symmetric* (in the subscripts) *part of* $[\hat{\sigma}^\mu{}_{\nu_1 \ldots \nu_p}]$ *equals* 0 *whenever* $p \neq 1$. Considering derivatives of (8.31), one concludes in the same way that *the totally symmetric part of* $[I_{\mu_1 \ldots \mu_p}]$ *is* 0 *whenever* $p \neq 0$. These facts about the diagonal values of derivatives of σ and I also follow easily from the explicit calculations in the next section; reversal of the present argument then provides a more concrete proof of (8.30) and (8.31) [cf. Widom 1980, reinterpreted following Drager 1978].

Recursive calculations of the Synge–DeWitt tensors

The objects $[\hat{\sigma}^\lambda{}_{\mu\nu}\ldots]$, $[\sigma_{\mu\nu}\ldots]$, and $[I_{;\mu\nu}\ldots]$ are of the utmost importance in computing asymptotic expansions of Green functions of partial differential operators, both by the classical HaMiDeW methods described in the next chapter and by the more modern and more general method of pseudodifferential operators [Widom 1980, Fulling & Kennedy 1988]. As we've seen in (8.29), the first class of these is easy to obtain from the second. Let us refer to the other two classes as the *Synge–DeWitt tensors*. DeWitt 1965 gave an algorithm for calculating them from the basic properties (8.26) and (8.27). It has been pushed to high order, with various degrees of computer assistance, by Schimming 1981, Rodionov & Taranov 1987, Fulling 1989 and unpublished, and Christensen & Parker 1989.

Remark: Widom 1980 gives a different algorithm, which, unlike DeWitt's, applies to connections which are not metric-compatible (or have incompletely antisymmetric torsion). (In that context σ is not defined, but $\hat{\sigma}^\mu$ is.) However, that approach requires consideration of many more index permutations than does DeWitt's ($p!$ versus p).

We already know that

$$[\sigma] = 0, \qquad [\sigma_\mu] = 0, \qquad [\sigma_{\mu\nu}] = g_{\mu\nu}. \qquad (8.32)$$

To obtain the next item in the list by DeWitt's method, differentiate (8.26) three times:

$$\sigma_{\mu\nu\rho} = \sigma^\alpha{}_{\nu\rho}\sigma_{\alpha\mu} + \sigma^\alpha{}_{\nu}\sigma_{\alpha\mu\rho} + \sigma^\alpha{}_{\rho}\sigma_{\alpha\mu\nu} + \sigma^\alpha\sigma_{\alpha\mu\nu\rho}.$$

Take the coincidence limit, using (8.32):

$$[\sigma_{\mu\nu\rho}] = [\sigma_{\mu\nu\rho}] + [\sigma_{\nu\mu\rho}] + [\sigma_{\rho\mu\nu}].$$

Subtract $3[\sigma_{\mu\nu\rho}]$ and use the Ricci identity:

$$\begin{aligned}
-2[\sigma_{\mu\nu\rho}] &= ([\sigma_{\nu\mu\rho}] - [\sigma_{\mu\nu\rho}]) + ([\sigma_{\rho\mu\nu}] - [\sigma_{\mu\nu\rho}]) \\
&= [\sigma_{\nu\mu\rho}] - [\sigma_{\mu\nu\rho}] + ([\sigma_{\rho\mu\nu}] - [\sigma_{\mu\rho\nu}] + [\sigma_{\mu\rho\nu}] - [\sigma_{\mu\nu\rho}]) \\
&= [(T^\alpha_{\mu\nu}\sigma_\alpha)_{;\rho}] + [(T^\alpha_{\mu\rho}\sigma_\alpha)_{;\nu}] + [T^\alpha_{\nu\rho}\sigma_{\mu\alpha}] - [\sigma_\alpha R^\alpha{}_{\mu\nu\rho}] \\
&= T_{\mu\nu\rho} + T_{\mu\rho\nu} + T_{\nu\rho\mu} \\
&= T_{\mu\nu\rho},
\end{aligned}$$

since T must be assumed antisymmetric. Therefore,

$$[\sigma_{\mu\nu\rho}] = -\tfrac{1}{2}T_{\mu\nu\rho}. \qquad (8.33)$$

In particular, $[\sigma_{\mu\nu\rho}] = 0$ in a torsion-free theory, the case of usual interest.

Exercise 35:

(a) Show that if there is no torsion, then

$$[\sigma_{\mu\nu\rho\tau}] = \tfrac{1}{3}(R_{\rho\mu\nu\tau} + R_{\tau\mu\nu\rho}). \qquad (8.34)$$

(b) Find the torsion terms left out of (8.34). *Answer:*

$$-\tfrac{1}{3}(T_{\mu\nu\rho;\tau} + T_{\mu\nu\tau;\rho}) + \tfrac{1}{4}T^\alpha_{\mu\nu}T_{\alpha\rho\tau} - \tfrac{1}{12}(T^\alpha_{\mu\rho}T_{\alpha\nu\tau} + T^\alpha_{\mu\tau}T_{\alpha\nu\rho}).$$

Exercise 36:

(a) Show that the pth derivative of (8.26) is

$$\sigma_{\mu_1\ldots\mu_p} = \sum \sigma^\alpha{}_{\nu_1\ldots\nu_k}\sigma_{\alpha\mu_1\rho_2\ldots\rho_{p-k}},$$

where the sum is over all 2^{p-1} partitions of the formal index sequence $(\mu_1, \ldots, \mu_p)$ into two subsequences $(\nu_1, \ldots, \nu_k)$ and $(\mu_1, \rho_2, \ldots, \rho_{p-k})$ with $\mu_1 \equiv \rho_1$ belonging to the second sequence. (Both subscript lists (not including α) remain in their natural order within $(\mu_1, \ldots, \mu_p)$.)

(b) Assuming no torsion, show in analogy with our treatment of the case $p = 3$ that

$$[\sigma_{\mu_1 \ldots \mu_p}] = \frac{1}{p-1} \left\{ \sum_{i=2}^{p} \sum_{j=1}^{i-1} \right.$$

$$\sum_{\substack{\text{subseqs.,} \\ 0 \le k \le p-j-1}} \sum_{l=1}^{j-1} [\sigma_{\mu_1 \ldots \mu_{l-1} \alpha \mu_{l+1} \ldots \mu_{j-1} \nu_1 \ldots \nu_k}] R^{\alpha}{}_{\mu_l \mu_j \mu_i; \rho_1 \ldots \rho_{p-j-1-k}}$$

$$\left. - \sum_{\substack{\text{subseqs.,} \\ 2 \le k \le [n/2]}} [\sigma_{\alpha \mu_1 \rho_2 \ldots \rho_{p-k}}][\sigma^{\alpha}{}_{\nu_1 \ldots \nu_k}] \right\}.$$

In the first term the ν's are a subsequence (of length k) of the sequence $(\mu_{j+1}, \ldots, \mu_p)$ with μ_i omitted. In the other term the ν's are any subsequence of the original subscript list of length $k \le [n/2]$, with the proviso (to prevent double counting) that $\nu_1 = \mu_1$ if $k = n/2$, and the ρ's are the complementary sequence.

The formula in part (b) of this exercise is not yet a *solution* of the recursion relation for $[\nabla^p \sigma]$, because it still involves $[\nabla^q \sigma]$ with $q < p$ on the right-hand side. But it is a start! The number of terms in the explicit formula for $[\nabla^p \sigma]$ is extremely large if $p \ge 8$. Research is continuing on the use of computers to evaluate quantities formed from $[\nabla^p \sigma]$ (such as the terms of the HaMiDeW series in the next chapter) in an efficient way.

The derivatives of I can be computed similarly. We already know that

$$[I] = 1 \quad \text{(the identity operator in the fiber).} \tag{8.35}$$

By differentiating (8.27), setting $\underline{x}' = \underline{x}$, taking (8.32) into account, and using the Ricci identity to isolate the derivative of highest order, one finds

$$[I_\mu] = 0, \qquad [I_{;\mu\nu}] = -\tfrac{1}{2} Y_{\mu\nu}, \tag{8.36}$$

and higher-order formulas involving the Riemann (and torsion) tensors as well as Y and its derivatives.

The formulas for the tensors $[\nabla^p \sigma]$ and $[\nabla^p I]$ have the following important qualitative properties (which can easily be proved by induction, absent an actual solution of the recursions):

(1) Each term is a product of covariant derivatives of the tensors R, T, and (in the case of I) Y.

(2) Each term in $[\nabla^{p+2}\sigma]$ or $[\nabla^p I]$ has *order* exactly p, in the following sense: Count 1 for each explicit covariant differentiation, 2 for each R, 1 for each T, and 2 for each Y, and add up. (Recall that T and R respectively involve 1 and 2 differentiations of the metric tensor.) The order of a quantity is the same as its physical dimension in powers of [mass] or [length]$^{-1}$, if $\hbar = c = 1$.

(3) There are no internal contractions (e.g., yielding Ricci tensors). In fact, there are not even any closed loops of contractions involving more than one factor. On the other hand, all the factors are linked by contractions into a single structure, if the entire product of Y-factors found in a term of $[\nabla^p I]$ is regarded as a single factor. That is, in the terminology of graph theory [Harary 1969], the contraction structure of a term is a *tree* — a graph which is both acyclic and connected. The number of contractions is exactly one fewer than the number of factors, as is clear from the way the terms are built up [see Exercise 36(b)]. (According to Harary, this is precisely the condition under which acyclicity and connectedness become equivalent.)

Exercise 37: Show that any term of the tree form has the correct relationship between *order* and *number of free indices* required for it to be a potential term in the Synge–DeWitt tensor of the appropriate order. (This does not mean that all possible tree terms actually appear.)

Chapter 9
Renormalization of the Stress Tensor

In Chapter 5 we looked at systems of the Casimir type, wherein space-time is locally flat, but its global structure is reflected nontrivially in the ground state of the quantum field. There we were able to calculate the energy-momentum (stress) tensor by investigating the limit $\underline{y} \to \underline{x}$ in

$$\langle T_{\mu\nu}(\underline{x}, \underline{y}) \rangle \equiv (\text{differential operator}) G^{(1)}(\underline{x}, \underline{y}).$$

For all quantum states considered, the singularity of the Green function was the same. Therefore, we could take the usual vacuum of the field in Minkowski space, $|0_{\text{Mink}}\rangle$, as a fiducial state, whose renormalized energy-momentum tensor can be regarded as 0 for sound physical reasons. Then

$$\langle T_{\mu\nu}(\underline{x}, \underline{y}) \rangle - \langle 0_{\text{Mink}} | T_{\mu\nu}(\underline{x}, \underline{y}) | 0_{\text{Mink}} \rangle$$

had a finite limit, to be interpreted as the physical, renormalized stress tensor in the state in question. In other words,

$$T_{\mu\nu}(\underline{x}) \equiv \lim_{\underline{y} \to \underline{x}} \left[T_{\mu\nu}(\underline{x}, \underline{y}) - \langle 0_{\text{Mink}} | T_{\mu\nu}(\underline{x}, \underline{y}) | 0_{\text{Mink}} \rangle \right]$$

defines a renormalized stress-tensor *operator* for such a system.

In curved space, calculations show that the short-distance singularity of $\langle T_{\mu\nu}(\underline{x}, \underline{y}) \rangle$ does not match that of $\langle 0_{\text{Mink}} | T_{\mu\nu}(\underline{x}, \underline{y}) | 0_{\text{Mink}} \rangle$. (This problem arises even for the "vacuum" state ($|0\rangle$) of a static system.) What, then, should one subtract from the naive, divergent expression to get the true stress tensor? In static theories we have no reason to assume that $\langle 0 | T_{\mu\nu}(\underline{x}, \underline{y}) | 0 \rangle = 0$ — indeed, we have just concluded that it is nonzero in Casimir-type problems. For time-dependent systems we don't even have a preferred state $|0\rangle$ to contemplate.

This chapter presents an answer, now widely accepted, to this question. It also offers an introduction to the rather complicated and delicate calculations which are needed to put the solution into practice.

WALD'S AXIOMATIC ANALYSIS

The mid-1970s abounded with *ad hoc* prescriptions for producing finite stress tensors, with little understanding of whether they would give the same answers, and few criteria for choosing one over another. Wald 1977, 1978a showed that a renormalized stress tensor satisfying certain reasonable physical requirements is essentially unique, in the sense that its ambiguity can be absorbed into redefinition of the coupling constants in the gravitational field equation ("renormalization" in its original meaning). He also showed that this $T_{\mu\nu}(\underline{x})$ can indeed be calculated by an appropriate subtraction of singular terms in the asymptotic expansion of the differentiated Green function, $T_{\mu\nu}(\underline{x}, \underline{y})$.

Besides Wald 1977 and 1978a, relevant papers include Davies 1977, Wald 1978b, Castagnino & Harari 1984, Brown 1984, Brown & Ottewill 1986, Allen et al. 1988. Expository summaries appear in Fulling 1983 and in Birrell & Davies 1982, Chap. 6.

Wald propounded five "axioms" to be satisfied by the stress tensor, of which three have stood the test of time. (The significance of the other two will be discussed later.) One anticipates an operator-valued distribution, $T_{\mu\nu}(\underline{x})$, such that $\langle A|T_{\mu\nu}(f)|B\rangle$ is defined (finite!) for all "physically reasonable" states $|A\rangle$ and $|B\rangle$ and all test functions f. (From experience with the free field and the Casimir effect, and also with the solvable time-dependent models of Chapter 7, one knows that "unreasonable" states can include both states associated with unphysical representations of the commutation relations and states which belong to a physical Fock space of the theory but are not in the domain of the unbounded operator $T_{\mu\nu}(f)$; and conversely that in models of infinite volume there are states of finite density which do not belong to the vacuum's Fock space but still appear "reasonable".) Wald argues that one expects this operator to have the following three properties:

(I) **Conservation:** $\nabla_\mu T^\mu{}_\nu = 0.$

Axiom I is certainly true in the classical field theory, and it is true "formally" in the quantized theory. (For instance, in models which can be solved by separation of variables, it is true in the sense of mode-by-mode differentiation of the divergent expression for $T_{\mu\nu}$.) The property is needed for consistency of Einstein's gravitational field equation, $R_{\mu\nu} - \frac{1}{2} R g_{\mu\nu} = G T_{\mu\nu}$. (From time to time it has been suggested that the phenomenon of particle creation by the gravitational field (see

Chap. 7) should be expected to disrupt the energy conservation law. This argument is unconvincing, however, since the particle creation arises from the same field dynamics which gives rise, through Noether's theorem, to the formal conservation law. It should be noted that parti- cle (electron-positron pair) creation by external electromagnetic fields does not violate the electromagnetic charge-current conservation law, $\partial_\mu J^\mu = 0$.)

The statement of the second axiom involves a whole class of asymp- totically static space-time metrics with the same initial spatial geom- etry, (M, γ). The initial states of the field within them can be un- ambiguously identified. (That is, the Fock spaces $\mathcal{F}^{\text{in}}$ for all these models are naturally isomorphic.) Similarly, one can consider the class of space-times with the same final geometry and the associated Fock space, $\mathcal{F}^{\text{out}}$.

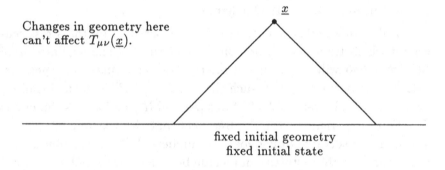

Changes in geometry here can't affect $T_{\mu\nu}(\underline{x})$.

fixed initial geometry
fixed initial state

(II) **Causality**: For a fixed in-state, $\langle T_{\mu\nu}(\underline{x}) \rangle$ is independent of varia- tions of $g_{\alpha\beta}(\underline{y})$ outside the past light cone of $\underline{x}$. For a fixed out- state, $\langle T_{\mu\nu}(\underline{x}) \rangle$ is independent of metric variations outside the fu- ture light cone of $\underline{x}$.

Before stating Axiom III, let's observe that in model calculations the infinities in $T_{\mu\nu}$ always come from terms of the type $a_j a_{j'}^\dagger$ in the naive formula; in fact, they come from the *commutator* term left over when the a operators are rearranged into their "normal" order. These divergent terms are proportional to $\delta(j - j')$ or $\delta_{jj'}$. The important point is that they are no longer operators, but just numbers (more pre- cisely, numerical multiples of the identity operator). (This would not be true in a nonlinear (interacting) field theory.) The other terms, of the types $a_j^\dagger a_{j'}$, $a_j a_{j'}$, and $a_j^\dagger a_{j'}^\dagger$, give no divergences for states in a suitable dense domain in the Fock space associated with the a and $a^\dagger$

operators; this is why *normal ordering* (simply discarding the offending commutator terms) with respect to any "reasonable" set of creation-annihilation operators yields a finite result (albeit nonunique and of dubious physical meaning). It therefore seems that the "infinite part" of $T_{\mu\nu}$ is a multiple of the identity. (This is all the more reason why it should be possible to throw it away with impunity, provided the act is done in a consistent, physically understood manner. Recall the arbitrariness in the zero point of the energy scale in ordinary mechanics.)

Exercise 38: Let T be an operator in Hilbert space (possibly unbounded). Show that the difference of the expectation value of T with respect to two normalized states,

$$\langle A|T|A\rangle - \langle B|T|B\rangle \qquad (\langle A|A\rangle = 1 = \langle B|B\rangle),$$

can be expressed in terms of (unsubtracted) matrix elements of T with respect to orthogonal pairs of states,

$$\langle C|T|D\rangle \qquad (\langle C|D\rangle = 0),$$

and conversely.

(III) **Consistency with the formalism**: The formal expression for $T_{\mu\nu}$ is valid for calculating $\langle A|T_{\mu\nu}|B\rangle$ whenever $\langle A|B\rangle = 0$. Equivalently, it is valid for calculating $\langle A|T|A\rangle - \langle B|T|B\rangle$ when $|A\rangle$ and $|B\rangle$ are normalized.

Wald's uniqueness theorem: Let $T_{\mu\nu}$ and $\overline{T}_{\mu\nu}$ be operators satisfying Axioms I–III. These objects are functionals of the space-time metric; we assume them to be defined for all globally hyperbolic models, or at least for all models that are asymptotically static in either the past or the future, so that Axiom II is meaningful. Define

$$U_{\mu\nu} = T_{\mu\nu} - \overline{T}_{\mu\nu}.$$

Then

(A) $U_{\mu\nu}$ is a multiple of the identity operator.

(B) $\nabla_\mu U^\mu{}_\nu = 0$.

(C) $U_{\mu\nu}(\underline{x})$ is a *local* functional of $g(\underline{x})$. That is, it depends only on the metric and its derivatives *at the same point* $\underline{x}$ (as opposed to nonlocal functionals of the sort $\int g_{\alpha\beta}(\underline{x}')B^{\alpha\beta}{}_{\mu\nu}(\underline{x},\underline{x}')\,d\underline{x}'$). Since the construction is completely covariant (intrinsic), derivatives of $\underline{g}$ can enter only through the Riemann curvature tensor and its covariant derivatives.

Remarks:

(1) Conclusion (C) has been stated for a field equation, such as $(\Box + m^2 + \xi R)\phi = 0$, whose coefficients are constructed entirely out of the metric. If an external gauge field or an external potential is present, the extended statement of (C) is that U can depend on these and their covariant derivatives in a gauge-covariant manner: The field operator $g^{\mu\nu}\nabla_\mu\nabla_\nu + V$, where ∇_ν equals $\partial_\nu + iA_\nu$ when acting on scalars (and involves Christoffel symbols more generally), will lead to terms such as $F^{\alpha\beta}F_{\alpha\beta}\,g_{\mu\nu}$ and $(\nabla_\mu\nabla_\nu + \nabla_\nu\nabla_\mu)V$ in the stress tensor (in conserved combinations). Cf. (9.8) below.

(2) Tensors with the properties (A), (B), and (C) are precisely those which can be absorbed by renormalization of the coupling constants in a suitably generalized gravitational field equation. This is what makes the theorem a uniqueness statement.

Outline of proof: Axiom III implies (A). Then the causality axiom implies (C): $U_{\mu\nu}$ can arise either as a $\langle U_{\mu\nu}\rangle_{\text{in}}$ or as a $\langle U_{\mu\nu}\rangle_{\text{out}}$, hence it must be independent of the metric to both the past and the future. Clearly, the conservation axiom implies (B).

To get an existence theorem and a calculational procedure for $T_{\mu\nu}$, we shall investigate the "point-split" object $T_{\mu\nu}(\underline{x},\underline{y})$. We need more mathematical information about the short-distance singularity of the latter. This is bound up with the general topic of the singular behavior of Green functions of linear partial differential equations with variable coefficients, to which we now turn.

HaMiDeW theory

The acronym "HaMiDeW" for this branch of asymptotic analysis was coined by G. W. Gibbons. It refers to the work of Hadamard 1952, who constructed fundamental solutions of hyperbolic and elliptic equations on manifolds [see Garabedian 1964 and Friedlander 1975], Minakshisundaram & Pleijel 1949 and Minakshisundaram 1953, who applied similar methods to the fundamental solutions of the associated parabolic equations in one higher dimension [see Friedman 1964 and Eidel'man 1969], and DeWitt 1965, who (extending the work of Schwinger

1951 in electrodynamics) independently developed the calculational aspects of the method in the context of operators of Schrödinger type with hyperbolic "Hamiltonians".

In its application to our subject, the HaMiDeW theory comes in a rigorous spatial version and a less rigorous space-time version. The former provides, in principle, a complete solution of the renormalization problem for *static* linear field theories. It also has important implications for the time-dependent case. Its space-time analogue provides a more direct and manifestly covariant method of renormalization in the general case. The following table lists the counterpart elements of the two developments. (Many of the notational conventions stated here are not universal, of course.)

Space	Space-time
Riemannian metric on M, signature $(+++)$, dimension d	Lorentzian metric on $\mathcal{M}$, signature $(+---)$, dimension $N \equiv d + 1$
Latin tensor indices	Greek tensor indices
Elliptic operator, K	Hyperbolic operator, $\square + \cdots$; becomes $\frac{\partial^2}{\partial t^2} + K$ in ultrastatic case.
Frequency $=$ (eigenvalue)$^{\frac{1}{2}} = \omega_j$	Mass, m
Fundamental solution, $G_\omega(x,y) =$ kernel of $(K - \omega^2)^{-1}$; e.g., Coulomb or Yukawa potential	Hadamard elementary solution, $G^{(1)}(\underline{x}, \underline{y})$; Retarded and advanced Green functions, $G_{\text{ret/adv}}(\underline{x}, \underline{y})$
Heat operator, $\frac{\partial}{\partial t} + K$ (parabolic)	Schwinger operator, $-i\frac{\partial}{\partial s} + \square + \cdots$ (of Schrödinger-hyperbolic type)

THE HEAT KERNEL

I shall denote by $H(t, x, y)$ the fundamental solution of a heat equation on a d-dimensional manifold M. It may be defined by any of the following essentially equivalent characterizations, in which K is a formally self-adjoint, elliptic, second-order operator on the scalar functions on M.

(1) For each $f \in C_0^\infty(M)$, the function $\Psi = e^{-Kt}f$ solves the initial value problem

$$\frac{\partial \Psi}{\partial t} = -K\Psi \quad \text{for } t > 0, \qquad \Psi(0, x) = f(x).$$

Then H is defined to be the integral kernel of e^{-Kt}:

$$\Psi(t, x) = \int H(t, x, y) f(y) \sqrt{\gamma(y)} \, d^d y.$$

(2) As a function of x, H satisfies the equation

$$\frac{\partial H}{\partial t} = -K_{(x)} H \quad \text{for } t > 0 \tag{9.1}$$

and the initial condition

$$H(t, x, y) \to \delta(x, y) \equiv \frac{\delta(x-y)}{\sqrt{\gamma}} \quad \text{as } t \downarrow 0. \tag{9.2}$$

(The covariant (scalar) Dirac distribution is represented in any particular chart by the ordinary delta function (a density) divided by the square root of the metric determinant.)

(3) For all t,

$$\frac{\partial H}{\partial t} + K_{(x)} H = \delta(x, y) \delta(t),$$

and

$$H(t, x, y) = 0 \quad \text{for } t < 0.$$

(4) For each $f \in C_0^\infty(\mathbf{R} \times M)$,

$$\Psi(t, x) \equiv \int H(t - u, x, y) f(u, y) \sqrt{\gamma} \, d^d y \, du$$

solves

$$\frac{\partial \Psi}{\partial t} + K\Psi = f,$$

with null initial data to the past of supp f.

We have no time to dwell on the analytic niceties here. (E.g., existence of a solution of the Cauchy problem for the heat equation does not obviously guarantee existence of a kernel.) Refer to Folland 1976, Chap. 4, for a textbook treatment of the basic case

$$K = -\nabla^2, \qquad H(t,x,y) = H_0(t, x - y) \equiv (4\pi t)^{-\frac{d}{2}} e^{-|x-y|^2/4t}, \quad (9.3)$$

to the books of Friedman and Eidel'man for more general theory, and to the sequence of papers Osborn et al. 1987 and Papiez et al. 1988 for a detailed, self-contained treatment of the heat kernel and its asymptotics when K contains external scalar and vector potentials (but not a gravitational field).

The t in the heat equation is not to be identified with the time coordinate of our relativistic space-time. We are interested in the heat kernel purely as a technical tool. In particular, we shall soon see that it is intimately related to the fundamental solution, or resolvent kernel, associated with any of the operators $(K - \omega^2)^{-1}$.

Exercise 39: In Exercise 2 you showed that any formally self-adjoint, elliptic, second-order differential operator on scalar functions takes the form

$$K\phi = \frac{1}{\rho} (D_j + A_j) \left[\rho \gamma^{jk} (D_k + A_k)\phi \right] + V\phi, \qquad (9.4)$$

where $D_j \equiv -i\partial_j$ and $\rho\, d^d x$ is the measure defining the scalar product; $\rho, \gamma, \mathbf{A}$, and V are real-valued functions of x, and γ is positive definite. Show that (9.4) is the expression, in a local coordinate system and local choice of gauge, of a formally self-adjoint operator acting on the sections of an Hermitian vector bundle (see definition below) of fiber dimension 1, equipped with a connection, ∇. In fact,

$$K\phi = \gamma^{jk} \mathcal{D}_j \mathcal{D}_k \phi + V\phi, \qquad (9.5)$$

where $\mathcal{D}_j \equiv -i\nabla_j$ and the indices are meant abstractly; in any particular coordinate system, and any gauge such that $\rho = \sqrt{\gamma}$, one has, for certain connection coefficients A_j,

$$\mathcal{D}_k \phi = (D_k + A_k)\phi$$

and

$$\mathcal{D}_j \mathcal{D}_k \phi = (D_j + A_j)\mathcal{D}_k \phi + i\Gamma^l_{jk}\mathcal{D}_l \phi.$$

(If the normalization of the sections is changed so that $\rho \neq \sqrt{\gamma}$, then A_k will acquire an imaginary part. In the notation of Chapter 8, $A_k = -iw_k$.)

Definition: A vector bundle is *Hermitian* if

(1) the base manifold is Riemannian or pseudo-Riemannian (with metric $\underline{g}$);

(2) each fiber is equipped with an inner product;

(3) an inner product on sections is defined by

$$\langle \phi, \psi \rangle \equiv \int_M \langle \phi(x), \psi(x) \rangle \sqrt{\gamma}\, d^d x.$$

Remarks: (1) Physicists would more likely call this a $U(r)$ bundle (where r is the fiber dimension), because the unitary group $U(r)$ is the natural symmetry group of the complex inner product in each fiber. (2) Some mathematicians reserve the term "Hermitian" for cases where the base manifold itself possesses a complex structure.

The operator $\gamma^{jk}\nabla_j\nabla_k$ appearing in (9.5) is the *covariant Laplacian* associated with the given metric, Hermitian bundle, and connection. When there is no gauge field in the problem, it is just the Laplace–Beltrami operator for the given metric. We shall always assume that the gravitational connection, Γ^l_{jk}, is the symmetric, metric-compatible connection (8.21) determined by $\underline{\gamma}$.

Remarks:

(A) Our treatment can be generalized to operators on bundles with fiber dimension greater than 1 (i.e., to nonscalar fields, in the broad sense of "scalar") if the operator is still of type (9.5) with γ^{jk} (for each index pair) equal to a multiple of the identity operator in the fiber. (V and A_j may be matrix-valued — i.e., nontrivial operators in the fiber.) Such an operator is said to have *scalar principal symbol*.

(B) Spinor fields ordinarily are taken to satisfy first-order field equations $A\psi = 0$, where A is not necessarily self-adjoint. In such cases one studies the self-adjoint, positive, second-order operators $A^\dagger A$ and $AA^\dagger$.

Theorem: Let K be as specified in Exercise 39 and Remark (A). Assume also that K is bounded below. Then

(1) $H(t, x, y)$ is nonsingular for $t > 0$ and also for $x \neq y$. (Obviously the degree of smoothness of H will depend on that of the coefficients in the differential operator. For simplicity let's assume that everything is C^∞.)

(2) Locally (that is, for x near y and t near 0) there is an asymptotic expansion

$$H(t, x, y) \sim (4\pi t)^{-\frac{d}{2}} e^{-\sigma(x,y)/2t} \sum_{n=0}^{\infty} a_n(x, y) t^n. \qquad (9.6)$$

(Compare (9.2), recalling that σ is half the square of the distance from x to y.) In particular,

$$H(t, x, x) \sim (4\pi t)^{-\frac{d}{2}} \sum_{n=0}^{\infty} a_n(x) t^n. \qquad (9.7)$$

(3) $a_n(x) \equiv a_n(x, x) \equiv [a_n]$ is independent of coordinate system and (as an operator in the fiber) of gauge. (Its matrix representation experiences a local similarity transformation when the gauge is changed: $a \mapsto U(x) a U(x)^{-1}$.)

(4) $a_n(x)$ depends *locally* on the coefficient functions in K — viz., γ^{jk}, A_j, V. In fact, $a_n(x)$ is a *polynomial* in the coefficients and their derivatives (at x), except for the appearance of $\gamma_{jk}(x)$ in lowering indices. For a fixed x, it is possible to choose coordinates so that $\gamma_{jk}(x) = \delta_{jk}$. It follows from (3) that $a_n(x)$ is then a polynomial in the curvature tensors $R^\alpha{}_{\beta\gamma\delta}$ and $Y_{\mu\nu}$, the potential V, and the covariant derivatives of these objects.

(5) $a_n(x)$ is "homogeneous of degree n" in the sense of dimensional analysis, where one counts 1 for each differentiation, 1 for each A, 2 for each V, and 0 for a γ. (This implies 2 for each R or Y. Compare the property (2) at the end of Chapter 8.) As a corollary, there is only a certain finite basis of possible terms for each n. (In other words, the degree of the polynomial grows with n in a very precise way.)

(6) The explicit form of $a_n(x)$ can be calculated recursively. One finds

$$[a_0] = 1, \qquad (9.8a)$$

$$[a_1] = -V + \frac{1}{6} R \qquad (R \equiv R^j{}_j \equiv R^{kj}{}_{kj}), \qquad (9.8b)$$

$$[a_2] = \frac{1}{2}V^2 + \frac{1}{6}\mathcal{D}^j\mathcal{D}_jV - \frac{1}{12}F_{jk}F^{jk} - \frac{1}{6}RV$$

$$+ \frac{1}{72}R^2 - \frac{1}{180}R_{jk}R^{jk} + \frac{1}{180}R^{jklm}R_{jklm}$$

$$+ \frac{1}{30}\nabla^j\nabla_jR, \tag{9.8c}$$

where

$$F_{jk} \equiv \partial_j A_k - \partial_k A_j + i[A_j, A_k] = -iY_{jk}$$

(the bundle curvature or gauge field strength), and

$$\mathcal{D}_jV \equiv D_jV + [A_j, V]$$

(cf. Exercise 31). Gilkey 1975 calculated $[a_3]$; there are 46 possible terms, of which 43 have nonzero coefficients (when the most obvious, monomial basis is used for the 46-dimensional vector space).

I shall not prove this theorem here, except to outline how the calculations in (6) are done. In addition to the works of Minakshisundaram and others mentioned previously, I recommend Gilkey 1975, 1979, Sakai 1971, and Wald 1979a, along with Gilkey's books 1974, 1984. For the calculational aspects and the application to renormalization, see DeWitt 1965, Christensen 1976, 1978, Fulling et al. 1981, and the Cargèse lectures of Boulware 1979 and Parker 1979. Many of these references also discuss various aspects of the connection between the heat kernel and the elliptic and hyperbolic Green functions, to which we turn below.

There is a variety of ways of computing the coefficients in (9.8) and, in principle, the higher-order terms:

(1) At least partial information can obtained with relatively little calculational labor by matching the general formula (with undetermined numerical coefficients) against known results for special cases of $(M, \gamma, \mathbf{A}, V)$ (such as spheres, ...); see McKean & Singer 1967 and Gilkey 1979.

(2) The most powerful and modern method is that of pseudodifferential operators, which was used by Gilkey 1975. [See also Seeley 1967, Greiner 1971.] In its conventional form the pseudodifferential method deals with Fourier transforms with respect to local coordinates; consequently, considerable labor or trickery is needed to rearrange the results into geometrically meaningful form (covariant derivatives and

curvature tensors). In recent years an "intrinsic" or "manifestly co-variant" approach to pseudodifferential operators has been developed [Bokobza-Haggiag 1969, Widom 1978, 1980, Drager 1978, Fulling & Kennedy 1988], which brings the calculations into the geometrical framework expounded in Chapter 8 and used in the traditional DeWitt method demonstrated below. The great promise of the pseudodifferential method is its applicability to a very wide class of linear differential operators, not just those of the form (9.5). Examples are second-order operators with nonscalar principal symbol and fourth-order operators. Such operators appear in the generalized quantum theories of gravity with fourth-order field equations [Barth & Christensen 1983].

(3) The time-honored method is to substitute (9.6) and (9.5) into (9.1) to derive recursion relations for the quantities $a_n(x, y)$:

Note first, by comparison with (9.3), that $[a_0] = 1$ [(9.8a)] is necessary and sufficient to guarantee (9.2). This is the base of the recursion.

Differentiate (9.6):

$$\frac{\partial H}{\partial t} = -\frac{d}{2t} H + \frac{\sigma}{2t^2} H + (4\pi t)^{-\frac{d}{2}} e^{-\sigma/2t} \sum_{n=0}^{\infty} n a_n t^{n-1};$$

$$\nabla_j H = -\frac{\nabla_j \sigma}{2t} H + (4\pi t)^{-\frac{d}{2}} e^{-\sigma/2t} \sum_{n=0}^{\infty} (\nabla_j a_n) t^n,$$

hence

$$-\nabla^j \nabla_j H = \frac{\nabla^j \nabla_j \sigma}{2t} H - \frac{\nabla^j \sigma \nabla_j \sigma}{4t^2} H$$
$$+ \frac{1}{t} \nabla^j \sigma (4\pi t)^{-\frac{d}{2}} e^{-\sigma/2t} \sum_{n=0}^{\infty} (\nabla_j a_n) t^n$$
$$- (4\pi t)^{-\frac{d}{2}} e^{-\sigma/2t} \sum_{n=0}^{\infty} (\nabla^j \nabla_j a_n) t^n.$$

(Note that covariant derivatives on a_n act upon its first variable and first gauge-tensor index only. The reader who is squeamish about this should think of the second variable and second index as being saturated by a vector-valued test function, or test section, as would be the case when H is used to form a solution of a heat equation with smooth data.) Therefore, (9.1) is

$$0 = (4\pi t)^{-\frac{d}{2}} e^{-\sigma/2t} \sum_{n=0}^{\infty} t^n \left[-\frac{d}{2} a_{n+1} + \frac{\sigma}{2} a_{n+2} + (n+1)a_{n+1} \right.$$

$$\left. + \frac{\nabla^j \nabla_j \sigma}{2} a_{n+1} - \frac{\nabla^j \sigma \nabla_j \sigma}{4} a_{n+2} + \nabla^j \sigma \nabla_j a_{n+1} - \nabla^j \nabla_j a_n + V a_n \right].$$

Since $\sigma = \frac{1}{2} \nabla^j \sigma \nabla_j \sigma$ [(8.26)], the terms in a_{n+2} cancel.

Each term in the power series must separately vanish, so we arrive at the recursion relation

$$\nabla^j \sigma \nabla_j a_{n+1} + (n+1)a_{n+1} + \frac{1}{2} \left(\nabla^j \nabla_j \sigma - d \right) a_{n+1} = \nabla^j \nabla_j a_n - V a_n$$

$$= -K a_n . \quad (9.9a)$$

Of course, a_n is interpreted as 0 if $n < 0$. It is useful to write this equation in the semicolon notation:

$$\sigma^j a_{n+1;j} + (n+1)a_{n+1} + \tfrac{1}{2}(\sigma_j{}^j - d)a_{n+1} = a_{n;j}{}^j - V a_n . \quad (9.9b)$$

Note that the first term in (9.9) is equal to the arc length, s, times the absolute derivative of a_{n+1} along the geodesic with respect to s [cf. (8.27)]. Thus (9.7) is really an ordinary differential equation; it describes the deviation of a_{n+1} from parallel transport as one moves along the geodesic. Hadamard, Minakshisundaram, Sakai, and others studied the solutions of this (or a counterpart) equation, starting from the fixed point, y, where $s = 0$. (See also Remark (4) below.) When $n + 1 > 0$ it turns out that all solutions of the related homogeneous equation are singular at y, so the admissible solution of the inhomogeneous equation is unique. This analysis is a step in one proof of the existence of the asymptotic approximation (9.6).

However, if one is interested merely in calculating the coincidence limits $[a_n]$, it is easier to follow a procedure demonstrated by DeWitt:

Note first, from (8.32), that

$$\sigma_{jk} = \gamma_{jk} + \text{higher-order terms};$$

therefore, since $\gamma_j{}^j = d$, the combination of terms

$$\frac{1}{2}(\sigma_j{}^j - d)a_{n+1}$$

vanishes when $x = y$. (Of course, it also vanishes identically if the metric is flat, even if there is an external gauge field.)

The coincidence limit of (9.9) is

$$(n+1)[a_{n+1}] = [a_{n;j}{}^j] - V[a_n]. \tag{9.10}$$

Consequently, we must find $[a_{n;j}{}^j]$ in order to find $[a_{n+1}]$ recursively. To this end, we differentiate (9.9) twice and form the covariant Laplacian:

$$\sigma^{jk}{}_k a_{n+1;j} + 2\sigma^j{}_k a_{n+1;j}{}^k + \sigma^j a_{n+1;jk}{}^k + (n+1)a_{n+1;k}{}^k$$
$$+ \tfrac{1}{2}\sigma^j{}_j{}^k{}_k a_{n+1} + \sigma_j{}^{jk} a_{n+1;k} + \tfrac{1}{2}(\sigma_j{}^j - d)a_{n+1;k}{}^k$$
$$= a_{n;j}{}^{jk}{}_k - V_{;k}{}^k a_n - 2V_;^k a_{n;k} - V a_{n;k}{}^k. \tag{9.11}$$

Now we take the coincidence limit of this, recalling that $[\sigma_{jkl}] = 0$ [(8.33)] and $[\sigma_{jklm}]$ can be expressed in terms of Riemann tensors [(8.34)], so that

$$[\sigma_j{}^{jk}{}_k] = -\frac{2}{3}R.$$

The result is

$$(n+3)[a_{n+1;j}{}^j]$$
$$= \tfrac{1}{3}R[a_{n+1}] + [a_{n;j}{}^{jk}{}_k] - V[a_{n;k}{}^k] - 2V_;^k[a_{n;k}] - V_{;k}{}^k[a_n]. \tag{9.12}$$

In particular,

$$[a_{0;j}{}^j] = \tfrac{1}{6}R[a_0] = \tfrac{1}{6}R.$$

So, from (9.10),

$$[a_1] = \tfrac{1}{6}R - V,$$

which is (9.8b). To obtain $[a_2]$ one needs $[a_{1;j}{}^j]$ and hence [for use in (9.12)] $[a_{0;j}{}^{jk}{}_k]$. Therefore, one must take the Laplacian of (9.11). Continuing in this way — repeatedly differentiating the recursion relation, and *afterwards* setting the two arguments equal — one can get all the $[a_n]$ and, as a byproduct, the coincidence limits of the derivatives of the a_n.

FURTHER REMARKS ON THE HEAT KERNEL

(1) The formula for $[a_n]$ is independent of the manifold dimension, d, and the fiber dimension, r — except insofar as the terms may become

linearly dependent in low dimensions, making the formula nonunique. (For example, when $d = 2$, the quantities R^2, $R^{jk}R_{jk}$, and $R^{jklm}R_{jklm}$ are all proportional.)

In particular, the term $\frac{1}{6}R$ in (9.8b) is the same in all dimensions and has nothing to do with the $\frac{1}{6}R$ in the four-dimensional conformally invariant wave equation (see (6.13) and surrounding discussion). Purely by accident, the conformally invariant modification of the Laplace–Beltrami equation (wherein $V = \frac{1}{6}R$) has $[a_1] = 0$, in dimension $d = 4$ only.

The numerical coefficients in the $[a_n]$ formulas do become dimension-dependent if the basic objects V, F_{jk}, and R_{jklm} are expressed in terms of other objects via dimension-dependent definitions. This rather obvious observation is a needed warning, because the Riemann tensor is often eliminated in terms of the *Weyl tensor*, its traceless part. Introducing the concept of "trace" inevitably produces a dependence on dimension.

(2) If m^2 is a constant (in practice, the mass of the field), then the heat kernel of $K + m^2$ equals $e^{-m^2 t}$ times the heat kernel of K. The mass is often inserted into the expansion through such a factor, rather than included as part of V. The relation between the resulting two different definitions of a_n is easily obtained by multiplying two power series together.

(3) Let K_1 and K_2 be operators on two manifolds, M_1 and M_2. Then $K \equiv K_1 + K_2$ is an operator on $M \equiv M_1 \times M_2$. [Familiar examples: (a) the total Hamiltonian of two noninteracting particles, M being the total configuration space of the system; (b) a partial differential operator that is split into ordinary differential operators by separation of variables.] Moreover, each K_i can be regarded in a natural way as an operator on M (i.e., in $\mathcal{L}^2(M)$), and these two operators commute. It follows that $e^{-tK} = e^{-tK_1}e^{-tK_2}$, and consequently that the heat kernel of K is the product of those of the K_i (which have t but no other variable in common).

This fact is a useful tool for deducing properties of the a_n without detailed calculations or tedious proofs by induction [e.g., Gilkey 1979]. For instance, let K_2 be simply $-d^2/dx_2{}^2$ on a one-dimensional manifold. Then the kernel $H(t, x_1, x_2, y_1, y_2)$ is

$$H_1(t, x_1, y_1) \times (4\pi t)^{-\frac{1}{2}} e^{-(x_2 - y_2)^2/4t},$$

according to (9.3). This multiplication converts fractional powers of t in (9.6) for H_1 into integral powers in the expansion for H, or vice versa. This phenomenon provides an "explanation" for the fact that the entire expansion of a heat kernel, including terms which are smooth functions of t and the other variables, is determined by the *local* behavior of the coefficients in the operator. Asymptotic or quasiasymptotic expansions of other Green functions typically are locally determined only in their singular terms. (For example, if the smooth part of the two-point function, or Hadamard elementary solution of the field equation, did not depend on the global structure of the manifold, there could be no Casimir effect!) A fractional power of t counts as a singularity in this sense, and the factorization of the heat kernel over product manifolds relates such terms to the integer powers.

I stress that this factorization property is a special feature of the heat kernel, traceable to the elementary identity $e^{a+b} = e^a e^b$. Factorization does not occur for other Green functions, such as the resolvent kernel, which are related to nonexponential functions of self-adjoint operators.

(4) If the metric is flat, equation (9.9) for a_0 reduces to $\sigma^j a_{0;j} = 0$, which is the same as (8.27). The initial value is also the same (namely, the identity matrix). Therefore, $a_0(x,y) = I(x,y)$, the parallel-transport matrix. It is common to write the series (9.6) with an explicit overall factor of I; this makes $a_0(x,y)$ identically equal to 1 and redefines the higher a_n's accordingly. (The $[a_n]$ are unchanged.) Since I, like H, is a map from the fiber F_y to the fiber F_x, this maneuver gives the a_n the relatively simple gauge-transformation properties of an operator within F_y, as opposed to an operator from one fiber to another. Furthermore, I serves as an integrating factor facilitating the recursive solution of the inhomogeneous equations for the a_n with $n > 0$.

Similarly, when the metric is curved, our a_0 can be expressed in terms of the *Van Vleck–Morette determinant*, a quantity which describes the manner of focusing of the geodesics emanating from y as they pass near x. (See Friedlander 1975, Sec. 4.2; DeWitt & Brehme 1960.) It essentially provides the natural Jacobian relating integration with respect to x with integration in the tangent space at y. In most of the published literature this solution for a_0 is factored out of the basic series (9.6), redefining the a_n and changing the appearance of the recursion relation, (9.9).

In the practical calculation of the $[a_n]$ by DeWitt's algorithm, separating out the geodesic determinant and the parallel-transport matrix is actually disadvantageous, since separate power series for those quantities must be calculated along the way. In the most recent calculational work, therefore, that practice has been abandoned.

(5) If the principal symbol of K is nonscalar or the order of the operator is other than 2, the ansatz (9.6) does not work. The essential singularity of H can't be concentrated into a known prefactor like the Gaussian in (9.6). Such problems can, in principle, be handled by the pseudodifferential method. See also Lee et al. 1986, 1987 for a purely coordinate-space algorithm for fourth-order operators.

(6) If A is a *non-self-adjoint* elliptic operator on a compact manifold of dimension $d = 2p$, then the integrand in the celebrated *index theorem* for A is $[a_p(A^\dagger A)] - [a_p(AA^\dagger)]$ (that is, the difference between the t-independent terms of the heat kernels of the two positive self-adjoint operators associated with A). The index theorem and related theorems describe important relationships among the operator theory, the global topology, and the local geometry of a manifold [Gilkey 1974, 1984; Eguchi et al. 1980; see Bonic 1969, Sec. 28, for background information on the space of Fredholm operators].

(7) The error in any truncation of the asymptotic series (9.7) may be nonuniform in x. In particular, if M has a boundary, integrating (9.7) over all of M will give an incorrect answer for

$$\int_M H(t,x,x)\sqrt{\gamma}\,d^dx \equiv \operatorname{Tr} e^{-tH} = \sum_{j=1}^{\infty} e^{-\lambda_j t}.$$

The corrections to the series due to the behavior of H near the boundary contain all integral powers of $t^{\frac{1}{2}}$, thus providing fractional powers of t when d is even and integral powers of t when d is odd, which heretofore were not present. See Kac 1966, Greiner 1971, Gilkey 1979, and also the papers of Candelas, Kennedy, and others cited in Chapter 5.

(8) In relativistic physics we would really like to carry out the foregoing analysis for a *hyperbolic* operator, $\Box + \cdots$, in the role of the semibounded elliptic operator K. The heat equation (9.1), however, is hopelessly pathological when the spectrum of K is not bounded below, a problem which is immediately encountered in the hyperbolic case. Schwinger and DeWitt avoided the worst aspect of this problem by

considering instead the "Schrödinger" equation

$$i\frac{\partial H}{\partial s} = K_{(x)}H. \tag{9.13}$$

Here s is an extra coordinate which can be identified with the *proper time*, or arc length, of relativistic particle mechanics. (The true time, t, is now one of the coordinates of K, which acts on a space-time $\mathcal{M}$.) The algorithm for calculating the coefficients $[a_n]$, or even $a_n(\underline{x}, \underline{y})$, is identical to the one we have discussed, except for stray factors of i. Unfortunately, however, clause (2) of the Theorem on the heat kernel is false for (9.13), unless the word "asymptotic" is replaced by something weaker. (That is, if the series is truncated after the term involving s^N, there is no guarantee that the error is $O(s^{N+1})$.) Nevertheless, I shall argue that the use which physicists have made of the formal expansion analogous to (9.6) in renormalization is justifiable.

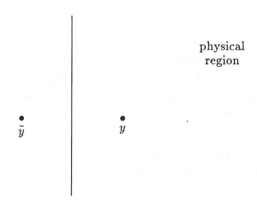

Before granting a pardon, however, let us substantiate the charge. A simple counterexample will suffice. Let the manifold first be $\mathbf{R}^d$, equipped with either a Euclidean or a Lorentzian metric, so that

$$K = -\nabla^2 \quad \text{(elliptic)} \qquad \text{or} \qquad \partial_t^2 - \nabla^2 \quad \text{(hyperbolic)}.$$

It is clear that, in analogy with (9.3), the fundamental solution of (9.13) is

$$H_0 \equiv (4\pi is)^{-\frac{d}{2}} e^{i\sigma/2s}.$$

Now consider restricting one spatial coordinate, say x^1, to positive values, and place a Dirichlet boundary condition on the hyperplane

$x^1 = 0$. The new problem is easily solved by the method of images: Let $\tilde{\underline{y}}$ be the image of $\underline{y}$ (namely, $(-y^1, y^2, \dots)$). Then

$$H(s, \underline{x}, \underline{y}) = H_0(s, \underline{x}, \underline{y}) - H_0(s, \underline{x}, \tilde{\underline{y}}).$$

Thus

$$H(s, \underline{x}, \underline{x}) = (4\pi i s)^{-\frac{d}{2}} \left[1 - e^{i(x^1)^2/s} \right].$$

As $s \downarrow 0$ with fixed $x^1 \neq 0$, the second term in the brackets oscillates furiously but always has modulus 1. So H simply does not possess an asymptotic expansion of the form

$$H(s, \underline{x}, \underline{x}) \sim (4\pi i s)^{-\frac{d}{2}} [a_0(\underline{x}) + a_1(\underline{x})s + a_2(\underline{x})s^2 + \cdots].$$

The Schwinger–DeWitt series for this system has only one nonzero term, H_0 itself. The error in the approximation is always of order unity, no matter how many of the terms $0 \times s^N$ are conceptually kept in the series!

A boundary is not necessary for this sort of phenomenon. If $K = -d^2/d\theta^2$, the Laplace–Beltrami operator on the circle, then the Schrödinger kernel has singularities as a function of s for positive s arbitrarily close to $s = 0$. (See the remark attributed to Atiyah on pp. 45–46 of Duistermaat & Guillemin 1975.)

On the other hand, under sufficiently strong conditions on the potential $V(x)$, the Schrödinger kernel of $-\nabla^2 + V$ does have a genuinely asymptotic Schwinger–DeWitt expansion [Osborn & Fujiwara 1983].

HYPERBOLIC AND ELLIPTIC GREEN FUNCTIONS

The heat kernel of a partial differential operator is very convenient to study, because of several special properties. These include the factorization discussed in Remark (3) above and the dimension independence discussed in Remark (1). Also, when $t > 0$ the heat kernel is a smooth function of x and y, whereas the other Green functions we're about to discuss have complicated singularities as $\sigma(x, y)$ approaches 0. Of more direct physical interest to us, however, are the resolvent kernel of an elliptic operator (the fundamental solution of the Helmholtz equation associated with that operator) and the two-point function of a quantum field and the other fundamental solutions associated with the field equation, which we first met in Chapter 4.

The heat-kernel expansion, fortunately, leads to expansions of the singular parts of these other Green functions. Here I'll briefly indicate how, and also summarize the information thereby obtained about these elliptic and hyperbolic kernels. Attention will be restricted to scalar fields.

For $z \in \mathbf{C}$, a *resolvent kernel* $G_z(x, y)$ satisfies

$$K_{(x)}G_z - zG_z = \delta(x, y). \qquad (9.14)$$

If $f \in C_0^\infty(M)$, then

$$u(x) \equiv \int G_z(x, y)f(y)\sqrt{\gamma}\,d^d y \qquad (9.15)$$

is a solution of $Ku = zu + f$. That differential equation, however, has several interpretations. (Similar remarks could have been made earlier about the heat equation.) First, if K is a self-adjoint operator in $\mathcal{L}^2(M)$ and z is not in the spectrum of K, then $u = (K-z)^{-1}f$ is well-defined as the unique solution of $Ku = zu + f$. In that case, (9.14) is supplemented by whatever boundary conditions or decay conditions are involved in the definition of K as a self-adjoint operator in $\mathcal{L}^2$ (cf. Chap. 2), and the integration in (9.15) is over all of M. But although that situation may be in the back of one's mind as an eventual application, our principal concern now is with the local, classical PDE theory of the elliptic operator K, and what I have to say is independent of boundary conditions and of whether $z \in \sigma(K)$. Then G_z is not unique and may, in a given context, be only locally defined; the point is that its *local singularity is* uniquely determined. A standard step in proving the existence of a local solution of (9.14) is to construct an approximate solution, or *parametrix*, as a series analogous to (9.6), which exhibits this characteristic singularity.

$G_z(x, y)$ may be obtained from $H(t, x, y)$. The key ingredient here is the numerical fact

$$\int_0^\infty e^{-t\lambda}\,dt = \frac{1}{\lambda}.$$

The spectral theorem allows this to be applied to each eigenvalue of a self-adjoint operator, yielding the operator identity

$$\int_0^\infty e^{-t(K-z)}\,dt = (K - z)^{-1}.$$

Formally, one has a similar identity for hyperbolic operators:

$$i \int_0^\infty e^{-is(\Box + \cdots + m^2)} \, ds = (\Box + \cdots + m^2)^{-1}.$$

Conversely,

$$\frac{1}{2\pi i} \int_C e^{-tz} (K - z)^{-1} \, dz = e^{-tK},$$

if C is a contour which surrounds all of $\sigma(K)$ clockwise.

These identities carry over to the kernel level and even to the local parametrices: One has

$$G_z(x, y) = \int_0^\infty H(t, x, y) e^{tz} \, dt.$$

And from the Green function of the Schwinger–DeWitt equation one would expect to get via

$$i \int_0^\infty H(s, \underline{x}, \underline{y}) e^{-im^2 s} \, ds \tag{9.16}$$

a Green function for the hyperbolic field equation; for the free field and other static models, the result is the Feynman propagator (4.3). (In both these cases the extra exponential factor may be absorbed into H according to Remark (2) above.)

If we plug the Minakshisundaram–DeWitt expansion (9.6) into one of these integrals and formally calculate, the result is the *Hadamard expansion* for the parametrix, which is independently derivable. [References include the books of Hadamard, Garabedian, and Friedlander already cited, and the papers of DeWitt & Brehme 1960 and Adler et al. 1977 (corrected by Wald 1978a and Adler & Lieberman 1978).] Incidentally, I should remark that in pseudodifferential treatments one usually *starts* from the symbol of $(K - z)^{-1}$ (that is, the Fourier transform of G_z) and derives H from it [e.g., Gilkey 1975]. However, the symbol of e^{-tK}, which is the Fourier transform of H, can also be obtained more directly [Schrader & Taylor 1984].

The structure of the Hadamard series depends on whether the dimension of the manifold is even or odd. The case of greatest interest to us is dimension 4, so I state the details only for it (and revert to a space-time notation). The general case is treated in the final chapter of Friedlander 1975. In this discussion the symbol "$\sim$" will mean that the indicated series correctly reproduces the singular part of the indicated function, even if not necessarily asymptotic to its smooth part. The four-dimensional series is

$$G(x,y) \stackrel{.}{\sim} \frac{U}{\sigma} + V \ln \sigma + W, \qquad (9.17a)$$

where U, V, and W are smooth functions of $\underline{x}$ and $\underline{y}$. V has an expansion

$$V(\underline{x},\underline{y}) \sim \sum_{n=0}^{\infty} V_n(\underline{x},\underline{y})\sigma^n, \qquad (9.17b)$$

where V_n satisfies essentially the same recursion relation as a_n. There is a standard way of constructing a similar series for W, the nonuniqueness of W being reflected in the freedom to choose W_0 arbitrarily. This function and its physical significance are discussed by Wald 1978a, Brown 1984, Castagnino & Harari 1984, Brown & Ottewill 1986. Of course, starting from the Schwinger–DeWitt series will yield a particular result for W (modulo the effect of the cutoff functions mentioned in the next paragraph).

The series (9.17) is not guaranteed to converge, but by multiplying its terms by a sequence of cutoff functions with rapidly shrinking support, one can obtain a convergent series with the same partial derivatives of all orders at the origin as the original series. (This is the standard *Borel construction* for finding a function with a given series as its asymptotic expansion. In the present context it is described in detail in Friedlander 1975, Lemma 4.3.2.) The resulting function, $G(\underline{x},\underline{y})$, is a parametrix for the partial differential equation under study.

In the elliptic case, the true Green function G_z is equal to G as here constructed plus some smooth correction to accommodate the particular boundary conditions or global manifold structure under study. The correction term, unlike the series (9.17), will be nonlocal in its dependence on the coefficients of K and will also depend on the boundary conditions and the topology of M. This is simply to reiterate the point that the series correctly reproduces the singularity (the U and V terms) of G_z, but is irrelevant to the smooth part of any specific solution. Note that by elliptic regularity (see Chapter 2) the difference between any two solutions of (9.14) is a smooth function (where the meaning of "smooth" may be weakened if the coefficients in K are not in C^∞).

In H the smooth terms (positive integral powers of t) were also correctly given by the parametrical series, (9.6). So in some sense we have lost something in the passage from the parabolic to the elliptic problem. In contrast, in the Lorentzian domain we *gain*. Although (Remark (8)) the Schwinger series is not an asymptotic series at all, the Hadamard series is, at least for the singular terms. Here the discussion is complicated by the fact that there are so many different types of Green functions for a hyperbolic equation (see Chap. 4).

First, using the V_n one can construct asymptotic (as $\underline{x} \to \underline{y}$) series and hence parametrices for G_{ret} and G_{adv}. This (which is the more important and natural problem in the study of classical wave propagation) is the main subject of Friedlander's book. The result, in the four-dimensional case, is

$$G_{\text{ret}}(\underline{x}, \underline{y}) \sim U\delta_+ + V\theta_+ , \tag{9.18}$$

where δ_+ is a delta function concentrated on the future light cone and θ_+ is a step function concentrated inside the future light cone (cf. point 4 under "Mathematical interpretation of singularities" in Chapter 4 above). G_{adv} is given by the analogous construction in the past light cone. The U and V are the same as in (9.17), up to normalization factors. These expansions are genuinely asymptotic; there is no freedom to add an arbitrary smooth solution of the homogeneous equation, since no such has the right support. Here *finite propagation speed* is taking the place of elliptic regularity.

Of greater interest in quantum field theory are solutions of the type $G^{(1)}$ and G_{F}, which do not have such support properties. Recall from Chapter 4 the identity

$$\frac{1}{x - i\epsilon} = \pi i \delta(x) + \mathcal{P}\frac{1}{x} \quad \text{as } \epsilon \downarrow 0.$$

A related equation is

$$\lim_{\epsilon \downarrow 0} \ln|x - i\epsilon| = \pi i \theta(x) + \ln|x| + \text{const.}$$

[Gel'fand & Shilov 1964, p. 336]. The first terms of these equations indicate how (9.18) can arise out of (9.17), and they suggest that the second terms might yield an interesting solution of the homogeneous equation. Recall that

$$\overline{G} \equiv \frac{1}{2}(G_{\text{ret}} + G_{\text{adv}}) \qquad \text{and} \qquad G_{\text{F}} = \overline{G} + \frac{i}{2}G^{(1)}.$$

We can identify G_{F} with i times the G given by the series (9.17). The resulting series $G^{(1)}$ might be anticipated to satisfy

$$\langle\Psi|\{\phi(\underline{x}), \phi(\underline{y})\}|\Psi\rangle \sim G^{(1)}$$

for any "physically reasonable" state Ψ — in particular, the vacuum two-point function $\langle 0|\{\phi, \phi\}|0\rangle$ of a static model, which is what we previously called $G^{(1)}$.

Schwinger, DeWitt, Christensen, and others in effect *assumed* that

$$\frac{\langle 0^{\text{out}}|\mathcal{T}(\phi(\underline{x})\phi(\underline{y}))|0^{\text{in}}\rangle}{\langle 0^{\text{out}}|0^{\text{in}}\rangle} \sim G_{\text{F}} + \text{smooth function},$$

and hence that

$$\frac{\langle 0^{\text{out}}|\{\phi(\underline{x}), \phi(\underline{y})\}|0^{\text{in}}\rangle}{\langle 0^{\text{out}}|0^{\text{in}}\rangle} \sim G^{(1)} + \text{smooth function},$$

and hence, under the assumption of finite particle creation, that

$$\langle\Psi|\{\phi(\underline{x}), \phi(\underline{y})\}|\Psi\rangle \sim G^{(1)} + \text{smooth function}. \qquad (9.19)$$

(Here G_{F} and $G^{(1)}$ mean the series to which we have just now temporarily assigned those labels, rather than any particular functions with those expansions.) Given (9.19), one can calculate a series for

$$\langle\Psi|T_{\mu\nu}(\underline{x}, \underline{y})|\Psi\rangle,$$

to find the singularity which must be subtracted off to get the expectation value of the renormalized operator,

$$\langle\Psi|T_{\mu\nu}(\underline{x})|\Psi\rangle.$$

This program was initiated by DeWitt 1965, 1975 and carried out in the paper of Christensen 1976 (extended to fields with spin in Christensen 1978). The input to the calculation is (9.16) with H given by the Schwinger–DeWitt expansion; all the integrals can be evaluated in terms of Bessel functions, which can then be expanded in power series.

Christensen's result for the singular terms of $T_{\mu\nu}(\underline{x}, \underline{y})$ is very long, filling a page plus one line of the article. He finds a leading term that is inversely quartic in the distance $\sqrt{2\sigma}$,

$$-\frac{1}{2\pi^2} \frac{1}{(\sigma^\rho \sigma_\rho)^2} \left[g_{\mu\nu} - 4 \frac{\sigma_\mu \sigma_\nu}{\sigma^\rho \sigma_\rho} \right];$$

a term inversely quadratic in the geodesic distance and linear in the Riemann tensor; a term inversely linear in the distance (which would disappear if the expansion were about the midpoint of the geodesic from $\underline{y}$ to $\underline{x}$ instead of an endpoint — see Davies et al. 1977); a term logarithmic in the distance and quadratic in the Riemann tensor; and a very long "finite" but still singular term, quadratic in the curvature, which remains bounded as the distance approaches 0, but depends on the direction of approach because it involves tensors such as

$$\frac{\sigma_\mu \sigma_\nu}{\sigma^\rho \sigma_\rho}.$$

Attempts to carry the expansion of (9.16) beyond this point lead to terms of increasing order in m^{-2} instead of increasing order in σ; this is a reflection of the fruitlessness of attempting to get the smooth part of $\langle T_{\mu\nu} \rangle$ out of the parametrical series, and it also indicates that the series is relevant to the large-mass "adiabatic vacuum" defined in Chapter 7 (see Bunch et al. 1978).

Once again, let me emphasize the existence of a finite, smooth part of $\langle T_{\mu\nu} \rangle$ which contains information that is *not* contained in the Christensen series. The ultimate aim of the calculation is to *subtract* the singular terms of the series, thereby revealing the physically signficant remainder. (I shall be more precise about this, presently.) This renormalized expectation value depends on the state Ψ; when the state is determined by the manifold (for instance, as the vacuum of a static model or the in-vacuum of an asymptotically static one), the expectation value depends on the global structure of the manifold, and therefore is inherently beyond the reach of any local asymptotic expansion. (The Casimir energy of Chapter 5 is the prototypical example.)

VINDICATION: THE UNIVERSALITY THEOREM

It remains to demonstrate that the two-point functions of quantum fields do have singularities of precisely the Hadamard form, so that renormalization based on the Schwinger–DeWitt series is possible and correct. From our experience with "infinite particle creation" in Chapter 7, we do not expect the two-point function of *every* quantum state to consist of a Hadamard singularity plus something finite; if the two-point function describes a state of genuinely infinite physical density of energy, then its renormalized stress tensor will not be finite. Thus the establishment of the correct renormalization procedure is inextricably bound up with the identification of the nonsingular, physically acceptable states.

The argument has two steps. Following Fulling et al. 1981, we first investigate the singularity of the two-point function of a general ultrastatic model, and then extend the conclusion to time-dependent models. The proof has been written out only for a scalar field, but there is no reason to doubt that the ideas extend to vector bundles. For ease of presentation let us take the dimension N of space-time to be 4 (hence $d = 3$), although the ideas are independent of dimension.

Definitions: Riemannian manifolds M_1 and M_2 are *locally equivalent* at $x \in M_1$ if there is an isometry Φ of a neighborhood N_1 of x onto a set $N_2 \subset M_2$. Two functions on the respective manifolds are locally equivalent if they coincide under the obvious identification: $f_1|_{N_1} = f_2 \circ \Phi$. Two operators, K_1 and K_2, are locally equivalent if their manifolds and coefficient functions are locally equivalent.

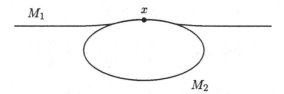

For notational ease the identification will henceforth be tacit: we shall write "$x \in N$" in place of the more pedantic "$x_1 \in N$ and the corresponding $x_2 \in \Phi[N]$", etc.

Universality theorem: Let K_1 and K_2 be locally equivalent at x_0. Assume that the ultrastatic space-times $\mathcal{M}_j \equiv \mathbf{R} \times M_j$ are globally hyperbolic and that the self-adjoint operators K_j are strictly positive, so that the quantum field theories described by K_j, and their vacuum states, are defined. Then there are a neighborhood N of x_0 (possibly smaller than the N_1 of the definition) and a $\delta > 0$ such that for all

$x, y \in N$ and all $|t| < \delta$,

$$G_1^{(1)}(t, x, y) - G_2^{(1)}(t, x, y) \in C^\infty\big((-\delta, \delta) \times N \times N\big),$$

where $G_j^{(1)}$ is the symmetrized vacuum two-point function (4.4), (4.6) for $(\mathcal{M}_j, K_j)$.

In words, the singularity of the ultrastatic two-point function near the diagonal is *locally* determined by the geometry of the manifold if the external field is purely gravitational; in the more general case where an external gauge field or scalar potential is present, it is locally determined by the symbol (all the coefficient functions) of the operator K.

Sketch of proof: An ultrastatic $G^{(1)}$ is the kernel of an operator $K^{-\frac{1}{2}} \cos(\sqrt{K}\, t)$ (4.4). The kernel of $K^{-\frac{1}{2}}$, like that of $(K - z)^{-1}$, has a (three-dimensional) Hadamard expansion; in particular, this kernel is smooth except at $x = y$, and its singularity there is locally determined in the sense just discussed. [The kernel of a power of K can be obtained from the heat kernel via

$$K^{-s} = \frac{1}{\Gamma(s)} \int_0^\infty e^{-tK} t^{-s-1}\, dt.$$

See Wald 1979a and ultimately Seeley 1967.] $G^{(1)}$ at a fixed t is obtained from that kernel by the action of

$$\cos(\sqrt{K}\, t) = \frac{d}{dt} \frac{\sin(\sqrt{K}\, t)}{\sqrt{K}} = \frac{d}{dt} G.$$

We recognize from Chapter 4 that we're dealing with the solution of a Cauchy problem with the kernel of $K^{-\frac{1}{2}}$ as initial data. The finite propagation speed of the Cauchy evolution then guarantees that the singularity of the result is locally determined. (The singularity will, however, extend over the light cone, rather than be concentrated on the diagonal.)

What is this universal, local singularity? By a roundabout argument, Fulling et al. 1981 show that it coincides with the (four-dimensional) Hadamard series. This conclusion can be extended to static (not necessarily ultrastatic) $\mathcal{M}$. A more direct proof of this **static Hadamard-form theorem** from the universality theorem and the asymptotics of eigenfunction expansions should be possible, but has not yet been achieved.

Now consider the general, time-dependent case. We consider a globally hyperbolic space-time and a field theory therein, possibly involving a time-dependent external potential in the place of the usual m^2.

Lemma [Fulling, Sweeny, & Wald 1978]: Let

$$G_{\Psi}^{(1)} \equiv \langle \Psi | \{\phi(\underline{x}), \phi(\underline{y})\} | \Psi \rangle$$

have the Hadamard form for $\underline{x}$ and $\underline{y}$ in a neighborhood of some Cauchy surface for $\mathcal{M}$. (That is, the singularity near the diagonal matches the Hadamard series, and the function is smooth elsewhere on the surface.) Then $G_{\Psi}^{(1)}$ is of Hadamard form throughout $\mathcal{M}$.

Main idea of proof: $G_{\Psi}^{(1)}$ satisfies the field equation in each variable.

General Hadamard-form theorem: In a globally hyperbolic $\mathcal{M}$, there is a class of states, forming a dense subspace of a Hilbert space, whose two-point functions have the Hadamard singularity structure.

Main idea of proof: Embed the region of interest in an initially ultrastatic $\mathcal{M}$ (of the same spatial topology); similarly extend the scalar potential to an asymptotically static and sufficiently positive mass, so that the static operator will have strictly positive spectrum. Then look at $\mathcal{F}^{\text{in}}$.

As a corollary, these states have finite renormalized $\langle \Psi | T_{\mu\nu} | \Psi \rangle$ (after subtraction of the Christensen series). We therefore identify these states as the "physically reasonable" ones.

Remark: Najmi & Ottewill 1985 have observed that it is possible for a state to have finite renormalized $\langle T_{\mu\nu} \rangle$ on a Cauchy surface although its two-point function has singularities at widely separated $\underline{x}$ and $\underline{y}$ (so it does not have the Hadamard form in the strong sense defined in the Lemma). Such a state may develop singularities in its $\langle T_{\mu\nu} \rangle$ at later times.

RENORMALIZATION PRESCRIPTIONS; UNIQUENESS ISSUES

Let Ψ be any physically reasonable state. Then $\langle \Psi | T_{\mu\nu}(\underline{x}, \underline{y}) | \Psi \rangle$, which can be formed by acting on $G_\Psi^{(1)}$ with a certain differential operator, can be decomposed as

$$\langle\Psi|T(\underline{x},\underline{y})|\Psi\rangle = T_{\text{sing}} + T_{\text{smooth}}\,, \qquad (9.20)$$

where T_{sing} is independent of Ψ, is locally determined by the symbol of K, and has the singularity structure described by the Christensen series. Unfortunately, the decomposition (9.20) is not unique, because a singular function plus an arbitrary smooth function is still singular. Naturally, we will consider only smooth functions which are independent of Ψ and depend locally on K, but that still leaves some room. Once this ambiguity has been resolved, we can define the renormalized $\langle\Psi|T(\underline{x})|\Psi\rangle$ by subtracting T_{sing} and taking the limit $\underline{y} \to \underline{x}$. Since the subtraction is independent of Ψ, this defines a renormalized energy-momentum tensor *operator*. We must also verify that the result satisfies Wald's axioms; in fact, that requirement serves to cut down the ambiguity. What is primarily at issue here is the conservation law, $\nabla_\mu T^\mu{}_\nu = 0$. The validity of the other two axioms can't be affected by any modification of T_{sing} of the types we've agreed to consider, and Wald's 1977 and 1978 papers, together with the Hadamard-form theorem, establish that they are indeed satisfied by our renormalized T.

The literature contains several ultimately equivalent prescriptions for carrying out the correct subtraction. The most straightforward is to follow the foregoing discussion literally, working with the point-split stress tensor. See, for example, Davies et al. 1977 and Bunch et al. 1978. Some other methods seek to cut down the calculational labor by performing the subtraction at the level of the Green function, $G_\Psi^{(1)}$, before differentiating to form the stress tensor. This proposal requires some ad hoc modifications to avoid producing a tensor that violates the conservation law. Two variations on this theme are given by Wald 1978a and by Bunch & Davies 1978 (repeated in Birrell & Davies 1982, p. 195). Brown & Ottewill 1986 advocate avoiding the Schwinger–DeWitt series altogether, working more directly with the Hadamard series (9.17).

Another way of implementing the subtraction is to form a linear combination of the $\langle T(\underline{x},\underline{y})\rangle$ of the physical field with the corresponding quantities for three fictitious fields having large masses. The mass ratios are adjusted so that all the singularities cancel in the total, and then the three auxiliary masses are taken to infinity. In that limit the auxiliary fields give no contribution to the finite remainder, so that the correct renormalized $\langle T \rangle$ is obtained. This method, which is akin to a technique called *Pauli–Villars regularization* in quantum electrodynamics, was introduced in our context by Vilenkin 1978.

In models where the field equation is solved by separation of variables, it should be possible to perform the subtraction "mode-by-mode under the integral sign" so that it is never necessary to calculate $\langle \Psi | T(\underline{x}, \underline{y}) | \Psi \rangle$ with the points separated (a very tedious task in practice). Unfortunately, it is not always obvious what mode-by-mode subtraction reproduces exactly the effect of the covariant, physically justified point-splitting subtraction. See Anderson & Parker 1987 for a recent addressing of such an issue in Robertson–Walker models. Generally speaking, the technology of doing such subtractions efficiently is still not sufficiently developed.

I close with a series of remarks which address the remaining questions of uniqueness.

(1) *The trace anomaly:* If $m = 0$ and $\xi = \frac{1}{6}$, the stress tensor of the classical field theory satisfies $T^{\mu}{}_{\mu} = 0$. Indeed, this is the Noether identity associated with the conformal invariance of the theory. The generalization to the massive case is a proportionality between the trace of T and the square of the field. In the quantum theory, one finds that if the renormalized stress tensor is forced to satisfy the conservation law, then it cannot also satisfy the trace identity. Instead, the trace contains an additional term, $a_2/16\pi^2$ (where a_2 is the second Schwinger-DeWitt coefficient, (9.8c)). This is the reason why the renormalization prescription needs to be artificially corrected when the subtraction is performed on the Green function before the differentiation. Historically, the trace anomaly was the battleground between adherents of various renormalization schemes and general approaches to quantum field theory in curved space-time. Within the present approach, informative references include Wald 1978a, Bunch et al. 1978, Vilenkin 1978, Brown 1984, Castagnino & Harari 1984.

(2) Wald's uniqueness theorem admits a certain nonuniqueness in the stress tensor: any term with the properties (A)–(C) could be added to it without violating the axioms. In the terms of our more recent discussion, any term with those properties could be moved from one term of (9.20) to the other. Two obvious examples of such terms are (arbitrary numerical multiples of) the metric tensor itself, $g_{\mu\nu}$, and the

Einstein tensor, $G_{\mu\nu}$. Additional examples are

$$\frac{\delta}{\delta g_{\mu\nu}} \int L \sqrt{g}\, d^N x, \tag{9.21}$$

where L is any scalar function formed covariantly from the metric and its derivatives. Taking $L = 1$ and $L = R$ yields $\underline{g}$ and $\underline{G}$. Taking $L = R^2$ and $L = R^{\mu\nu}R_{\mu\nu}$ yields two tensors involving the Riemann tensor quadratically and its second derivatives linearly. Let us call them $A_{\mu\nu}$ and $B_{\mu\nu}$. See Birrell & Davies 1982, Secs. 6.2 and 6.4, for the details. (Their notation for A and B is $^{(1)}H$ and $^{(2)}H$.)

The seemingly independent object $R^{\mu\nu\rho\tau}R_{\mu\nu\rho\tau}$ gives nothing new in the local equation of motion, because of the four-dimensional analogue of the Gauss–Bonnet theorem, which states that

$$\int \left(R^{\mu\nu\rho\tau}R_{\mu\nu\rho\tau} - 4R^{\mu\nu}R_{\mu\nu} + R^2 \right) \sqrt{g}\, d^4 x \tag{9.22}$$

is a topological invariant. [See DeWitt 1975, Sec. 6.15. The formula (9.22) is hard to find in the mathematical literature, but it can be shown equivalent to the more fundamental expressions found in Allendoerfer 1940 and Chern 1944, 1945, 1962. (The last of these papers specifically refers to Lorentzian manifolds.)] This term can't be ignored in the quantization of the gravitational field itself, however, because there the global topology becomes significant (see, for instance, the papers cited on pp. 97–98 of Duff 1981).

Clearly, a term proportional to $G_{\mu\nu}$ is indistinguishable from a change in the other side of the Einstein equation — a *renormalization of the gravitational constant*. The gravitational field equation is a subject for experimental investigation; only the *total* numerical strength of the term proportional to $\underline{G}$ can be determined by experiment, and the division of that term into a gravitational term on the left-hand side and a matter term on the right-hand side is an arbitrary convention. (See Fulling & Parker 1974 for further discussion.) Furthermore, it has long been known that Einstein's equation can be generalized by the addition of a term $\Lambda g_{\mu\nu}$, where Λ is called the *cosmological constant*; astronomical evidence is that Λ is very small. The ambiguous term in the stress tensor proportional to the metric can be absorbed by a renormalization of the cosmological constant. (Wald originally stated a separate axiom to set this term to zero, so that the stress tensor would reduce to its usual value in flat space. I prefer to consider this a convention for defining the cosmological constant, not an axiom.)

(3) In the same way, one could modify the gravitational field equation by terms proportional to $A_{\mu\nu}$ and $B_{\mu\nu}$. The equation of motion of $\underline{G}$ thereby becomes of *fourth order*. (This raises some dynamical worries — see Horowitz & Wald 1978 and Horowitz 1981.) The necessity for these terms is already visible in the logarithmic term of the Christensen series, which contains precisely these tensors, multiplied by $\ln \sigma$. Since σ is a quantity with physical dimensions, such a term is not covariant under changes of one's unit of length. A rescaling will cause the logarithmic term to spit out a smooth, finite term proportional to some linear combination of the four conserved tensors so far discussed. Originally Wald proposed to eliminate this ambiguity in the renormalized stress tensor by a fifth axiom, requiring the field equation to have the dynamical properties of a second-order equation. However, later [Wald 1978a, last section] he discovered an argument that this is not possible: Whatever term proportional to $A_{\mu\nu}$ or $B_{\mu\nu}$ is included in the subtraction T_{sing}, it can be absorbed into a suitable logarithmic term by modifying its logarithm to read $\ln(\sigma/l^2)$, where l is some length. (Thus, although the coupling constants multiplying the fourth-order tensors in the gravitational equation are dimensionless numbers, they may be replaced by quantities with physical dimension. Such *dimensional transmutation* arises also in other field theories; see Coleman & Weinberg 1973.) But at most *one* possible term of each of these two types is acceptable according to the proposed axiom, because $A_{\mu\nu}$ or $B_{\mu\nu}$ itself explicitly contains fourth derivatives. If the fifth axiom could be implemented, therefore, it would necessarily introduce one or two length scales into the theory. Now suppose that we start from a conformally invariant theory; it contains no length scales at all. By pure reasoning, unaided by experiment, we cannot possibly get a length scale out if it was never put in. Therefore, it must be the case that *no* choice of the terms in T_{sing} proportional to $A_{\mu\nu}$ and $B_{\mu\nu}$ can do the job. (It should be clear that there is nothing paradoxical about fixing such length scales *experimentally*, however. Indeed, precisely this is believed to happen in quantum chromodynamics.)

(4) What about higher-degree polynomials in curvature as prospective gravitational Lagrangians? Must we be prepared to admit them all, with resulting terms of sixth and higher order in the gravitational

equation? No, at least not within the framework of a quantum field without self-interaction in a classical gravitational background. Within the point-splitting method there are no singular terms of such high order in the curvature, and therefore no imperative to consider terms of those orders as possible subtractions. From the deeper standpoint of uniqueness, suppose that a rival renormalization scheme produced a result which differed from ours by terms of this nature. For dimensional reasons, the discrepant terms would have to have numerical coefficients proportional to negative powers of the mass of the field. (Here I assume that m^{-1} is the only length scale in the classical theory.) The rival theory would then be quite singular in the massless limit, and therefore implausible.

(5) Brown 1978 questioned whether there might be covariant, conserved tensors formed *nonpolynomially* out of the metric and curvature. He believed that such terms could be used to remove the trace anomaly without spoiling conservation. His specific prescription involved the quantity $\ln(C^{\mu\nu\rho\tau}C_{\mu\nu\rho\tau})$ ($C \equiv$ Weyl tensor), which is singular in the limit of flatness or conformal flatness. His proposal was therefore rejected by reasoning similar to that in (4). However, as far as I know, there is no rigorous theorem stating the nonexistence of such nonpolynomial objects with nicer properties. (I regret that in Fulling 1983 the Wald uniqueness theorem is misstated so as to imply that this problem has been solved.) One's feeling is that the stress tensor in a fixed state (in the sense of Wald's Axiom II) ought to be an "analytic" functional of the curvature; therefore, it should be expandable in a power series, and the reasoning in (4) comes into play. (See Epstein 1975, including the cautious remark at the end.)

Conclusion

The central problem of fundamental theoretical physics today is reconciling general relativity with quantum theory. These two profound insights of the early twentieth century have never really fit together into a consistent picture of the world. The theory of quantum fields in curved space-time, satisfying linear equations of motion, is a small, preliminary skirmish in this campaign. It has provided at least its fair share of conceptual problems.

One of the striking things about the subject is the intertwining of the conceptual issues with the mathematical tools. Historically there has been a close association between relativity and differential geometry, while rigorous research in quantum theory has looked more toward functional analysis. Field quantization in a gravitational background brings these two alliances into head-to-head confrontation: A field is a function of a time and a space variable,

$$\phi(t, \mathbf{x}).$$

Relativity and modern geometry persuade one with an almost religious intensity that these variables must be merged and submerged; the true domain of the field is a space-time manifold:

$$\phi(\underline{x}), \quad \underline{x} \mapsto \{x^\mu\}.$$

But quantum theory and its ally, analysis, are constantly pushing in the opposite direction. They want to think of the field as an element of some function space, depending on time as a parameter:

$$\phi_t(\mathbf{x}).$$

This complementarity provides the subject with much of its difficulty and much of its interest.

Progress has been made by abandoning the naive extrapolation of the "particle" concept into a context where it has no clear meaning, and concentrating instead on observables built out of the field itself, such as the electric current and the energy-momentum tensor. As we've seen in Chapters 5, 7, and 9, studying the stress tensor reveals the physical significance of a field model and the various quantum states within it. The condition of renormalizability of the stress tensor (and other observables) helps determine which of the mathematically possible states are physically relevant. As for the particles, they are still there in regimes of sufficient symmetry or approximate symmetry (such as the present-day slowly expanding, very flat universe). They are an epiphenomenon, like the phonons, magnons, and so on of condensed-matter physics: They are excitations of a more fundamental substrate, and when external conditions make the latter too turbulent, they fade away.

This new understanding provides the base for calculations of physical quantities in an unlimited supply of problems. As an example of the application of the renormalization theory to a particular problem, I recommend a series of papers on the stress tensors of various quantum states in the Schwarzschild metric: Candelas 1980, Candelas & Howard 1984, Howard & Candelas 1984, Howard 1984.

Nevertheless, one cannot yet feel confident that this theory rests on a sufficiently solid physical base. One wants to understand how it emerges as an approximation to a more fundamental theory in which gravity itself is quantized. Recently there has been an upsurge of interest in this question [e.g., Banks 1985, Halliwell 1987, Padmanabhan 1989]. These authors argue that a fully quantized theory of gravity and other fields has a semiclassical limit, in which the matter fields obey a quantum field theory in a classical gravitational background. They have more difficulty, however, in reproducing the dynamical effect of the matter on the gravitational field; usually they arrive at a metric satisfying the *vacuum* Einstein equation (with no stress tensor). If one attempts to go beyond the lowest asymptotic order, Padmanabhan argues, quantum excitations of the gravitational field will become equally as significant as those of the matter fields, and the background-field picture will break down. (This same point has long been made by M. J. Duff [private communications].) The manifest cosmological relevance of the *classical* Einstein equation with matter source suggests to me that we simply don't yet know how to take the correct limit. The solution of this problem will demand strong physical insight combined with sound asymptotic analysis.

Other criticisms of quantum field theory in curved space-time are that the separation of a nonlinear theory into classical and quantized parts is inherently ambiguous [Duff 1981], and that the fourth-order terms (9.21) in the gravitational equation of motion resulting from renormalization give rise to physically implausible instabilities in the time evolution of the gravitational field [Horowitz & Wald 1978; Horowitz 1981]. (These instabilities presumably will not be cured by trying to quantize the gravitational field itself, since they are an aspect of the nonrenormalizability of the gravitational interaction. They will, however, be absorbed into that broader problem.) In any event, one could never rest content with such a theory as an ultimate description of the world, even if it should turn out that there are many situations in which it gives an adequate quantitative model of reality.

In Fulling 1984 I have speculated on some ways in which we might move rather directly from curved-space quantum field theory into some kind of true quantum gravity. Another possibility is that the rigorous study of *interacting* fields in curved space will lead to a return to the Schrödinger picture (in an infinite-dimensional configuration space!) and the introduction of the methods of constructive quantum field theory [e.g., Glimm & Jaffe 1981] into the gravitational context. (See also Fredenhagen & Haag 1987.) This is much easier said than done! Of course, it is also possible that the new ideas of string theory [Green et al. 1987], the old idea of discretization of space and time themselves, or some other conceptual revolution will sweep the problems away.

In the meantime, there is no shortage of physical scenarios (the early or late universe, black holes, Kaluza–Klein (extra-dimensional) models) in which the theory of field quantization and renormalization on manifolds could be put into practice. Surprisingly few concrete calculations have yet been carried out on this relatively sound basis.

There are interesting mathematical points still to be settled, as well. The reader has surely noticed gaps, well labelled and otherwise, in this exposition. J. Dimock [private communications] has pointed out that there is still no mathematically respectable theory of the electromagnetic field in curved space-time, because the need to choose a gauge interferes with the geometrically covariant construction of the field algebra in analogy with the scalar and spinor fields [Dimock 1980, 1982]. This book will have served its purpose if its very deficiencies recruit a new generation of investigators to finish the job.

Appendix
Varieties of Instability
of a Boson Field in an External Potential
and Black Hole Klein Paradoxes

This document was originally distributed in 1975 by the Mathematics Department of King's College, University of London, as a technical report. (The research was supported by the Science Research Council.) A brief account of its most novel conclusions was published as Fulling 1976.

It is reproduced here verbatim, except for certain improvements connected with the revolution in scientific typography, and the updating of references to some journal articles that were not yet in print at that time.

Analogous studies of the Klein effect for fermions have since been conducted by Bilodeau 1977 for the neutrino field and by Manogue 1988 for the massive Dirac field. Ambjorn & Wolfram 1983 investigate the Schiff–Snyder–Weinberg scenario further; they present evidence that the reaction of the quantized field on the electric field suffices to suppress the instabilities.

Recent years have seen considerable attention to the implications of strong-field effects (on fermions, primarily) for realistic nuclear physics. I understand that the experimental evidence is still inconclusive. From this literature I will cite only these reviews: Rafelski et al. 1978; Soffel et al. 1982; Greiner et al. 1985.

Abstract

Part One

A relativistic scalar field is quantized in a one-dimensional "box" comprising two broad electrostatic potential wells. As the potential difference increases, the phenomena found by Schiff, Snyder, and Weinberg in such a model occur: merging of mode frequencies and disappearance of the vacuum as a discrete state, followed by appearance of complex frequencies and unboundedness below of the total energy. However, a new effect appears for some values of the potential: The discrete vacuum (with the associated particle interpretation) reappears, but the energy remains unbounded below because some negative-norm modes have greater frequencies than some positive-norm modes. That is, a particle-antiparticle pair can have energy less than that of empty space. As the outer walls of the box approach infinity, this situation goes over into the boson Klein "paradox", marked by nonuniqueness of the vacuum and coexistence of positive- and negative-norm continuum modes at the same frequency.

Part Two

Particle creation in quantum field theory near a time-independent rotating or charged black hole is an instance of the Klein "paradox"; in the rotating case a second Killing vector is used to reduce a gravitational external-potential problem to an equivalent electrostatic one. The essential geometrical features responsible for the effect are brought out here by constructing a simple rectilinear model of a rotating black hole. As a special case one has the original Klein "step", the infinite-volume limit of the Schiff–Snyder–Weinberg square well, which can be solved exactly. Here, in the quantized scalar field theory, there is a spontaneous breaking of time-reversal symmetry, and a particular choice of "vacuum" state leads to the prediction of particle creation. (The calculation is done here for a massive field.) Alternative choices are discussed, as are the physical reality of the field quanta, and the questionable relevance of such an "unstable" field theory in a real physical situation.

Part One

I. Introduction

Nearly fifty years have passed since Klein discovered the mathematical phenomenon which bears his name[1,2] in the solutions of a relativistic wave equation featuring a static electric potential with behavior such as

$$A_0(\mathbf{x}) \to 0 \qquad \text{as } z \equiv x^3 \to \infty,$$
$$A_0(\mathbf{x}) \to \text{constant} \neq 0 \quad \text{as } z \to -\infty. \tag{1.1}$$

It has been thirty-five years since Schiff, Snyder, and Weinberg[3] discovered related peculiarities in the solutions of the Klein–Gordon equation in a potential well. It was quickly realized that the Klein "paradox" indicates the impossibility of interpreting the solutions of the wave equation as quantum-mechanical wave functions of a single particle, and consequently the need for a field theory. The Klein phenomenon is generally understood to represent a continuous creation of charged-particle pairs by the strong electrostatic field. Detailed explanations in the context of a quantized field theory are strangely lacking in the literature, however.[4] The Schiff–Snyder–Weinberg effects have been even

---

[1]   O. Klein, *Z. Physik* **53**, 157 (1929). Klein was studying the Dirac equation, but a similar situation arises for the Klein–Gordon equation, which is the only case considered in the present paper.

[2]   A list of later references is given by A. I. Nikishov, *Nucl. Phys.* B **21**, 346 (1970). Note in particular the famous calculation of J. Schwinger, *Phys. Rev.* **82**, 664 (1951) for a potential linear in z.

[3]   L. I. Schiff, H. Snyder, and J. Weinberg, *Phys. Rev.* **57**, 315 (1940).

[4]   For example, the basic algebraic facts of the Klein phenomenon for the Dirac equation are presented in J. D. Bjorken and S. D. Drell, *Relativistic Quantum Mechanics* (McGraw–Hill, New York, 1964), pp. 40–42. It is pointed out that single-particle relativistic quantum mechanics is inadequate in this context. However, one searches in vain the index of J. D. Bjorken and S. D. Drell, *Relativistic Quantum Fields* (McGraw–Hill, New York, 1965) for a resolution of the paradox. (See, however, p. 75 of the first book.) The Klein effect for boson fields is almost never discussed.

more neglected. Except for an important paper of Schroer and Swieca,[5] recent work has concentrated on showing that the most disturbing aspect of the situation, unboundedness of the total field energy from below, can be prevented by including in the model a nonlinear self-interaction of the quantized field.[6,7] The precise relationship of the Schiff–Snyder–Weinberg effects to the Klein paradox (for bosons) has never been made clear. Nor is there any consensus on whether all these effects represent merely a breakdown of the single-particle interpretation of the wave equation, a breakdown of the particle interpretation of the second-quantized field theory, or a breakdown of the field theory itself.

Recently the subject has been made much less academic by calculations of an outward flux of energy and angular momentum or charge in the quantum theory of a scalar or spinor field in the (time-independent) region surrounding a rotating or charged black hole.[8] These effects have been recognized as instances of the Klein paradox (in a generalized sense).

In attempting to understand the black hole results better, the author has reinvestigated the simplest models in which the Klein and Schiff–Snyder–Weinberg effects appear. A potential step exhibiting the Klein paradox should be obtainable as the infinite-volume limit of a square potential well, in which the Schiff–Snyder–Weinberg effects arise. Following Ref. 3, we shall study a square well enclosed in a box (at one end) in order to make the spectrum of modes discrete; such a model may also be described as consisting of two adjoining square wells with infinitely high outer walls (Fig. 1). Whereas Schiff et al. kept one well constant in size while the other expanded to infinity, we shall here keep the ratio of well lengths constant as they both tend to infinity. Under those circumstances a new effect not reported by Schiff et al. is observed in the solutions of the canonical boson field theory. (De-

---

[5]   B. Schroer and J. A. Swieca, *Phys. Rev. D* **2**, 2938 (1970). See also B. Schroer, *Phys. Rev. D* **3**, 1764 (1971). We consider only the "no-vacuum" theory developed by these authors, since it seems more justified physically than their alternative quantization involving an indefinite metric in the state space of the field theory.

[6]   A. B. Migdal, *Zh. Eksp. Teor. Fiz.* **61**, 2209 (1971) [*Sov. Phys.–JETP* **34**, 1184 (1972)].

[7]   A. Klein and J. Rafelski, *Phys. Rev. D* **11**, 300 (1975).

[8]   See Refs. 5, 11, and 1 of Part Two.

scriptions of the various effects are deferred to the body of the paper.) It will be seen that such a modification of the Schiff–Snyder–Weinberg scenario is in fact necessary if the Klein type of behavior is to arise as a limiting case.

II. PRELIMINARIES

A. Quantization of a Scalar Field in an Electrostatic Potential.

Let us begin by reviewing the necessary formalism.[9] We consider the Klein-Gordon equation with minimal electromagnetic coupling,

$$\left(\frac{\partial}{\partial t} + ieA_0\right)^2 \phi - \nabla^2\phi + m^2\phi = 0 \qquad (\hbar = c = 1). \qquad (2.1)$$

The potential $A_0(\mathbf{x})$ is independent of t, and the vector potential $\mathbf{A}$ is zero. Naturally, one separates variables, studying (c-number) solutions of the form[10]

$$\phi_{(j)}(t, \mathbf{x}) = \phi_j(\mathbf{x})e^{-i\omega_j t}, \qquad (2.2)$$

for which

$$(\omega_j - eA_0)^2\phi_j = (-\nabla^2 + m^2)\phi_j \qquad (2.3)$$

must hold. This equation is invariant under the trivial gauge transformation of addition of a constant V_0 to both ω and eA_0 [corresponding to multiplication of $\phi(t, \mathbf{x})$ by a phase, $\exp(-iV_0 t)$]. The phases of solutions may be chosen so that ϕ_j is real when ω_j is real, and

$$\phi_k(\mathbf{x}) = \phi_j(\mathbf{x})^* \quad \text{if } \omega_k = \omega_j{}^*. \qquad (2.4)$$

Equation (2.3) does not display the squares of the frequencies ω as eigenvalues of an Hermitian operator, in contrast to the result of separation of variables in the free Klein-Gordon equation, or in the presence of a time-independent vector potential, an $\mathbf{x}$-dependent mass

---

[9]   W. Pauli and V. Weisskopf, *Helv. Phys. Acta* **7**, 709 (1934); H. Snyder and J. Weinberg, *Phys. Rev.* **57**, 307 (1940). See also Refs. 5 and 7.

[10]   The index j (and later k) will be used to label the distinct solutions (ω, ϕ) of the problem (2.3). Note that when $A_0 = 0$, the same function $\phi_j \propto e^{i\mathbf{p}\cdot\mathbf{x}}$ appears for two values of j, corresponding to $\omega = \pm(\mathbf{p}^2 + m^2)^{\frac{1}{2}}$.

term (Ref. 5), or a static[11] gravitational field.[12] A consequence is that the solutions need not be orthogonal in the usual (L^2) sense. Instead, one finds from the Hermiticity of $-\nabla^2 + m^2$ that

$$(\omega_k - \omega_j{}^*) \int d^3x \left[\omega_k + \omega_j{}^* - 2eA_0(\mathbf{x})\right] \phi_j{}^*(\mathbf{x})\phi_k(\mathbf{x}) = 0 \qquad (2.5)$$

for any two solutions of Eq. (2.3) satisfying suitable boundary conditions. This equation yields a nontrivial relation of quasi-orthogonality except when (1) ω_j is real and $\omega_k = \omega_j$, or (2) ω_j is complex and $\omega_k = \omega_j{}^*$. For instance, for real and distinct frequencies we have

$$\int d^3x \, (\omega_j + \omega_k - 2eA_0) \, \phi_j \phi_k = 0, \qquad (2.6)$$

and for complex $\omega_j = \omega_k$,

$$2 \int d^3x \, (\mathrm{Re}\ \omega_j - eA_0)|\phi_j|^2 = 0. \qquad (2.7)$$

For real ω Schiff et al. define

$$\epsilon_j = 2 \int d^3x \, (\omega_j - eA_0) \, \phi_j{}^2, \qquad (2.8)$$

which may be positive, negative, or zero. (We assume for the moment that it is finite.) Equation (2.5) for $\omega = \omega_j{}^*$ [with the convention (2.4)] suggests extending this definition to complex ω.

For technical reasons it is often convenient to replace the second-order equation (2.1) by a system of two equations of first order in time.[13] Let

$$\pi = \left(\frac{\partial}{\partial t} - ieA_0\right) \phi^* \qquad (2.9)$$

---

[11]   This term means that $g_{0j} = 0$ for $j \neq 0$ (in a suitable coordinate system), as well as that the metric tensor is independent of time. The formalism of second quantization for this case has been given by S. Bonazzola and F. Pacini, *Phys. Rev.* **148**, 1269 (1966); S. Fulling, *Phys. Rev. D* **7**, 2850 (1973); B. S. Blum, Ph.D. Thesis, Brandeis University (1973); A. Ashtekar and A. Magnon, *Proc. Roy. Soc. A* **346**, 375 (1975).
[12]   Rigorous mathematical treatments of the self-adjoint operators which arise in all these classes of problems are: T. Ikebe and T. Kato, *Arch. Rat. Mech. Anal.* **9**, 77 (1962); P. R. Chernoff, *J. Func. Anal.* **12**, 401 (1973).
[13]   Cf. H. Feshbach and F. Villars, *Rev. Mod. Phys.* **30**, 24 (1958). There are many equivalent first-order formalisms, related by local or nonlocal transformations in the two-component space. The most convenient or elegant choice depends on context.

and consider the two-component object

$$\Phi = \begin{pmatrix} \phi \\ \pi* \end{pmatrix}.$$ (2.10)

Then Eqs. (2.1) and (2.3) are respectively equivalent to

$$\frac{d}{dt}\Phi(t,\mathbf{x}) = -iH\Phi(t,\mathbf{x}),$$ (2.11a)

$$H\Phi_j(\mathbf{x}) = \omega_j\Phi_j(\mathbf{x}),$$ (2.11b)

where

$$H = \begin{pmatrix} eA_0(\mathbf{x}) & i \\ -i(-\nabla^2 + m^2) & eA_0(\mathbf{x}) \end{pmatrix}$$ (2.12)

The elementary normal-mode solution (2.2) is the first component of a solution

$$\Phi_{(j)}(t,\mathbf{x}) = \Phi_j(\mathbf{x})e^{-i\omega_j t}$$ (2.13)

of Eq. (2.11a), the second component being

$$\pi_{(j)}*(t,\mathbf{x}) = \pi_j*(\mathbf{x})e^{-i\omega_j t},$$
$$\pi_j*(\mathbf{x}) = -i(\omega_j - eA_0)\phi_j(\mathbf{x}).$$ (2.14)

In the space of two-component functions $\Phi(\mathbf{x})$ the expression

$$\langle\Phi_1,\Phi_2\rangle = i\int d^3x\,(\phi_1*\pi_2* - \pi_1\phi_2)$$ (2.15)

$$= i\int d^3x\left\{\phi_1*\left(\frac{\partial}{\partial t} + ieA_0\right)\phi_2 - \left[\left(\frac{\partial}{\partial t} + ieA_0\right)\phi_1\right]*\phi_2\right\}$$

has all the properties of a scalar product except positive definiteness. Furthermore, H is Hermitian with respect to it:

$$\langle H\Phi_1,\Phi_2\rangle = \langle\Phi_1,H\Phi_2\rangle;$$ (2.16)

this implies that if Φ_1 [$= \Phi_1(t,\mathbf{x})$] and Φ_2 are solutions of Eq. (2.11a), then $\langle\Phi_1(t),\Phi_2(t)\rangle$ is independent of the time t at which the integration is performed. For the eigenvectors $\phi_j(\mathbf{x})$ one has

$$\langle\Phi_j,\Phi_k\rangle = \int d^3x\,(\omega_k + \omega_j* - 2eA_0)\,\phi_j*\phi_k.$$ (2.17)

Hence Eqs. (2.6)–(2.8) can be restated as

$$\langle \Phi_j, \Phi_k \rangle = \epsilon_j \delta_{\omega_j \omega_k} \quad \text{for real } \omega_j ; \qquad (2.18a)$$

$$\langle \Phi_j, \Phi_j \rangle = 0, \quad \langle {}^*\Phi_j, \Phi_j \rangle = \epsilon_j \quad \text{for complex } \omega_j, \qquad (2.18b)$$

where ${}^*\Phi_j$ is the solution of the form (2.10) or (2.13) whose first component is $\phi_j{}^*$ at $t = 0$. For lack of a better word, $\langle \Phi, \Phi \rangle$ will be called the *norm* of Φ.

To construct the quantum theory of the charged field ϕ one expects to start from the *general* solution of Eq. (2.1), from which the field operator will be built. One would like to argue, first, that the Cauchy problem for this system is well-posed, so that its solutions are in one-to-one correspondence with choices of (sufficiently well behaved) initial values $\phi(0, \mathbf{x})$ and $\pi^*(0, \mathbf{x})$, and, second, that each set of initial data can be expanded in terms of the eigenvectors $\Phi_j(\mathbf{x})$. These properties obviously are closely related (cf. Chernoff, Ref. 12). In this paper a favorable answer to these technical questions surrounding the classical theory will have to be assumed. (Also, the treatment of continuous spectra will be intuitive and cavalier.)

In all the models considered, it will be physically obvious that the Cauchy problem is well posed, because proper boundary conditions are imposed whenever the field is in danger of propagating "off an edge" or "into a hole" or "to infinity within a finite time". Unfortunately the standard methods of proving completeness of eigenfunctions apply to the Klein–Gordon electrostatic case only when the potential is so weak that the phenomena of greatest interest to us here do not occur. The spectral theorem cannot be applied to the operator H unless H is self-adjoint with respect to some *positive definite* scalar product. Since H generates the time evolution of the system, sesquilinear forms in which it is Hermitian are associated with conserved quantities, such as charge or energy. The charge form (2.15) is never positive definite. However, if the potential is sufficiently weak, the total field energy [cf. Eq. (2.28)] yields a positive definite form with respect to which H can be proved self-adjoint.[14] The failure of this proof for a strong potential is not merely a technicality. If H is self-adjoint with respect to *any* (positive definite) scalar product, then it has no complex eigenvalues and no Jordan associated eigenvectors [see Eqs. (3.24–26)]; but it is

---

14   L.-E. Lundberg, *Commun. Math. Phys.* **31**, 243 (1973).

precisely these conditions which cease to hold when the Schiff–Snyder–Weinberg effects arise. The theory of operators in spaces with indefinite scalar products[15] is in its infancy, and it is not clear that the problem of completeness of eigenfunctions can even be posed uniquely, since the scalar product does not determine a unique topology with respect to which the space ought to be completed. Schroer and Swieca (Ref. 5) have sketched an argument to the effect that a suitably modified spectral theorem should be true for any operator satisfying Eq. (2.16) in an indefinite metric. [Their proof, unfortunately, does not appear to apply to the Klein situation (1.1), since it assumes vanishing of $A_0(\mathbf{x})$ at spatial infinity.]

We now reinterpret $\phi(t,\mathbf{x})$ as an operator-valued distribution, identifying π of Eq. (2.9) as its canonical conjugate,

$$[\phi(t,\mathbf{x}), \pi(t,\mathbf{y})] = i\delta(\mathbf{x}-\mathbf{y}),$$
$$[\phi,\phi] = [\pi,\pi] = [\phi,\phi^\dagger] = [\phi,\pi^\dagger] = [\pi,\pi^\dagger] = 0 \quad \text{at equal times,}$$

(2.19)

and replacing complex conjugation by Hermitian conjugation where necessary. Consider a mode $\Phi(j)$ for which ω_j is real. We anticipate that the field will contain a term proportional to $\phi(j)$, with an operator coefficient:

$$\phi(t,\mathbf{x}) = c_j \phi_j(\mathbf{x})e^{-i\omega_j t} + \cdots \qquad (2.20)$$

[and similarly for π and Φ with Eqs. (2.13)–(2.14)], and that this term is orthogonal to the remainder of the field in the sense of Eq. (2.15). Then the orthogonality relations (applied at $t=0$ for simplicity) imply

$$\epsilon_j c_j = \langle \Phi_j, \Phi \rangle = i\int_{t=0} d^3x \, (\phi_j{}^*\pi^\dagger - \pi_j \phi)$$
$$= \int d^3x \, \phi_j{}^* \left[i\frac{\partial\phi}{\partial t} + (\omega_j^{(*)} - 2eA_0)\phi \right], \qquad (2.21)$$

where the symbols without subscripts refer to the quantized field. Equations (2.19) now yield[16]

$$\epsilon_j\epsilon_k[c_j, c_k^\dagger] = \epsilon_j\delta_{\omega_j\omega_k}, \qquad \epsilon_j\epsilon_k[c_j,c_k] = 0. \qquad (2.22)$$

15   J. Bognár, *Indefinite Inner Product Spaces* (Springer, Berlin, 1974).
16   The reverse implication from Eqs. (2.22) to Eqs. (2.19) is equivalent to certain completeness relations for the functions ϕ_j [Ref. 3, Eqs. (34)–(36)] or Φ_j [Ref. 7, Eq. (2.11)].

If $\epsilon_j \neq 0$, it is natural to normalize ϕ_j so that $\epsilon_j = \pm 1$. Moreover, if there are two or more modes with the same value of ω, one can choose basis vectors Φ_j which are orthogonal to each other. (If modes of frequency ω occur with both signs of the norm ϵ, this last step introduces an ambiguity which is characteristic of the Klein paradox — see Sec. VI.) Equation (2.18a) can then be written

$$\langle \Phi_j, \Phi_k \rangle = \epsilon_j \, \delta(j,k); \qquad (2.23)$$

the right-hand side has been written as a generalized Dirac δ function[17] to allow for a continuous spectrum [thus generalizing Eq. (2.8)].

With these conventions the c_j for positive-norm modes can be interpreted as annihilation operators for particles, while the c_j for negative-norm modes are creation operators for antiparticles. Accordingly, define[18]

$$a_j = c_j \quad \text{for } \epsilon_j = 1, \qquad a_j = b_j = c_j^\dagger \quad \text{for } \epsilon_j = -1. \qquad (2.24)$$

Then Eqs. (2.22) take the form

$$[a_j, a_k^\dagger] = \delta(j,k), \qquad [a_j, a_k] = 0. \qquad (2.25)$$

The appropriateness of the "particle" terminology is supported by the forms assumed in terms of the a's by the principal conserved quantities of the field theory: the total charge,

$$eQ = e\langle \Phi, \Phi \rangle \int d^3x \, e\rho, \qquad (2.26)$$

$$e\rho(\mathbf{x}) = \tfrac{1}{2}ie\,(\phi^\dagger \pi^\dagger + \pi^\dagger \phi^\dagger - \phi\pi - \pi\phi) \qquad (2.27)$$

[cf. Eq. (2.15)], and the total energy or Hamiltonian,

$$E = \int d^3x \, (\pi^\dagger \pi + \nabla \phi^\dagger \cdot \nabla \phi + m^2 \phi^\dagger \phi + e\rho A_0). \qquad (2.28)$$

---

[17]   That is, $f(j) = \int d\mu(k)\, f(k)\, \delta(j,k)$ for all functions f, where μ is some measure on the index space, which may include both a discrete sum and an integration over a continuum.

[18]   When $\epsilon_j < 0$, b_j will be used to emphasize that an antiparticle mode is being considered, but a_j will be used when it is convenient to treat the two kinds of modes on the same footing. With the notational convention of Ref. 10, this introduces no ambiguity.

[Note that a gauge transformation of the type described below Eq. (2.3) changes E to $E + QV_0$.] Routine calculations show that (for a discrete mode)

$$eQ = \tfrac{1}{2}e\,\epsilon_j(c_j^\dagger c_j + c_j c_j^\dagger) + \cdots = e\epsilon_j(a_j^\dagger a_j + \tfrac{1}{2}) + \cdots, \qquad (2.29)$$

$$E = \epsilon_j\omega_j(a_j^\dagger a_j + \tfrac{1}{2}) + \cdots, \qquad (2.30)$$

where the dots indicate the contributions from all the other modes. Thus $a_j^\dagger a_j$ is a number operator for quanta of charge $e\epsilon_j$ and energy $\epsilon_j\omega_j$, and this mode also contributes $\tfrac{1}{2}e\epsilon_j$ and $\tfrac{1}{2}\epsilon_j\omega_j$ to the (indeterminate) charge and (infinite) energy of the vacuum in the standard Fock representation. (If the mode is part of a continuous spectrum, these statements remain true when suitably reformulated.) It turns out that $\epsilon_j\omega_j$ is positive for all sufficiently large $|\omega_j|$, but it may be negative for certain modes, as will be seen in Secs. IIIA and IV.

In summary, despite the complications introduced by the term linear in ω in Eq. (2.3), each normal mode of a quantized complex scalar field in an external electrostatic potential possesses a conventional particle interpretation, provided that its frequency is real and its norm is not zero. The failure of either of these conditions is a Schiff–Snyder–Weinberg effect; these will be treated in Sec. III. Even if all the modes have real frequency and nonvanishing norm, three other "pathologies" are possible: (1) The normal-mode solutions may not form a complete set, as when the operator in a conventional eigenvalue problem is not self-adjoint. (This case will not be discussed further here.) (2) The vacuum state and associated Fock representation may not be unique, because positive- and negative-norm modes coexist at the same frequency ω (see Sec. VI). (3) The vacuum state may be unstable, because a negative-norm mode has a higher frequency than some positive-norm mode (see Secs. IV and V).

B. A Family of Models.

From now on we shall concentrate on a class of explicitly solvable models in space-time dimension 2. Let

$$-eA_0(x) = 0 \quad \text{for } 0 < x < L_+,$$
$$-eA_0(x) = V \quad \text{(constant)} \quad \text{for } -L_- < x < 0. \qquad (2.31)$$

The possibilities $L_+ = \infty$ and $L_- = \infty$ are to be included. If $L_\pm$ is finite, the boundary condition

$$\phi(\pm L_\pm) = 0 \qquad (2.32)$$

is imposed, corresponding to complete reflection of the scalar waves at the walls of the "box" (Fig. 1). Because of the gauge invariance, there is no loss of generality in taking the potential to be 0, rather than some constant, on the right side. Furthermore, we may assume that $V \geq 0$ and $L_+ \geq L_-$, since the other cases may be reduced to this one by space reflection and charge conjugation.

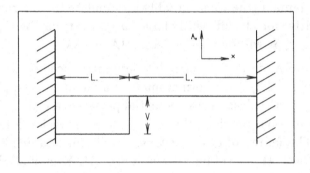

Fig. 1.   The general Schiff–Snyder–Weinberg square-well potential.

An additional reduction of the independent cases comes from using one of the dimensioned parameters of the model as the unit of length (or energy). Define[19]

$$\psi = L_- L_+^{-1} \qquad (0 \leq \psi \leq 1),$$

$$\theta = m \min(L_+, L_-) = mL_- \qquad (0 \leq \theta \leq \infty), \qquad (2.33)$$

$$Z = \max(m, L_-^{-1}, L_+^{-1}) = \max(m, L_-^{-1}).$$

---

[19]   These variables have been chosen to parametrize the limits $m \to 0$ or $L_\pm \to \infty$ in the least singular manner. The first forms given for θ and Z apply also when $L_- > L_+$.

Except for the one case with $Z = 0$, every choice of (m, L_+, L_-) is essentially equivalent to one with $Z = 1$. In the space of the parameters m, L_+^{-1}, L_-^{-1} we may thus take the independent cases to comprise one face of the unit cube ($L_- = 1$, $1 \leq L_+ = \psi^{-1} \leq \infty$, $0 \leq m = \theta \leq 1$), half of an adjoining face ($m = 1$, $1 \leq L_- = \theta \leq \infty$, $L_- \leq L_+ = \theta/\phi \leq \infty$), and the origin ($m = 0$, $L_+ = L_- = \infty$). Within each case one must consider the effect of varying V from 0 to ∞.

The normal-mode solutions of the form (2.13), (2.2), (2.14) must obey Eq. (2.3), which becomes

$$-\frac{d^2}{dx^2}\phi_j(x) = (\omega_j^2 - m^2)\phi_j(x) \quad \text{for } x > 0, \qquad (2.34a)$$

$$-\frac{d^2}{dx^2}\phi_j(x) = [(\omega_j + V)^2 - m^2]\phi_j(x) \quad \text{for } x < 0, \qquad (2.34b)$$

and must be continuous and have continuous first derivative at $x = 0$, and must satisfy Eq. (2.32) if applicable. A solution for $x \geq 0$ which satisfies the boundary condition at L_+ is, in a notation close to that of Ref. 3,

$$\phi_j(x) = (\sinh \eta)^{-1} \sinh[\eta(L_+ - x)/L_+], \qquad (2.35a)$$

where

$$\eta = L_+(m^2 - \omega^2)^{\frac{1}{2}} = (\theta/m)\,\psi^{-1}(m^2 - \omega^2)^{\frac{1}{2}} \qquad (2.35b)$$

and we may take Im $\eta > 0$ if $|\text{Im } \eta| > |\text{Re } \eta|$, but Re $\eta > 0$ if $|\text{Re } \eta| \geq |\text{Im } \eta|$; i.e., $-\pi/4 \leq \arg \eta \leq 3\pi/4$. If ω is real and $|\omega| < m$, η is real and positive and we have the ordinary sinh function; if ω is real and $|\omega| > m$, then $\eta = i\mu$ where $\mu > 0$, and

$$\phi_j(x) = (\sin \mu)^{-1} \sin[\mu(L_+ - x)/L_+], \qquad (2.36a)$$

$$\mu = L_+(\omega^2 - m^2)^{\frac{1}{2}}. \qquad (2.36b)$$

If $\omega = \pm m$, we have

$$\phi_j(x) = 1 - \frac{x}{L_+}, \qquad \eta = 0. \qquad (2.37)$$

If ω is complex, the function (2.35a) becomes complex (unless ω is imaginary); Eq. (2.4) is fulfilled. For convenience the functions have been arbitrarily and temporarily normalized so that $\phi_j(0) = 1$; the cases $\mu = n\pi$ [for which $\phi_j(0) = 0$] can be treated separately.

For $x \leq 0$ the solution is

$$\phi_j(x) = (\sinh \lambda)^{-1} \sinh[\lambda(x + L_-)/L_-], \qquad (2.38a)$$

$$\lambda = L_-[m^2 - (\omega + V)^2]^{\frac{1}{2}} = (\theta/m)[m^2 - (\omega + V)^2]^{\frac{1}{2}} \qquad (2.38b)$$

with sign conventions analogous to those for η. If ω is real and $|\omega+V| > m$, then in analogy to Eqs. (2.36) we set $\lambda = i\xi$, $\xi > 0$. If $\omega + V = \pm m$, we have $\lambda = 0$ and $\phi_j(x) = 1 + x/L_-$.

All that remains in order to determine which values of ω actually occur for finite $L_\pm$ is to equate the logarithmic derivatives of the expressions (2.35a) and (2.38a) at $x = 0$. The general formula is

$$L_-^{-1}\lambda \coth \lambda = -L_+^{-1}\eta \coth \eta. \qquad (2.39)$$

Let us investigate first the real values of ω. To simplify the formulas we consider only the case $Z = m = 1$, $\theta = L_- \geq 1$. (The other face of the cube is equally easy.) Then Eq. (2.39) is

$$-F_-(\omega + V) = F_+(\omega), \qquad (2.40)$$

$$F_+(z) = -\psi\eta \coth \eta$$
$$= \begin{cases} -\theta(1 - z^2)^{\frac{1}{2}} \coth[\theta\psi^{-1}(1 - z^2)^{\frac{1}{2}}] & \text{if } 0 \leq z < 1, \\ -1 & \text{if } z = 1, \quad (2.41a) \\ -\theta(z^2 - 1)^{\frac{1}{2}} \cot[\theta\psi^{-1}(z^2 - 1)^{\frac{1}{2}}] & \text{if } z > 1, \end{cases}$$

$$F_-(z) = -\lambda \coth \lambda$$
$$= \begin{cases} -\theta(1 - z^2)^{\frac{1}{2}} \coth[\theta(1 - z^2)^{\frac{1}{2}}] & \text{if } 0 \leq z < 1, \\ -1 & \text{if } z = 1, \quad (2.41b) \\ -\theta(z^2 - 1)^{\frac{1}{2}} \cot[\theta(z^2 - 1)^{\frac{1}{2}}] & \text{if } z > 1, \end{cases}$$

$$F_\pm(-z) = F_\pm(z). \qquad (2.42)$$

The roots ω can be read off from the superimposed graphs of the two sides of Eq. (2.40) for a given value of V (see Figs. 3, 4 and 7).

To distinguish between particle and antiparticle modes one must evaluate the norm (2.8):

$$\epsilon_j = 2 \int_{-L_-}^{0} dx\,(\omega + V)\,\phi_j(x)^2 + 2 \int_{0}^{L_+} dx\,\omega\phi_j(x)^2$$

$$\equiv \epsilon_- + \epsilon_+$$

$$= (\omega + V)\theta\,[\lambda^{-1}\coth\lambda - (\sinh\lambda)^{-2}]$$

$$+ \omega\theta\psi^{-1}[\eta^{-1}\coth\eta - (\sinh\eta)^{-2}]. \tag{2.43}$$

For imaginary $\eta = i\mu$, one has the analytic continuation

$$\epsilon_+ = \omega\theta\psi^{-1}[(\sin\mu)^{-2} - \mu^{-1}\cot\mu], \tag{2.44}$$

and if $\eta = 0$, $\epsilon_+ = \frac{2}{3}\omega\theta\psi^{-1}$; similarly for $\lambda = i\xi$ or 0. In fact, the analytic continuation of Eq. (2.43) correctly gives ϵ_j, as defined by Eq. (2.8), even for complex ω, although ϵ_j is then no longer the norm of ϕ_j [see Eqs. (2.18b)]. A useful observation of Schiff et al. (Ref. 3) can now be generalized:

$$\epsilon_+ = \psi\theta^{-1}\frac{d(-\eta\coth\eta)}{d\omega}, \qquad \epsilon_- = -\theta^{-1}\frac{d(+\lambda\coth\lambda)}{d\omega}, \tag{2.45a}$$

or

$$\epsilon_\pm = \frac{dF_\pm(z)}{d\omega}. \tag{2.45b}$$

It follows that when $F_+(\omega)$ and $-F_-(\omega + V)$ are plotted on the same graph, as in Figs. 3, 4 and 7, the norm of the mode corresponding to an intersection of the curves is determined by the slopes of the curves at that point. For one has

$$\epsilon_+ > 0 \quad \text{if and only if } F_+ \text{ is increasing } (\omega > 0), \tag{2.46a}$$

$$\epsilon_- > 0 \quad \text{if and only if } -F_- \text{ is decreasing } (\omega + V > 0). \tag{2.46b}$$

The "normal" situation is for the slopes to have opposite signs, so that ϵ_+ and ϵ_- have the same sign. If V is sufficiently large, however, ϵ_+ and ϵ_- may have opposite signs, and the sign of ϵ is determined by the steeper curve. When the curves are tangent, ϵ is 0. It is possible for the curves to cross at infinity in the vertical coordinate; the rules can then be applied to the slopes in the surrounding neighborhood. Tangency also can occur at infinity.

When the allowed values of ω are plotted against V, continuous curves result (Figs. 2, 5 and 6). The V–ω plane is divided into regions by the qualitative behavior of the corresponding functions ϕ_j for positive and negative x (Fig. 2). When $\omega > 1$ ($= m$) or $\omega < -(1 + V)$, the solution is oscillatory [like Eq. (2.36a)] in both regions. In the strip where $|\omega| < 1$ and $\omega > 1 - V$, the solution shows an exponential decay for $x > 0$, as one would expect for the wave function of a particle of charge $+e$ bound in the attractive potential well on the left side of Fig. 1. When $|\omega + V| < 1$ and $\omega < -1$, correspondingly, there is exponential behavior for $x < 0$, suggesting an antiparticle of charge $-e$, which sees the potential (2.31) as repulsive. If $\omega > -1$ and $\omega < 1 - V$, both parts of $\phi_j(x)$ are exponential, so the condition of a smooth derivative at the origin cannot be satisfied and no solution exists. Finally, when $\omega < -1$ and $\omega > 1 - V$, both parts of the solution are again oscillatory. The existence of this last region is central to the Klein paradox, although the paradox itself does not arise until L_+ and L_- are increased to infinity, so let us call it the "Klein region".

It is easy to see [by inspection of Fig. 3, etc., in the light of the criteria (2.46)] that $\epsilon_\pm$ are both positive when $\omega > 1$ and both negative when $\omega + V < -1$. Hence the solutions in these regions are unambiguously interpretable as the wave functions of free particles and free antiparticles, respectively ("free" because of the everywhere oscillatory behavior, allowing construction of wave packets which propagate). In the two bound-state regions and the Klein region, on the other hand, ϵ_+ and ϵ_- may have different signs (and in the Klein region always do). Consequently, a root ω in these domains may correspond to a particle mode, an antiparticle mode, or a mode with $\epsilon = 0$. In what follows we shall examine in detail the significance of solutions in these regions.

Of course, if no complex frequencies or solutions with $\epsilon = 0$ exist, the system may be quantized as described earlier (the solutions ϕ_j being renormalized so that $\epsilon_j = \pm 1$).

III. The Schiff–Snyder–Weinberg effects

A. Normal Modes.

In Ref. 3 the potential (2.31) was thought of as a model of an atomic or nuclear potential of radius L_-. Therefore, the cases studied had $L_+ \gg L_-$ and L_- comparable to m^{-1}, with $L_+ = \infty$ being the situation of ultimate interest. Under those conditions the authors found[20] that the mode frequencies have the behavior in the $V - \omega$ plane sketched qualitatively in Fig. 2.

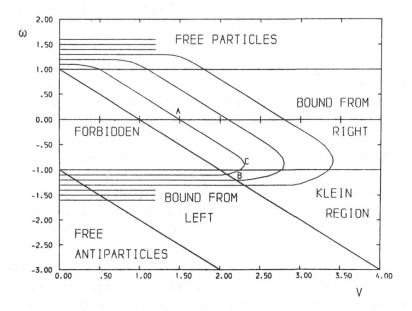

Fig. 2. Frequency levels computed by Schiff et al. for $m = 1$ and $L_+ \gg L_-$ (schematic). The characterizations of the solutions in various regions are valid for any L_+ and L_-. For the significance of points A, B, and C see Sec. IIIA.

At $V = 0$ the spectrum of ω is, of course, that of a free boson field quantized in a finite box. As L_+ increases, the spacing between the frequency levels decreases, resulting in a continuous spectrum when $L_+ = \infty$. The particle frequencies remain separated from the antiparticle frequencies, however, by a mass gap of width $2m$ ($= 2$ in the chosen units).

As the potential is turned on, the mode frequencies change. In particular, "bound states" grow out of the particle continuum into the mass gap. Before we turn to the Schiff–Snyder–Weinberg effects proper,

---

[20]   Reference 3 also treats the case of a quantized fermion (Dirac) field, for which the results are qualitatively different. Up-to-date treatments include: Ya. B. Zel'dovich and V. S. Popov, *Usp. Fiz. Nauk* **105**, 403 (1971) [*Sov. Phys.–Usp.* **14**, 673 (1972)]; L. Fulcher and A. Klein, *Phys. Rev. D* **8**, 2455 (1973); *Ann. Phys.* (N.Y.) **84**, 335 (1974).

two features which appear in Fig. 2 at relatively small V deserve discussion, because they appear strange at first sight.

First, at the point labelled A the frequency ω of the lowest-lying particle mode becomes negative. According to Eq. (2.30), this means that a particle can exist with negative energy (relative to the vacuum state); in fact, since arbitrarily many particles can be present in that mode, the total energy operator of the second-quantized theory is not bounded below. No catastrophic instability of the system thereby results, however, since the absolute law of charge conservation prevents the vacuum from decaying into such a state, even under the influence of any physically permissible external perturbation (contrast Sec. IV). In fact, the negativity of the energy can be removed simply by a gauge transformation which adds a constant to all the frequencies in the theory. To put it another way, it is charge conservation which allows the energy observable to be redefined by the addition of a term proportional to the total charge Q [see remark below Eq. (2.28)], because the energy difference between two states of different charge of a closed system cannot be measured. (A gauge transformation must be applied consistently to *all* charged fields, including any external perturbing field whose quanta might carry charge away from the subsystem of original interest.)

Second, at the point B of Fig. 2 one of the antiparticle modes moves into the region of wave functions bound into the left potential well. This phenomenon appears paradoxical, since one would expect a particle of charge $-e$ to be repelled, not attracted, by the negative potential. Klein and Rafelski (Ref. 7) have pointed out the physical explanation of this effect. For such a mode the oscillatory part of the wave function inside the well makes a positive contribution ϵ_- to the norm, but the exponential part on the right-hand side contributes a negative term, ϵ_+, with $|\epsilon_+| > \epsilon_-$ so that the total norm, ϵ, is negative — corresponding to an antiparticle mode. However, the energy of a state contains a term proportional to $\int dx\, e\rho A_0$ [see Eq. (2.28)], where $\rho(x)$, for a one-particle state, is just the integrand of the integral which defines ϵ [see Eqs. (2.8), (2.15), (2.20), (2.26), (2.27), (2.29)]. In the present case, with the chosen gauge, only the part of the function inside the potential well contributes to this integral; the result is $-\epsilon_- V$, which is negative. This term makes the total energy of the antiparticle less than m (i.e., $\omega > -m$), so that such an antiparticle bound state really can exist. In other words, the charge density of the state is *polarized* by the potential, so that the total charge is negative, but the charge density near the potential is positive and hence contributes a negative (expectation value of) binding energy. The modes in the Klein region, with free-particle-like wave functions of opposite charge both inside and outside the wells, may be regarded as extreme instances of this polarization. (It should be understood that an extended charge structure is being attributed here not to the particle itself, but only to its wave function.)

The really significant effects begin at the value V_c of V corresponding to the point labelled C in Fig. 2. There the bound antiparticle mode appears to coalesce with the lowest bound particle mode. The graphs of the two frequencies as functions of V form a single smooth curve with a vertical tangent at C. Call the corresponding frequency ω_c. At that point there is, of course, only one solution of Eq. (2.3) satisfying the boundary conditions (2.32). This situation arises because two points of intersection of the graphs of $F_+(\omega)$ and $-F_-(\omega + V)$ have flowed together into a single point of tangency (cf. Fig. 3). It follows that $\epsilon = 0$ for such a solution, which we shall call a *singular mode*. As V increases beyond V_c, there is no intersection at all, and hence no real roots ω, corresponding to the pair of roots that previously existed. A conjugate pair of complex roots, however, can be found growing out from ω_c into the complex ω plane.

Qualitatively similar behavior is observed at larger V for successive higher-lying pairs of particle and antiparticle modes. The occurrence of singular modes and of complex frequencies for sufficiently deep external potentials is the important discovery of Schiff, Snyder, and Weinberg. These phenomena require an extension of the quantization procedure described in Sec. IIA, and also a physical interpretation.

First we analyze the generality of the features observed in Fig. 2. Equation (2.40) implicitly defines ω as a local function of V, or vice versa; from it one can derive a differential equation for the curves in Fig. 2 and their extensions into the complex ω plane. Differentiating Eq. (2.40), solving for $d\omega/dV$, and using Eq. (2.45), one obtains

$$\frac{d\omega}{dV} = -\frac{\epsilon_-}{\epsilon}. \qquad (3.1)$$

A vertical tangent in the graph of $\omega(V)$ therefore occurs when and only when $\epsilon = 0$. [Note that ϵ and ϵ_- can never vanish simultaneously, since Eq. (2.40) is never satisfied when the derivatives of both sides are zero (at $z = 0$). Also, when $\epsilon_- = \pm\infty$, ϵ_+ for a solution is also infinite ($\mu = n\pi$, etc.), and the right side of Eq. (3.1) can be defined by continuity.] This equation is equivalent to a special case of Eq. (2.17) of Klein and Rafelski (Ref. 7), which was derived for an arbitrary potential, provided only that the integrals $\int |\phi_j|^2 \, dx$, $\int |\phi_j|^2 A_0 \, dx$, and $\int |\phi_j|^2 A_0^2 \, dx$ converge for the mode in question.

The behavior of the frequencies in the immediate vicinity of a singular mode may be found be regarding V as a function of ω. With $dV/d\omega = 0$,

$$\frac{d^2V}{d\omega^2} = [F'_-(\omega + V)]^{-2}[F''_-(\omega + V)F'_+(\omega) - F'_-(\omega + V)F''_+(\omega)], \quad (3.2)$$

$$\Delta V = V - V_c, \qquad \Delta\omega = \omega - \omega_c, \quad (3.3)$$

we have

$$\Delta V = \frac{1}{2}\frac{d^2V}{d\omega^2}(\Delta\omega)^2 \quad (3.4)$$

in lowest order, and hence a square-root branch point of $V(\omega)$. If $d^2V/d\omega^2 < 0$, then real $\Delta\omega$ of either sign corresponds to real $\Delta V < 0$, but a real $\Delta V > 0$ arises from imaginary $\Delta\omega$. Thus a singular mode always marks the conversion of a pair of real frequencies into a pair of complex frequencies. Since the solution of Eq. (3.1) can otherwise be continued indefinitely, and since there are no complex frequencies when $V = 0$, the only complex frequencies at a given V are those which have been created in this way as V increases from 0.

For the real frequencies in the vicinity of a singular mode, the sign of $d\omega/dV$ is the opposite of the sign of ϵ; consequently, whether a mode of the field corresponds to particles or antiparticles depends only on the slope of the curve through it in the V–ω plane. (This criterion is not valid far away from a singular point, since the curve may have passed through a point where its tangent is horizontal — see Fig. 5. If $V < 0$, then $d\omega/dV$ and ϵ have the same sign.) This can be seen by following the intersections of the $F_\pm$ curves as V varies — keeping in mind the discussion surrounding Eqs. (2.46). The analogous statement for a general potential can be derived from the Klein–Rafelski equation mentioned above, which can be written

$$\left.\frac{d\omega}{d\lambda}\right|_{\lambda=1} = e\langle A_0\rangle - \frac{e^2\langle(A_0 - \langle A_0\rangle)^2\rangle}{\omega - e\langle A_0\rangle}, \quad (3.5)$$

where the derivative is with respect to the overall strength of the potential λA_0, and[21]

$$\langle B \rangle = \frac{\int |\phi_j|^2 B\, dx}{\int |\phi_j|^2\, dx} \tag{3.6}$$

for any function $B(x)$. Since ϵ is given by Eq. (2.8), whenever ϵ is close to zero the second term in Eq. (3.5) dominates and is proportional to ϵ^{-1} with a negative coefficient.

B. Quantization.

It remains to show, following Schroer and Swieca (Ref. 5), how the construction of Fock space in Sec. IIA can be generalized to include singular and unstable (complex-frequency) modes. Each mode or pair of modes can be analyzed as an independent system of one or two degrees of freedom, and a notation adopted which permits the states of this system to be classified neatly by their contributions to the energy and charge of the whole scalar field system; this has already been done for ordinary modes at the end of Sec IIA. States of finite total charge and energy then exist in certain infinite-tensor-product representations[22] of the field algebra (2.19). If there are only finitely many "pathological" modes, an irreducible representation (of the type just mentioned) is uniquely determined by the requirement that the representation of the subalgebra of ordinary modes be a multiple of the Fock representation of the operators (2.25). If there are infinitely many discrete strange modes, or if they form a continuum, then any statement as to which is the "physical" representation (if any) is tentative, pending elucidation of the physical meaning of the field theory under such conditions.

Let us consider first a pair of complex frequencies, ω_j and $\omega_j{}^*$, where $\mathrm{Im}\,\omega_j > 0$. To simplify notation we assume that these modes are discrete and drop the index j in some contexts. The corresponding wave functions ϕ_j and $\phi_j{}^*$, or Φ_j and ${}^*\Phi_j$, satisfy Eqs. (2.4), (2.7), (2.8), and (2.18b) and are orthogonal to those of all other modes, and the normalization (including phase) can be chosen[23] so that $\epsilon = i$. In

---

[21] $\langle B \rangle$ must not be interpreted as the physical expectation value of B, since $|\phi(x)|^2$, for ϕ a solution of the Klein–Gordon equation, is not a probability density in x-space.

[22]   See, e.g., M. C. Reed, J. Func. Anal. **5**, 94 (1970).

[23]   If $\epsilon = 0$, Φ_j would be orthogonal to every vector in the space, which contradicts the known nondegeneracy of the sesquilinear form (2.15).

analogy to Eq. (2.20) we write the contributions to the field as

$$c\phi_j(\mathbf{x})e^{-i\omega t} + d\,\phi_j(\mathbf{x})^* e^{-i\omega^* t}, \tag{3.7}$$

where c and d are operators. Then one finds that

$$-i\,d = \langle \Phi_j, \Phi \rangle \tag{3.8a}$$

as given by Eq. (2.21), and

$$ic = \langle {}^*\Phi_{-j}, \Phi \rangle = i \int_{t=0} d^3x \, (\phi_j \pi^\dagger + \pi_j{}^* \phi), \tag{3.8b}$$

since the second component of ${}^*\Phi_j$ is $-\pi_j{}^*$ when $t = 0$. The commutation relations which follow are

$$\begin{aligned} [d, c^\dagger] &= i = -[c, d^\dagger], \\ [c, c^\dagger] &= [d, d^\dagger] = [c, d] = 0. \end{aligned} \tag{3.9}$$

The charge and energy, Eqs. (2.26)–(2.28), therefore contain the terms

$$eQ_j = ei(d^\dagger c - c^\dagger d) + e, \tag{3.10}$$

$$E_j = -i\omega^* c^\dagger d + i\omega d^\dagger c + \mathrm{Re}\,\omega. \tag{3.11}$$

Let

$$\begin{aligned} c &= \tfrac{1}{2}[(\mathrm{Im}\,\omega)^{\frac{1}{2}}(q_1 - iq_2) + (\mathrm{Im}\,\omega)^{-\frac{1}{2}}(p_1 - ip_2)], \\ d &= \tfrac{1}{2}[(\mathrm{Im}\,\omega)^{\frac{1}{2}}(q_1 - iq_2) - (\mathrm{Im}\,\omega)^{-\frac{1}{2}}(p_1 - ip_2)]. \end{aligned} \tag{3.12}$$

The canonical commutation relations for the Hermitian quantities q_1, p_1, q_2, p_2 are equivalent to Eqs. (3.9). One finds

$$Q_j = q_1 p_2 - q_2 p_1, \tag{3.13}$$

$$\begin{aligned} E_j &= \tfrac{1}{2}[p_1^2 + p_2^2 - (\mathrm{Im}\,\omega)^2(q_1^2 + q_2^2)] + (\mathrm{Re}\,\omega)Q_j \\ &\equiv E_j^0 + (\mathrm{Re}\,\omega)Q_j. \end{aligned} \tag{3.14}$$

The second term of E_j commutes with the first (and could be eliminated from this particular mode pair by a gauge transformation). Q_j and E_j^0 have the form of the angular momentum and the Hamiltonian of a two-dimensional "repulsive harmonic oscillator". [Recall that a mode with real ω gives rise to an ordinary attractive harmonic oscillator Hamiltonian, Eq. (2.30), which could be written as $\pm\tfrac{1}{2}(p^2 + \omega^2 q^2)$.]

Alternatively, they can be realized as the generators of rotations and dilations, respectively, in a two-dimensional Euclidean space (see Ref. 5). The spectra of such operators are known: Q_j has all the integers as eigenvalues (so charge is still quantized). E_j^0 has a continuous spectrum consisting of all the real numbers (without multiplicity), but has no (discrete, normalizable) eigenvectors.

We turn now to the problem of a mode with real ω but $\epsilon = 0$. As pointed out in Sec. IIIA, this can happen only when a pair of real frequencies is about to turn into a pair of complex frequencies as the strength of the potential changes. Expression (2.13) is still a solution of the field equation, but evidently a second solution has somehow been lost. To recover it we consider two solutions of the ordinary type, which are about to flow together:

$$(\omega_1 - eA_0)^2\phi_1 = (-\nabla^2 + m^2)\phi_1 , \tag{3.15a}$$

$$(\omega_2 - eA_0)^2\phi_2 = (-\nabla^2 + m^2)\phi_2 , \tag{3.15b}$$

$$\omega \equiv \tfrac{1}{2}(\omega_1 + \omega_2), \qquad \Delta\omega \equiv \tfrac{1}{2}(\omega_1 - \omega_2), \tag{3.16}$$

$$\epsilon_1 = \langle \Phi_1, \Phi_1 \rangle = 1, \qquad \epsilon_2 = \langle \Phi_2, \Phi_2 \rangle = -1. \tag{3.17}$$

All the quantities are functions of a parameter λ, the strength of the potential. Define (cf. Ref. 7)

$$\phi_e = (\Delta\omega/2)^{\frac{1}{2}}(\phi_1 + \phi_2), \qquad \phi_o = (2\Delta\omega)^{-\frac{1}{2}}(\phi_1 - \phi_2). \tag{3.18}$$

Adding and subtracting the equations (3.15) and passing to the limit $\Delta\omega \to 0$, $\omega_1 \to \omega \leftarrow \omega_2$, we obtain formally

$$[\nabla^2 - m^2 + (\omega - eA_0)^2]\phi_e = 0, \tag{3.19}$$

$$[\nabla^2 - m^2 + (\omega - eA_0)^2]\phi_o = -2(\omega - eA_0)\phi_e . \tag{3.20}$$

Consequently when a singular mode exists, one must anticipate the need to include a solution, ϕ_o, of an *inhomogeneous* separated wave equation in the complete set in which the field's initial data are expanded.

The solutions of the full field equation (2.1) corresponding (at $t = 0$) to ϕ_e and ϕ_o are, in the notation (2.2),

$$\begin{aligned}\phi_{(e)} &= (\Delta\omega/2)^{\frac{1}{2}}(\phi_{(1)} + \phi_{(2)}) \\ &= \phi_e e^{-i\omega t}\cos \Delta\omega T - i\Delta\omega\phi_o e^{-i\omega t}\sin \Delta\omega t \\ &\to \phi_e e^{-i\omega t} \quad \text{as } \Delta\omega \to 0\end{aligned} \tag{3.21}$$

and

$$\phi_{(o)} = (2\Delta\omega)^{-\frac{1}{2}}(\phi_{(1)} - \phi_{(2)})$$
$$= \phi_o e^{-i\omega t}\cos\Delta\omega t - i(\Delta\omega)^{-1}\phi_e e^{-i\omega t}\sin\Delta\omega t$$
$$\to \phi_o e^{-i\omega t} - i\phi_e\, t\, e^{-i\phi t} \quad \text{as } \Delta\omega \to 0. \tag{3.22}$$

The limits indicated are not uniform in t. Nevertheless, the final expressions in Eqs. (3.21) and (3.22), with ϕ_e and ϕ_o satisfying Eqs. (3.19) and (3.20), are easily seen to be solutions of the (homogeneous) space-time field equation (2.1). Two-component vectors Φ_e and Φ_o defined according to Eq. (2.10) satisfy Eq. (2.11a) and

$$\langle\Phi_e, \Phi_e\rangle = 0, \qquad \langle\Phi_o, \Phi_o\rangle = 0, \qquad \langle\Phi_o, \Phi_e\rangle = 1 \tag{3.23}$$

(for $\Delta\omega \neq 0$ or $= 0$). [If the normalization of ϕ_1 and ϕ_2 had been chosen so that ϵ passed smoothly through zero as one followed the curve of real frequencies ω_1 and ω_2 through the singular point in the $\lambda-\omega$ plane, then it follows from the discussion at the end of Sec. IIIA that $\epsilon \propto \Delta\omega \propto (\Delta\lambda)^{\frac{1}{2}}$ near the singular point. The normalization convention (3.17), therefore, forces ϕ_1 and ϕ_2 to grow as $(\Delta\omega)^{-\frac{1}{2}}$. Hence one expects ϕ_e and, in view of Eq. (3.20), ϕ_o to converge to finite, nonzero functions at the singular point; this is consistent with $\langle\Phi_o, \Phi_e\rangle = 1$.] Finally, the two-component eigenvalue equations equivalent to Eqs. (3.19), (3.20), and (2.9) are

$$H\Phi_e = \omega\Phi_e, \tag{3.24}$$

$$H\Phi_o = \omega\Phi_o + \Phi_e, \tag{3.25}$$

where H is given in Eq. (2.12). That is, Φ_o is an *associated eigenvector* of H, whose restriction to the space spanned by Φ_e and Φ_o has the Jordan canonical form

$$\begin{pmatrix} \omega & 1 \\ 0 & \omega \end{pmatrix}. \tag{3.26}$$

We now argue abstractly in the converse direction. Let Φ_e and Φ_o be an eigenvector and an associated eigenvector, with real eigenvalue ω, of an operator H which is Hermitian in the sense of Eq. (2.16). Then

$$\omega\langle\Phi_o, \Phi_e\rangle = \langle\Phi_o, H\Phi_e\rangle = \langle H\Phi_o, \Phi_e\rangle = \omega\langle\Phi_o, \Phi_e\rangle + \langle\Phi_e, \Phi_e\rangle,$$

and hence $\epsilon = \langle\Phi_e, \Phi_e\rangle = 0$. [Of course, solutions of Eq. (3.20) exist for all ω, but only when $\epsilon = 0$ does one of them satisfy the boundary conditions — such as Eq. (2.32) — which make it a normalizable vector in the domain of H. On nondiagonal Jordan structure in the continuous spectrum see Part Two, Sec. III.] Similar manipulations show that Φ_e and Φ_o are orthogonal to the eigenvectors and associated eigenvectors affiliated with other values of ω, and that $\langle \Phi_o, \Phi_e \rangle$ is real. Exploiting the freedom to multiply Φ_e by an arbitrary complex number and to add an arbitrary multiple of Φ_e to Φ_o, one can impose all of Eqs. (3.23). Writing out Eqs. (3.24) and (3.25) explicitly for H defined by Eq. (2.12), one arrives at Eqs. (3.19) and (3.20), and hence at the elementary solutions of the field equation presented as limits in Eqs. (3.21) and (3.22).

To quantize

$$f\phi_{(e)} + g\phi_{(o)} \tag{3.27}$$

one proceeds much as in the case of complex frequencies [Eq. (3.7)], obtaining

$$[f, g^\dagger] = [g, f^\dagger] = 1 \tag{3.28}$$

as the only nonvanishing commutators, and

$$eQ_j = e(f^\dagger g + g^\dagger f + 1), \tag{3.29}$$

$$E_j = g^\dagger g + \omega Q_j \equiv E_j^0 + \omega Q_j \tag{3.30}$$

for the charge and energy of the singular mode system. Let

$$f = -i\frac{q_1 - iq_2}{\sqrt{2}}, \qquad g = \frac{p_1 - ip_2}{\sqrt{2}}. \tag{3.31}$$

Then the q's and p's satisfy the canonical relations, Q_j takes the form (3.13), and the charge-independent part of the energy is

$$E_j^0 = \tfrac{1}{2}(p_1^2 + p_2^2). \tag{3.32}$$

E_j^0 is the Hamiltonian of a nonrelativistic free particle of unit mass in two dimensions (the obvious boundary case between attractive[24] and repulsive oscillators, but qualitatively different from both). Its spectrum is entirely continuous, but runs only from 0 to $+\infty$ [contrast the cases (2.30) and (3.14)].

---

[24] Note that the energy of the modes ϕ_1 and ϕ_2 of Eqs. (3.15)-(3.18) can be written as $\tfrac{1}{2}[p_1^2 + p_2^2 + (\Delta\omega)^2(q_1^2 + q_2^2)] + \omega(Q_1 + Q_2)$.

To summarize, gathering together Eqs. (2.20), (2.24), (3.7), and (3.27) for the field, Eqs. (2.29), (3.10), and (3.29) for the charge operator, and Eqs. (2.30), (3.11), and (3.30) for the energy or Hamiltonian, one has the complete field expansion[25]

$$\phi(t, \mathbf{x}) = \int d\mu_+(j)\, a_j \phi_j e^{-i\omega_j t} + \int d\mu_-(j)\, b_j^\dagger \phi_j e^{-i\omega_j t}$$

$$+ \int d\mu_s(j)\, [f_j \phi_{ej} e^{-i\omega_j t} + g_j(\phi_{oj} e^{-i\omega_j t} - i\phi_{ej}\, t\, e^{-i\omega_j t})]$$

$$+ \int d\mu_c(j)\, [c_j \phi_j e^{-i\omega_j t} + d_j \phi_j{}^* e^{-i\omega_j{}^* t}], \qquad (3.33)$$

with the conserved quantities

$$eQ = \int d\mu_+(j)\, a_j^\dagger a_j - e \int d\mu(j)\, b_j^\dagger b_j \qquad (3.34)$$

$$+ e \int d\mu_s(j)(f_j^\dagger g_j + g_j^\dagger f_j + 1) + e \int d\mu_c(j)(id_j^\dagger c_j - ic_j^\dagger d_j + 1)$$

and

$$E = \int d\mu_+(j)\, \omega_j a_j^\dagger a_j - \int d\mu_-(j)\, \omega_j b_j^\dagger b_j$$

$$+ \int d\mu_s(j)\, [g_j^\dagger g_j + \omega_j(f_j^\dagger g_j + g_j^\dagger f_j + 1)]$$

$$+ \int d\mu_c(j)\, [i(\omega_j d_j^\dagger c_j - \omega_j{}^* c_j^\dagger d_j) + \mathrm{Re}\ \omega_j]. \qquad (3.35)$$

(The vacuum charge and vacuum energy of the ordinary modes have been discarded.) Here μ_+, μ_-, μ_s, and μ_c are measures (see Ref. 17) on the ordinary modes with $\epsilon_j > 0$, ordinary modes with $\epsilon_j < 0$, singular modes, and modes with Im $\omega_j > 0$, respectively. The normalization conventions (2.23) (with $|\epsilon_j| = 1$), (2.18b) (with $\epsilon = i$), and (3.23), or their continuum generalizations, are understood. The contributions to the charge and energy from singular and complex modes can be put into the more transparent forms (3.13), (3.14), and (3.32) by the transformations (3.12) and (3.31).

A state vector of the subsystem labelled by an index value j is most conveniently specified in the spectral representation of its charge and

---

[25]   The contribution from associated eigenvectors appears incorrectly in Eq. (43) of Ref. 5.

energy operators. That is, a state of definite charge, Q_j , is represented by a "wave packet", $\psi(E_j)$ with $\int |\psi|^2 = 1$, where Q_j and E_j are now numerical variables. The expectation value of the energy is then, in the cases with continuous spectra,

$$\langle E_j \rangle = \int_{0 \text{ or } -\infty}^{\infty} dz \, z \, |\psi(z)|^2. \tag{3.36}$$

Consider now the entire field system. The most natural generalization of the usual "vacuum state" is constructed by choosing for each mode a state with $Q_j = 0$ (thus, in particular, the usual noparticle state for ordinary modes), such that the expectation value of the total energy,

$$\langle E \rangle = \int d(\mu_s + \mu_c)(j) \, \langle E_j \rangle, \tag{3.37}$$

is finite. The state of the whole system is the formal tensor product of these states of the subsystems. This vector obviously is not unique; it uniquely determines a representation of the canonical field algebra if and only if there are only finitely many strange modes.

What is the physical significance of singular modes and complex frequencies? First of all, when complex frequencies exist, the energy spectrum of the field theory is unbounded below, a condition which is generally believed to lead to instability. This subject will be discussed further in Sec. IV and in Part Two, Sec. IV. On the other hand, both cases are marked by the nonexistence of a discrete vacuum state and the impossibility of defining particle annihilation and creation operators. Intuitively, what happens as a particle and an antiparticle mode coalesce at a singular point is that the energy of an antiparticle becomes equal to the negative of the energy of a particle, so that arbitrarily many pairs can be added to the system without changing the total energy and charge. To have a finite energy $\langle E_j \rangle$ in a singular mode one must have, in some sense, infinitely many of these zero-energy pairs; every normalizable state of the field contains such a "cloud" of particles. The complex modes might be thought of in the same way as due to pairs with negative energy. These descriptions must not be taken literally, however. The vacuum does not merely coexist with the continuum of "cloud" states; it actually dissolves into it and disappears, except as a limit point of the continuum. (This situation is to be contrasted with the case discussed in the next section, where a discrete, normalizable vacuum state exists, even though pairs with negative energy are possible.) Under these conditions the field theory really has no particle interpretation. As in time-dependent external-potential problems, the field concept is seen to be fundamental, while particles are a secondary construct, meaningful only under certain conditions.

IV. NEGATIVE-ENERGY PAIRS

In the original work of Schiff et al. the potential (2.31) was regarded as a simple model of a realistic potential of subatomic size (radius L_-), sitting in empty space ($L_+ = \infty$). A finite value of L_+ was introduced only for convenience. Hence they naturally considered only $L_+ \gg L_-$. However, in connection with the Klein paradox and analogous phenomena near black holes, one is interested in the limits $L_+ \to \infty$, $L_- \to \infty$. These limits could be taken successively, or with the ratio L_+/L_- fixed, or in many other ways, with no guarantee of identical results (although one feels that all these limits ought to be equivalent in some loose physical sense). However, the most natural finite analogue of the case $L_+ = L_- = \infty$ would appear to be a situation in which the two ends of the "box" are treated on an equal basis. Consequently, the model (2.31) has been investigated with $L_+ = L_-$ ($\psi = 1$).

It is obvious from the beginning that the results must be qualitatively different from those shown in Fig. 2 in at least one respect. There the "free" states for both particles and antiparticles were essentially determined by the vast empty space outside the potential well; the paths of their frequencies in the V–ω plane were roughly parallel to the lines $\omega = \pm m$. But when $L_+ = L_-$, the quantity $\omega + V$ must play a role fully as important as that of ω in the qualitative description of the frequency levels. (This expectation is confirmed in Fig. 5.) The joining of positive- and negative-norm modes to form singular modes must occur in the Klein region, or else symmetrically in the "bound from left" and "bound from right" regions. In fact, a gauge transformation by the constant $V_0 = \frac{1}{2}V$ leads to a description of the system which is globally symmetrical with respect to particles and antiparticles, and with respect to left and right.

The normal-mode frequencies have been computed numerically for the case $\theta = 4$ ($L_+ = L_- = 4m^{-1}$). The $F_\pm$ functions for some interesting values of V are shown in Figs. 3 and 4. Because of the equality of L_+ and L_-, these curves can be tangent only when $F_\pm$ are

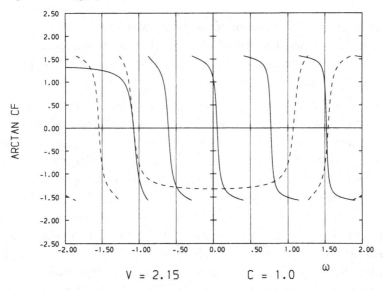

Fig. 3.   Graphical solution of Eq. (2.40) for $L_+ = L_-$ and $V = 2.15$. The dashed curve is the graph of the function $F_+(\omega)$, and the solid curve is that of $-F_-(\omega + V)$. The vertical axis has been mapped onto a finite interval by a monotonic function. At approximately this value of V the curves are tangent at $\omega = -1.1$ (a singular mode).

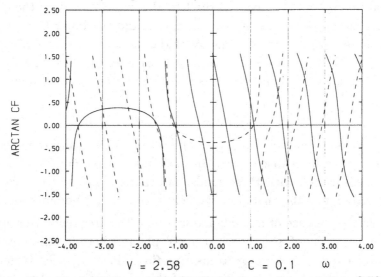

Fig. 4.   Graphical solution of Eq. (2.40) for $L_+ = L_-$ and $V = 2.58$. At this value of V the root $\omega = -1.4$ corresponds to a solution of positive norm, while the mode at $\omega = -1.2$ has negative norm; hence negative-energy pairs exist in the quantum theory. As V changes, the solid curve moves to left or right, and these roots flow into points of tangency and then disappear into the complex ω plane (see Fig. 6).

zero or infinity. The behavior of the first few modes in the $V-\omega$ plane is shown in Fig. 5, and the Klein region of that figure has been replotted in the symmetrical gauge in Fig. 6. Any other value of θ is expected to yield qualitatively similar results.

Figures 5 and 6 display an unexpected new feature. Every curve beyond the first (counting outward from the origin) has more than one point of vertical tangency. In fact, the nth curve has n such points (as far as the computations have been carried out, at least). It is clear from the discussion in Sec. IIIA that the pair of complex frequencies created at a singular point is not "permanent" in this case, but must rejoin at a singular point on an adjacent curve of real frequencies. This behavior has been indicated schematically in Fig. 6 by dashed lines. In particular, for certain values of V (e.g., $V = 2.6$ in Fig. 6; cf. Fig. 4) all complex frequencies disappear and the particle interpretation (discrete vacuum, Fock representation), which does not exist for some smaller V, is regained.

On the other hand, the remarks in Sec. IIIA concerning the sign of the norm remain valid here. Near a singular point at which the curve of real frequencies is convex to the left, the upper branch of the curve corresponds to solutions of negative norm, and the lower to solutions of positive norm. Since the energy of the particles or antiparticles in a mode with frequency ω and norm $\epsilon = \pm 1$ is $\epsilon\omega$ [see Eq. (2.30)], the energy spectrum associated with such an "inverted" pair of modes has the form

$$E = -\omega_1 N_1 + \omega_2 N_2 \qquad (\omega_1 > \omega_2), \qquad (4.1)$$

where N_j is the number of quanta present in mode j (and the vacuum energies have been discarded). In particular, if $N_1 = N_2 = N$, the energy associated with these modes is negative, and since N can be arbitrarily large, the energy of the field theory is not bounded below.

Unlike the case of a single mode of negative frequency (discussed at the beginning of Sec. IIIA), it is not possible here to eliminate the negative energy by a gauge transformation, since $\omega_1 - \omega_2$ is gauge-invariant. The physical significance of this observation is that a state with $N_1 = N_2$ (and, say, no other particles present) has the same charge as the vacuum. Consequently, the vacuum could conceivably decay into this state, with an associated release of energy to be carried away by some other physical system (such as the electromagnetic radiation field) weakly coupled to the quantized scalar field. Therefore, a quantum

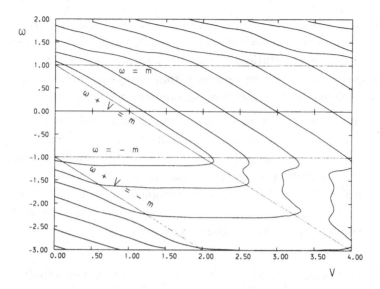

Fig. 5. Dependence of the frequency levels on V for $L_+ = L_-$ (plotted from the results of many computations like those presented in Figs. 3 and 4). Note that the inverted pair of roots in Fig. 4 and the two roots adjacent to them all lie on a smooth curve including the second particle and antiparticle eigenvalues of the free system $(V = 0)$. All solutions corresponding to points on this curve have exactly one node.

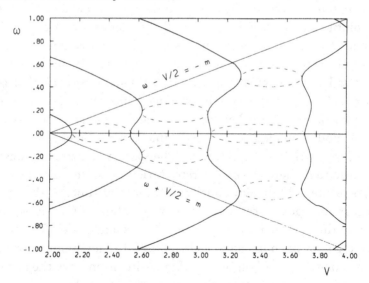

Fig. 6. Enlargement of the Klein region of Fig. 5, replotted in a gauge which displays the charge conjugation symmetry of the system. The dashed lines qualitatively represent extensions of the curves into the complex frequency plane, with the $\operatorname{Im}\omega$ axis perpendicular to the page.

system with energy unbounded below is expected to be catastrophically unstable against small perturbations, even if, when taken at face value, it possesses discrete stationary states. To what extent a theory with this property can be taken seriously from a physical point of view is a debatable question, to which we shall return in Part Two.

A final observation on the solutions of this model is that all the functions $\phi_j(x)$ corresponding to a continuous curve in the $V-\omega$ plane must have the same number of nodes, since the function satisfies a second-order equation and varies continuously as V varies [cf. the explicit forms (2.35)–(2.38)]. Thus, for instance, when $V = 2.6$ (see Figs. 6 and 4), there are two particle "eigenfunctions" with one node and none with no nodes. This contrasts with a familiar property of Sturm–Liouville operators.

Let us summarize the varieties of unstable or potentially unstable behavior which have been found in the simple square-well models, and which clearly will occur in other time-independent boson external-potential problems if the electrostatic potential is sufficiently deep.[26] We are assuming that the general solution of the field equation can be expanded in normal modes (including the inhomogeneous partners of any singular modes). In other words, the analogue of failure of self-adjointness is neglected here. If all the modes are of the ordinary type, and if all positive-norm modes have larger frequencies than all negative-norm modes, then no instability occurs. If modes occur in inverted order, however, the energy in the Fock representation is unbounded below, and the system is judged unstable against external perturbations. On the other hand, a singular mode (which is really a pair of modes tangled in an irreducible Jordan block) deprives the theory of its particle interpretation. The system is internally unstable in the sense that there are no discrete stationary states and the matrix elements of the field display runaway behavior in time. A theory with complex modes possesses both kinds of instability. Of course, mode pairs of all three types may be present in the same model. To complete the catalogue of phenomena one should add the nonuniqueness of the vacuum which occurs when modes of opposite norm have the same frequency; this can give rise to a flux of created "particles" in a putative

---

[26]   A model with a smooth potential has been solved by B. A. Arbuzov and V. E. Rochev, *Teor. Mat. Fiz.* **12**, 204 (1972) [*Theor. Math. Phys.* **12**, 761 (1972)].

"vacuum" state. This, the boson Klein paradox, will be discussed in Sec. VI and Part Two.

V. THE SCHIFF–SNYDER–WEINBERG MODEL REVISITED

Although the foregoing computations indicate a qualitative difference in the structure of the normal modes for the two cases $L_+ \gg L_-$ and $L_+ \approx L_-$, it is clear that there is no sharp dividing line between these regimes. Furthermore, it will be argued in Sec. VI that only the latter type of behavior is consistent with the "Klein paradox" as a limiting case; but if L_- is, say, 10^{10} meters, then the physics around the potential jump at $x = 0$ is surely very similar to that around an idealized Klein step ($L_\pm = \infty$), even if $L_+ \ggg L_-$. To the extent that the relevant parameter is the ratio of $\min (L_+, L_-)$ to the de Broglie wavelength, rather than the Compton wavelength m^{-1}, one must expect Klein behavior for fixed $L_\pm$ as ω, μ, and ξ increase. Thus the authors of Ref. 3 might have seen negative-energy pairs if they had closely examined the solution of their model at large quantum numbers. (This remark does not apply to the case $L_+ = \infty$, which engaged most of their attention.)

This conjecture has been confirmed by computations of the normal modes of the model with $L_- = 1$, $L_+ = 10$ ($\theta = 1$, $\psi = .1$). For small V the conventional Schiff–Snyder–Weinberg behavior (as in Fig. 2) was observed. When $V \approx 30$, however, the F_- curve becomes sufficiently steep to intersect several times with a single branch of the F_+ curve, and the corresponding modes exhibit the abnormal ordering of positive and negative norms. [See Fig. 7, and compare Eqs. (2.46) and surrounding discussion.] By imagining the F_- curve shifted left or right on Fig. 7, one can easily see that the qualitative behavior of the frequencies in this region is as sketched in Fig. 8. That is, there is a sequence of transitory pairs of complex-frequency modes (cf. Fig. 6), until finally a pair "escapes" permanently (cf. Fig. 2). For V slightly larger than the value at which a pair of complex modes disappears, the real frequencies appear in the abnormal order, so that negative-energy particle pairs occur in the second-quantized theory. Figure 8 emphasizes, incidentally, the point that there is, in general, no natural pairing of the positive- and negative-norm ordinary modes.

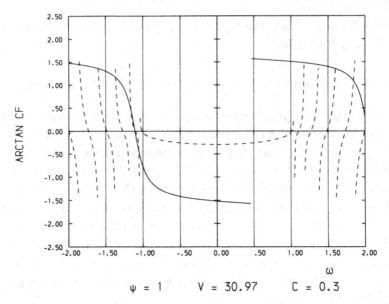

$$\psi = 1 \qquad V = 30.97 \qquad C = 0.3$$

Fig. 7.   Graphical solution of Eq. (2.40) for $L_+ = 10\,L_-$ and $V = 30.97$, exhibiting an inverted pair of roots near $\omega = -1.1$. *Note:* A solution falling on the left-hand branch of the solid curve has 9 nodes inside the potential well, while one on the right-hand branch has 10. The number of nodes outside the well is determined by the branch of the dashed curve, counting outwards from the broad central minimum as zero.

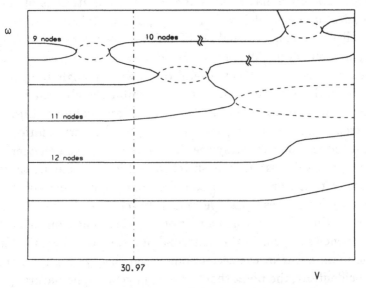

Fig. 8.   A highly schematic picture of the behavior of the frequency levels for the system of Fig. 7 as V varies in the neighborhood of 30.97.

VI. The Klein Limit

When $L_+ = \infty$ and $L_- = \infty$, the entire Klein region of the V–ω plane is filled with degenerate pairs of (continuum) modes, one of positive norm and one of negative norm. (See Part Two, Sec. III, for details). This situation clearly is not a limiting case of the old Schiff–Snyder–Weinberg picture (cf. Fig. 2), in which the modes with $\epsilon = +1$ are sharply separated in the V–ω plane from those with $\epsilon = -1$ by a horizontal curve, and also complex frequencies exist. The resolution of this discrepancy was a primary motivation of the present investigation.

The limiting Klein situation is, however, consistent with the new picture (Figs. 5 and 6), in which positive- and negative-norm modes are interspersed. As θ increases, the curves in Figs. 5 and 6 will get closer together, and a fixed value of V will find itself in a region of curves with many "wiggles" in the Klein region. The complex-frequency loops will shrink and disappear in the limit. For a fixed V one anticipates that in the limit adjacent solutions ϕ_j with odd numbers of nodes will become essentially indistinguishable, while the even-node functions alternating with them flow together into another continuum (distinguishable from the first). Thus one will obtain (outside the bound-state regions) the double continuum of solutions of the infinite-volume problem. [In a simpler analogue, the nonrelativistic Schrödinger equation with a potential symmetric under space reflection, the discrete solutions with odd numbers of nodes give rise in the infinite-volume limit to the continuum solutions of odd parity, while the solutions with even node numbers collapse into an even-parity continuum, degenerate in energy with the odd solutions. The conjecture of similar behavior in the present case is supported by the observation that the intersections in Figs. 3, 4, and 7 tend to form two alternating sequences (as characterized, e.g., by F_+).] For ω in the free particle and antiparticle regions, appropriate linear combinations of the two solutions constitute the conventional *in*-states of scattering theory (one representing a beam of particles coming from the left and the other a beam from the right), both having the same sign of ϵ. In the bound regions there is only one solution per value of ω, representing a beam which is totally reflected by the potential. In the Klein region there will again be a solution with the form of an *in*-state from each side, but these two solutions will have opposite values of ϵ. [See Eqs. (3.12)–(3.19) of Part Two.]

What are the consequences for field theory of this type of spectrum of modes? The energy will still be unbounded below, because in the Klein region one has negative-norm modes with frequencies greater than those of some positive-norm modes. In addition, there is a new phenomenon due to the existence of modes with opposite norm and precisely the same value of ω. Suppose that one had a pair of discrete modes[27] with that property:

$$\Phi = a\Phi_+ e^{-i\omega t} + b^\dagger \Phi_- e^{-i\omega t} + \cdots, \tag{6.1}$$

$$\langle \Phi_+, \Phi_+ \rangle = 1, \qquad \langle \Phi_-, \Phi_- \rangle = -1, \qquad \langle \Phi_+, \Phi_- \rangle = 0; \tag{6.2}$$

$$Q_+ + Q_- = a^\dagger a - b^\dagger b, \tag{6.3a}$$

$$E_+ + E_- = \omega(a^\dagger a - b^\dagger b) = \omega (Q_+ + Q_-) \tag{6.3b}$$

[see Eqs. (2.20), (2.23), (2.24), (2.29), (2.30)]. Consider the Bogolubov transformation

$$\Phi'_+ = \alpha\Phi_+ + \beta\Phi_-,$$
$$\Phi'_- = \beta^*\Phi_+ + \alpha^*\Phi_-, \tag{6.4}$$

where α and β are any complex numbers satisfying

$$|\alpha|^2 - |\beta|^2 = 1. \tag{6.5}$$

[It is not necessary, and turns out to be inconvenient, to adhere to the convention that $\Phi_\pm(x)$ are real.] Then Eqs. (6.2) remain valid for the new basis vectors $\Phi'_\pm$, and if a' and b' are defined by analogy with Eq. (6.1), one finds that

$$a = \alpha a' + \beta^* b'^\dagger,$$
$$b^\dagger = \beta a' + \alpha^* b'^\dagger. \tag{6.6}$$

A short calculation shows that the charge and energy expressions are invariant under this transformation:

$$Q_+ + Q_- = \omega^{-1}(E_+ + E_-) = a'^\dagger a' - b'^\dagger b'. \tag{6.7}$$

Nevertheless, from Eqs. (6.6) it follows that the original vacuum state (annihilated by the operators a and b) carries finite probability amplitudes for the presence of various numbers of pairs of a'-particles and b'-antiparticles. A state specified by a given "particle" structure will

---

[27]   The continuum case, which seems to be the only one that arises in practice, will be discussed in Part Two, Sec. III.

have different expectation values for observables defined in terms of the field operators, depending on which basis of normal modes is used to define "particle". There is no evident way to choose among these definitions, since the energy and charge are diagonalized in each case. Thus the particle terminology has become thoroughly ambiguous.[28]

VII. THE MASSLESS CASE

Finally, one should consider whether our conclusions must be modified when the scalar field has $m = 0$, which is the case most often studied in the literature oriented toward general relativity. Evidently the most important difference from the massive case is that when $m = -m = 0$ there are no bound and forbidden regions in the spectrum (cf. Fig. 2). Consequently, Klein and Schiff–Snyder–Weinberg phenomena will appear as soon as the external potential is switched on (no matter how weakly).[29]

Indeed, when a free massless scalar field is quantized in a finite box with periodic boundary conditions, with no external potential at all, the mode with $\omega = 0$, corresponding to a constant spatial solution $\phi_j(\mathbf{x})$, is already a singular mode in the sense of Sec. III. Since the right side of Eq. (3.20) vanishes, we have $\phi_o = \phi_e = \phi_j$, and the corresponding term in the field expansion (3.33) becomes

$$f\phi_j + g\phi_j\,(1 - it), \quad \phi_j = \text{constant}. \qquad (7.1)$$

Note that the second term really must be included in the most general solution of $\Box\phi = 0$ with periodic boundary conditions on a finite spatial interval.

---

[28]   Note that nothing similar can happen when two modes of the same norm are degenerate. In that case the scalar products and the charge and energy contributions are preserved by a unitary transformation on the vectors $\Phi_\pm$, but then the counterparts of Eqs. (6.6) will not mix creation operators with annihilation operators. Hence such a relabelling of modes induces merely a change of basis within each n-particle subspace of the Fock space. Any mixing of creation and annihilation operators would disrupt the diagonal forms of the charge and energy operators.

[29]   To be precise, in some models with entirely discrete modes a certain minimum strength of the potential is needed to bring the lowest particle mode and highest antiparticle mode into contact, so that the first Schiff–Snyder–Weinberg singular mode can form.

Part Two

I. DECAY OF A ROTATING OR CHARGED BLACK HOLE

Strong evidence has been presented recently that the gravitational
field near a black hole, or an astrophysical object collapsing to form a
black hole, will cause particle creation, or instability of the vacuum, in
the quantum theory of fields against the black hole background geometry.[1] The study of this topic goes back, on the astrophysical side, to
the observation by Penrose[2] that a classical particle can enter the ergosphere of a rotating black hole and come out again with more energy
than it had originally, the black hole losing energy and angular momentum in the process. The mechanism is the production of another
particle which falls into the black hole, carrying the opposite angular
momentum and also negative energy as defined by an asymptotic observer. There is a corresponding effect for classical waves, called *superradiance*.[3,4] A wave incident on the ergosphere will be reflected out, as
from a potential barrier. The interesting point is that in certain modes
of the wave field the reflected wave will have an amplitude greater than
that of the incident wave. Zel'dovich and Starobinsky (Refs. 3) then
argued, in analogy with photon emission in atomic physics, that in a
quantum theory there should be spontaneous creation of particles in

---

[1]   These and other topics in the quantum theory of fields in curved space-
time have been expounded by B. S. DeWitt, *Phys. Reports* **19**, 295 (1975);
a short preliminary version (in which the treatment of the Hawking effect
differs significantly from the final version) appears in *Particles and Fields —
1974 (APS/DPF Williamburg)* (AIP Conference Proceedings No. 23), edited
by C. E. Carlson (American Institute of Physics, New York, 1975), pp. 660–
688. The various effects occur in both boson and fermion theories, but we
shall consider only the former.

[2]   R. Penrose, *Rivista Nuovo Cimento* **1**, special number, 252 (1969),
especially pp. 270–272; R. Penrose and R. M. Floyd, *Nature Phys. Sci.* **229**,
177 (1971).

[3]   Ya. B. Zel'dovich, *ZhETF Pis. Red.* **14**, 270 (1971) [*JETP Lett.* **14**,
180 (1971)]; A. A. Starobinsky, *Zh. Eksp. Teor. Fiz.* **64**, 48 (1973) [*Sov.
Phys.–JETP* **37**, 28 (1973)].

[4]   C. W. Misner, unpublished; W. H. Press and S. A. Teukolsky, *Nature*
238, 211 (1972); J. D. Bekenstein, *Phys. Rev. D* **7**, 949 (1973).

these superradiant modes. This expectation was confirmed by explicit calculations in quantum field theory by Unruh.[5,6]

Unruh dealt with the Kerr metric,[7] which describes a stationary, rotating black hole. In the exterior region (including the ergosphere but not the space-time inside the horizon) this metric is independent of time in the conventional coordinate system,[8] and the scalar wave equation can be solved by separation of variables.[9] Therefore, there is a natural construction of the stationary states of the system, following the standard treatment of a static external potential (see later sections and Part One for details). There will be a "vacuum" state and states with various numbers of quanta or "particles" present. Unruh calculated that in his vacuum state there is a net flux at infinity of energy and angular momentum away from the black hole. Since the space is asymptotically flat, one expects that far from the black hole these currents of energy and angular momentum are associated with real particles. In this calculation the black hole is a time-independent external potential, posited once and for all. However, the principles of energy and angular momentum conservation suggest that, if the dynamics of the black hole itself were considered, it must be losing mass and angular momentum. Indeed, the mechanism of this "spin-down" appears to be visible in the external-potential theory as a compensating flux of negative energy and angular momentum inward over the horizon, corresponding to the transmitted wave in classical superradiance. The Unruh effect may thus be visualized as the Penrose process with the original incoming particle omitted.

Later, Hawking[10] announced that a more realistic, time-dependent

5   W. G. Unruh, *Phys. Rev. D* **10**, 3194 (1974).
6   This problem has also been studied by L. H. Ford, Ph.D. Dissertation, Princeton University (1974); *Phys. Rev. D* **12**, 2963 (1975) [erratum ibid. *D* **14**, 658 (1976)].
7   R. P. Kerr, *Phys. Rev. Lett.* **11**, 237 (1963); S. W. Hawking and G. F. R. Ellis, *The Large Scale Structure of Space-time* (Cambridge Univ. Press, Cambridge, 1973), Sec. 5.6.
8   R.H. Boyer and R.W. Lindquist, *J. Math. Phys.* **8**, 265 (1967).
9   B. Carter, *Commun. Math. Phys.* **10**, 280 (1968); *Phys. Rev.* **174**, 1559 (1968); D. R. Brill, P. L. Chrzanowski, C. M. Pereira, E. D. Fackerell, and J. R. Ipser, *Phys. Rev. D* **5**, 1913 (1972); Ref. 17; Ref. 6.
10 S. W. Hawking, *Nature* **248**, 30 (1974); in *Quantum Gravity: An Oxford Symposium*, edited by C. J. Isham, R. Penrose, and D. W. Sciama (Clarendon Press, Oxford, 1975), pp. 219–267; *Commun. Math. Phys.* **43**,

black hole (formed by the collapse of an ordinary body without a past horizon) will explosively decay into particles, even if it is not rotating. Hawking showed that Unruh's result for a preexisting rotating black hole is consistent with a limiting case of his.

An analogous effect involving a charged (Reissner–Nordström) black hole has been studied by Gibbons.[11,12] Here the classical electrostatic Klein phenomenon occurs in an astrophysical setting. It is energetically favorable to produce particle-antiparticle pairs, which are separated by the electric field, one member going to infinity and the other over the horizon. The black hole presumably loses its charge and part of its mass by this process. Gibbons applied his analysis also to a collapsing charged body, following Hawking.

The vacuum states in the works of Unruh and Gibbons were specified in terms of the absence of particles at past null infinity, $\mathcal{I}^-$. Here, however, we shall proceed as in ordinary quantum scattering theory, describing states by their properties at finite, but arbitrarily early or late, times. This approach has several advantages: (1) It is applicable equally to massive and massless fields. (2) The physical interpretation of various boundary conditions on the field, always a delicate matter, seems clearer on real spacelike hypersurfaces than on fictitious null hypersurfaces. (3) It provides an alternative conceptual framework, more familiar and convincing to those (e.g., the present author) whose backgrounds are in quantum theory, not general relativity. Its disadvantages are that one abandons an important tool of the modern relativist, and that, as we shall now see, certain conceptual problems arise that apparently are avoided by the other way of thinking. It is hoped that readers will find that the discussion of these paradoxes, even if they are regarded as straw men, leads to a deeper understanding of the theory.

The physical description of Unruh's process offered above is self-contradictory if all the references to "particles" are taken literally. One starts with a vacuum state, so that, by definition, no particles are present. The particle notion employed here is time-independent, since

199 (1975).

[11]   G. W. Gibbons, *Commun. Math. Phys.* **44**, 245 (1975).

[12]   A related prediction had been made qualitatively by L. Parker and J. Tiomno, *Astrophys. J.* **178**, 809 (1972), on the basis of the pair-creation formula of Schwinger (Ref. 2 of Part One). More recently the effects of vacuum polarization by charged and rotating black holes have been studied further by R. Ruffini and coworkers.

the creation and annihilation operators have been introduced in correlation with the stationary normal modes of the system. Nevertheless, one reaches the conclusion that particles are being continually created and radiated away. To clear up this inconsistency of language requires a deeper understanding of the physics of the situation.

Another surprising aspect is that the process is not invariant under time reversal, although the model itself is. Because rotation is involved, this point needs to be stated rather carefully. Figure 9(a) represents the black hole and the particles emitted from it, whose motion can be resolved into two components, one due to their ejection and the other to their orbital angular momentum. Reversing the directions of all motions in the picture yields Fig. 9(b), which portrays a different system, since the black hole (a prescribed external potential) is now rotating backwards. However, an additional space reflection in a plane containing the axis of rotation restores the original geometry, in Fig. 9(c). It is the combined transformation leading from (a) to (c) which can be regarded as the time-reversal symmetry of the Kerr geometry. How does this operation affect the state of the field? The particles in Fig. 9(c) carry angular momentum of the same sign as at first. However, now they are falling into the ergosphere, and they will increase the mass and angular momentum of the black hole. If state (a) exists in the theory, then by an elementary symmetry principle state (c) must exist also.

In the electromagnetic case (of Klein[13] or Gibbons) the electrostatic field is time-symmetric in the usual sense. If it creates pairs and shoots the particles off to infinity, then there must also be the possibility of beams of particles and antiparticles coming in from infinity and annihilating in the region of strong electric field.

It should be noted that this time asymmetry actually has nothing to do with a classical picture of the phenomenon, or even with a particle interpretation. It could have been discussed, not graphically, but directly in terms of the fluxes of energy and angular momentum computed by Unruh's method. Clearly, an initial condition has been somehow imposed to insure an outward flux. From the point of view of Refs. 5 and 11, this is the condition that the vacuum state is empty at $\mathcal{I}^-$, rather than, say, empty at $\mathcal{I}^+$. In the context of the present

---

[13]   See Refs. 1 and 2 of Part One.

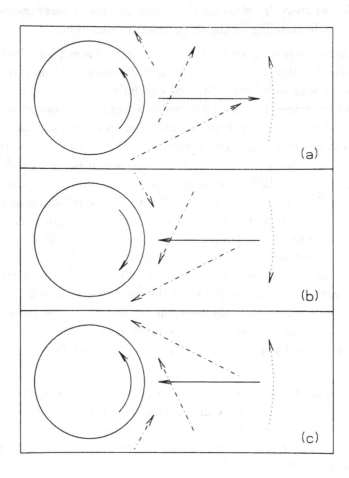

Fig. 9. (a) Particles expected to be ejected (solid arrow) from a rotating
black hole have orbital angular momentum (dotted arrow) of the same sign
as that of the hole. Hence they should leave the ergosphere in directions such
as those indicated by the dash-dotted arrows.   (b) Time reversal of the
situation in (a).   (c) Space reflection of (b) restores the original black hole
geometry, but in this state particles are carrying angular momentum into the
hole, so as to accelerate its rotation.

work, the apparent paradoxes posed here are manifestations of the freedom to make Bogolubov transformations on some of the normal modes, pointed out in Sec. VI of Part One. This will be explained in Sec. III, after some machinery has been set up.

II. A RECTILINEAR MODEL OF THE KERR METRIC

A. Geometry.

Let us consider a simple model which exhibits precisely those geometrical features of the Kerr black hole which are responsible for the superradiance effects (Refs. 2–5), without irrelevant complications. The metric is

$$ds^2 = dt^2 - [dx + V(z)dt]^2 - dy^2 - dz^2, \qquad (2.1a)$$

where $V(z)$ is an arbitrary function of z. Thus we have

$$g_{00} = 1 - V^2, \qquad g_{01} = -V, \qquad g_{11} = -1, \qquad \dots \qquad (2.1b)$$

We note the following properties of this space-time:

(1) The determinant of the metric is $g = -1$. It follows that the metric has a normal space-time signature $(+ - - -)$ at every point, and the scalar wave equation is thus hyperbolic.

(2) The hypersurfaces of constant t are spacelike. One expects, therefore, to have a well-posed Cauchy problem for initial data on one of these surfaces.

(3) The metric coefficients $g_{\mu\nu}$, and hence the dynamics of the field obeying the scalar wave equation, are independent of t. We shall call such a metric *manifestly time-independent* ("manifestly" because the property refers to the distinguished coordinate system as well as the intrinsic geometry).

(4) The Killing vector associated with translation in the coordinate t is timelike wherever $|V(z)| < 1$, but is spacelike where $|V(z)| > 1$. Therefore, the metric is *not* manifestly *stationary* if $|V(z)| \geq 1$ for some z.

The distinction made here between "time-independent" and "stationary" requires an explanation. The time independence described in points (2) and (3) above can be expressed in terms of the field of normal vectors to the equal-time hypersurfaces. The normal vector has

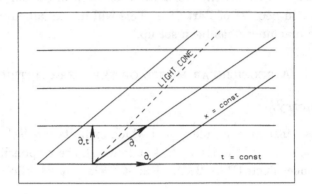

Fig. 10. A nonorthogonal coordinate system in Minkowski space.
The generator of time translations is a *spacelike* Killing vector.
The normal to the equal-time hypersurfaces (also a Killing vector
in this example) is timelike.

the *covariant* components

$$\partial_\mu t = (1,0,0,0), \qquad (2.2)$$

hence the norm[14] $g^{00} = 1$. The statement of time independence is that
this norm is greater than zero, and that t is an ignorable coordinate.
This property is the crucial one for the formalism of second quantization
(see below). On the other hand, the Killing vector generating time
translations has the *contravariant* components

$$\partial_t x^\mu = (1,0,0,0) \qquad (2.3)$$

and the norm $g_{00} = 1 - V^2$. Hence when $|V| > 1$ the Killing vector
becomes spacelike, and its existence does not qualify the metric as
stationary. An example of a spacelike time translation generator is
provided by Minkowski space equipped with a rectilinear coordinate
system whose time axis is so oblique that it falls outside the light cone
(Fig. 10).

(5) A "Galilean" coordinate transformation of the form

$$t' = t, \qquad x' = x - V_0 t, \qquad y' = y, \qquad z' = z \qquad (2.4a)$$

---

[14]   More pedantically, the square of the norm.

(V_0 a constant) leaves the metric (2.1) unchanged except for the replacement of V by

$$V'(z) \equiv V(z) + V_0. \tag{2.4b}$$

Thus it can be said to induce a *gauge transformation* on the potential V. In particular, if $V(z)$ is equal to a constant in some region, then the space-time is actually flat there, and the gauge transformation simply "straightens" the time axis, so that the associated Killing vector is the normal vector (see Fig. 10). It is important to note that such a Galilean transformation leaves the hypersurfaces of constant t invariant. Also, one can always make $|V(z)| < 1$ in the neighborhood of a given point by adding some V_0. Therefore, our metric is *locally stationary* (in the intrinsic sense). The new locally timelike Killing vector is

$$\partial_{t'} = \partial_t + V_0 \partial_x. \tag{2.5}$$

(6) Let

$$\Delta V = \sup_z V(z) - \inf_z V(z). \tag{2.6}$$

We shall be primarily interested in cases where V approaches its supremum as $z \to \infty$ and its infimum (taken as zero) as $z \to +\infty$. If $\Delta V < 2$, then one can choose $V_0 = -\frac{1}{2}(\sup V + \inf V)$ in order to make $|V'(z)| \leq \frac{1}{2}\Delta V < 1$. In this case the geometry is globally stationary. On the other hand, if $\Delta V \geq 2$, then there is no gauge in which $|V'(z)| < 1$ for all z at once.

Consequently, if $\Delta V \geq 2$, the space-time with metric (2.1) is (a) manifestly time-independent (b) everywhere locally stationary (c) *not* globally stationary. The first property makes the formalism of time-independent field quantization applicable. We shall see that (c) is a necessary condition for the appearance of instabilities and "paradoxes" precisely analogous to those of Klein and of Schiff, Snyder, and Weinberg. The condition (b), which is equivalent to the presence of the second Killing vector, ∂_x, makes the problem formally almost identical to the electrostatic theory treated at length in Part One. Presumably the instability effects can also arise in time-independent nonstationary gravitational models without this simplifying feature, which have not yet been analyzed.

B. Reduction of a Gravitational Problem to an Electromagnetic One.

The scalar field equation in the most general background metric is

$$(-g)^{-\frac{1}{2}}\partial_\mu[(-g)^{\frac{1}{2}}g^{\mu\nu}\partial_\nu\phi] + \mu^2\phi = 0. \tag{2.7}$$

(It has become customary in the present context to write μ for the mass of the field, since m and M are firmly entrenched as an angular momentum quantum number and the mass of the black hole, respectively.) The conformal term, $R\phi/6$, could be added (but would become singular in the step-function model treated below). If the metric is time-independent and $x^0 \equiv t$ is the time coordinate, the substitution

$$\phi(t,\mathbf{x}) = \phi_j(\mathbf{x})^{-i\omega_j t} \tag{2.8}$$

yields[15]

$$-\omega_j^2 g^{00}\phi_j - 2i\omega_j g^{0k}\partial_k\phi_j - i\omega_j(-g)^{-\frac{1}{2}}\partial_k[(-g)^{-\frac{1}{2}}g^{0k}]\phi_j$$
$$+(-g)^{-\frac{1}{2}}\partial_k[(-g)^{\frac{1}{2}}g^{kl}\partial_l\phi_j] + \mu^2\phi_j = 0. \tag{2.9}$$

Now suppose that ∂_1, as well as ∂_0, is a Killing vector, and that

$$g^{02} = g^{03} = 0. \tag{2.10}$$

Write

$$\phi_j(\mathbf{x}) = \phi_j(x^2, x^3)e^{ik_j x^1}, \tag{2.11}$$

obtaining

$$(\omega_j - k_j g^{01}/g^{00})^2 g^{00}\phi_j$$
$$= k_j^2(g^{01}/g^{00})^2 g^{00}\phi_j - k_j^2 g^{11}\phi_j + 2ik_j g^{1l}\partial_l\phi_j$$
$$+ik_j(-g)^{-\frac{1}{2}}\partial_l[(-g)^{\frac{1}{2}}g^{1l}]\phi_j + (-g)^{-\frac{1}{2}}\partial_i[(-g)^{\frac{1}{2}}g^{il}\partial_l\phi_j] + \mu^2\phi_j$$
$$(i,l = 2,3). \tag{2.12}$$

This is somewhat like the separated Klein-Gordon equation in spacetime dimension 3 for a scalar field of charge k_j in an electrostatic potential g^{01}/g^{00}, with additional magnetic and zeroth-order coupling terms. In particular, if

$$g^{12} = g^{13} = 0, \tag{2.13}$$

---

[15]   S. A. Fulling, Ph.D. Dissertation, Princeton University (1972), p. 266. This equation is equivalent to Eqs. (1.4) of Ref. 5 when some misprints in the latter are corrected and the conformal term is replaced by the mass term.

as is the case for the Kerr metric and our rectilinear model, then the equation is invariant under simultaneous change of sign of ω and k ("charge conjugation"). In the rectilinear model ∂_2 is also a Killing vector, so one may factor $e^{iq_j y}$ from ϕ_j and obtain

$$[\omega_j + k_j V(z)]^2 \phi_j(z) = -\phi_j''(z) + (k_j^2 + q_j^2 + \mu^2)\phi_j(z), \qquad (2.14)$$

which formally is precisely a two-dimensional electrostatic problem with charge k_j and mass

$$m = (k_j^2 + q_j^2 + \mu^2)^{\frac{1}{2}} \qquad (2.15)$$

[Part One, Eq. (2.3)]. Henceforth we omit the index j and take k and q to be fixed.

The Klein region, in the sense of Part One, of the eigenfrequency spectrum of Eq. (2.15) arises when $e\Delta V \geq 2m$, or

$$k\Delta V \geq 2(k^2 + q^2 \mu^2)^{\frac{1}{2}}. \qquad (2.16)$$

This inequality is satisfied quickest in the limit of small q and μ, where it takes the form $\Delta V \geq 2$ (and $k \neq 0$). The geometrical significance of the condition $\Delta V \geq 2$ has already been pointed out.

The quantization of the field is now routine. The field operator has an expansion of the form of Eq. (3.33) of Part One, where the functions ϕ_j are understood to include the factors $e^{ikx}e^{iqy}$ and the integrations include integrations over k and q.

The particle interpretation of the ordinary modes deserves some comment. In the terms of the expansion corresponding to a given k and q, the operators $a = a_{k,q,j}$ annihilate particles with transverse momentum (k, q). The $b^\dagger$ operators create antiparticles of momentum $(-k, -q)$ if the field is charged (complex). If the field (in the original gravitation-coupled theory) is neutral, however, then $b_{k,q,j}^\dagger$ creates *particles* of momentum $(-k, -q)$. In the latter case the full field expansion is Hermitian, because of the "charge conjugation" invariance of Eq. (2.14): Every solution with $\epsilon < 0$, labelled by (ω, k, q), is the complex conjugate of a solution with $\epsilon > 0$, quantum numbers $(-\omega, -k, -q)$, and the same dependence on z. Therefore, every term in the $b_j^\dagger$ sum is the adjoint of a term in the a_j sum for the inverted values of k and q. When attention is restricted to fixed $k \neq 0$, a neutral field behaves formally like a charged field, coupling nontrivially in the electrostatic analogy.

All the discussion of unstable and singular modes in Part One and its references carries over to the gravitational case. In particular, the explicit numerical solutions, which indicate the range of phenomena to be expected in general, apply to the case, rather unphysical, of a universe of the form (2.1) of finite extent in z, with reflection boundary conditions at the ends, and with $V(z)$ a step function.

Because of the freedom to make a Galilean gauge transformation (2.4) [which becomes a gauge transformation in the usual sense in the context of Eq. (2.14)], the frequency or energy of a mode, ω_j, is determined only up to a term linear in the x-momentum: $k_j V_0$. The energy difference between two states of different total x-momentum is ill-defined, but the conservation law for that quantity, which is valid because $\partial_1 \equiv \partial/\partial x$ is a Killing vector, assures that transitions between such states do not occur. Note that the analogous remarks in Part One (Sec. IIIA) for the electromagnetic case did not appeal to the superselection rule[16] for charge, but only to the conservation law.

C. Analogy with the Kerr Metric.

The separation of variables in the Kerr metric and consequent scalar field quantization (Refs. 9 and 3–6) lead to results very similar to those for the simpler rectilinear model. The space-like Killing vector analogous to $\partial/\partial x$ is $\partial/\partial\phi$, the generator of rotations about the symmetry axis of the black hole. The corresponding conserved quantity, or "charge" in the electrostatic reduction of the problem, is the (orbital) angular momentum along the symmetry axis, $m \equiv L_z$, rather than the linear momentum k. Thus in translating from electrostatic language to gravitational language after quantization, one must for "antiparticle" read "particle of the opposite angular momentum" if the scalar field is Hermitian, and "antiparticle of the opposite angular momentum" if the field is complex.

The polar angle θ is a trivial coordinate analogous to y. The factor $e^{ikx}e^{iqy}$ in the normal modes is replaced by the spheroidal harmonic

---

[16]   F. Strocchi and A. S. Wightman, *J. Math. Phys.* **15**, 2198 (1974). It is clear from Sec. 1 of this reference that the proof of the superselection rule requires a quantized electromagnetic field; in an external-potential theory the rule is at best an additional physical assumption. The authors point out in Sec. 4 that the question of a superselection rule for energy and momentum in quantized gravity theory is unsettled.

$S_{lm}(\omega, \theta, \phi)$, where l is a generalization of the quantum number of total orbital angular momentum of a spherically symmetric system.

The coordinate which enters the equation nontrivially, analogous to z, is the radial coordinate r, or a suitable monotonic function of it, r^*. The reduced wave equation does not have a purely electromagnetic form like (2.14); the coefficients have a more complicated geometrical dependence, which does not, however, affect the formalism of quantization.

In the rectilinear model one has natural inertial frames at $x = +\infty$ and $x = -\infty$ respectively, which are related to each other by a Galilean transformation. The analogue in the Kerr case is the passage from the rest frame at infinity ($r^* = \infty$) to the local inertial frame at the horizon of the black hole ($r^* = -\infty$), which is "dragged" by the gravitational influence of the rotating massive object (the Lense–Thirring effect). The corresponding potential difference, playing the role of $-\Delta V$, between $+\infty$ and $-\infty$ is

$$\Omega_H = (a/2M)[M + (M^2 - a^2)^{\frac{1}{2}}]^{-1}, \qquad (2.17)$$

called the angular velocity of the horizon. (M is the hole's mass and Ma is its angular momentum.) The radial wave equation reduces to

$$\frac{d^2 R}{dr^{*2}} = -(\omega^2 - \mu^2)R \qquad (2.18a)$$

in the limit $r^* \to \infty$, and to

$$\frac{d^2 R}{dr^{*2}} = -(\omega - m\Omega_H)^2 R \qquad (2.18b)$$

as $r^* \to -\infty$, where $R(r^*) \propto (r^2 + a^2)^{\frac{1}{2}} \phi_{\omega lm}(r)$. Thus the behavior of the coefficients in that equation is such that there is always a Klein region (i.e., there are always superradiant modes) even if μ is nonzero and a is small; there is no analogue to Eq. (2.16).

It is not yet known[17] whether the Kerr model exhibits Schiff–Snyder–Weinberg effects (unstable and singular modes) or negative-energy pairs (see Part One, Sec. IV), in addition to Unruh's Klein effect. A rigorous spectral theory for the operator appearing in the radial equation is needed to decide the question.

---

[17] See, however, S. L. Detweiler and J. R. Ipser, *Astrophys. J.* **185**, 675 (1973).

III. THE KLEIN PARADOX

It remains to discuss the solutions of the field equation (2.14) of our model and their significance for quantization in the circumstances where the Klein paradox, or particle creation effect, actually occurs. It suffices to consider a step-function potential $V(z)$ without a boundary at either end $(-\infty < z < \infty)$. This is simply the model of Part One, Sec. IIB, with $L_+ = L_- = \infty$. (As a matter of fact, most of the discussion of this section does not depend on the potential function, so long as it tends to constant values at $\pm\infty$.) Let us assume the condition (2.16) (which now reduces to $V \geq 2$), so that the middle of the frequency spectrum will be a Klein interval, not a forbidden interval (see Fig. 2).

Thus we are to solve Eqs. (2.34) of Part One, with $x = z$ and, without loss of generality, $m = 1$ (i.e., $Z = 1$, $\theta = \infty$, ψ undefined). [Most of the literature treats the case $m = 0$ $(Z = 0)$, which is similar and in some respects simpler.] The general solution for real ω satisfying

$$|\omega| > 1 = m, \qquad |\omega + V| > 1 \tag{3.1}$$

is

$$Ae^{i\mu z} + Be^{-i\mu z} \quad \text{for } z > 0, \tag{3.2a}$$

$$Ce^{i\xi z} + De^{-i\xi z} \quad \text{for } z < 0, \tag{3.2b}$$

$$\mu = (\omega^2 - 1)^{\frac{1}{2}}, \qquad \xi = [(\omega + V)^2 - 1]^{\frac{1}{2}}. \tag{3.3}$$

The continuity conditions at $Z = 0$ yield

$$A + B = C + D, \tag{3.4a}$$

$$(A - B)\mu = (C - D)\xi. \tag{3.4b}$$

If one of Eqs. (3.1) is violated, say the first, then there is only one linearly independent bounded solution, and Eq. (3.2a) is replaced by

$$A'e^{-\eta z}, \qquad \eta = (1 - \omega^2)^{\frac{1}{2}}, \tag{3.5}$$

whence

$$A' = C + D, \qquad i\eta A' = C - D. \tag{3.6}$$

To choose a convenient basis pair of the solutions (3.2), it is advisable to study the behavior of wave packets at large positive and negative times, as in ordinary quantum scattering theory. Let $g_1(p)$ be a square-integrable function, say a Gaussian, such that both it and its Fourier transform have reasonably well-defined peaks. Consider

$$\int_0^\infty d\mu\, A e^{i\mu z} e^{-i\omega t} g_1(\mu), \qquad (3.7)$$

where A may depend on μ. The standard stationary-phase argument says that this integral is large only near values of x, t, and μ related by

$$\frac{d}{d\mu}(\mu z - \omega t + \arg A + \arg g_1) = 0,$$

or

$$z = (\mu/\omega)t + \text{term independent of } z \text{ and } t.$$

If ω is positive, therefore, for any fixed g_1 there is a time before which the packet (3.7) is located almost entirely on the negative half of the z axis. Similarly, the packet

$$\int_0^\infty d\xi\, D e^{-i\xi z} e^{-i\omega t} g_2(\xi)$$

travels with group velocity $-\xi/(\omega + V)$, and hence is concentrated on the positive z axis at large negative t, provided that $\omega + V$ is positive. If $\omega + V$ is negative, on the other hand, a packet formed from $e^{+i\xi z}$ has this type of motion.

Now, for each $\mu > 0$, define a solution ${}^+\phi^{\text{in}}_{-\mu}(z)$ as follows. Define $\omega > 0$ and ξ as functions of μ by Eqs. (3.3). Let ${}^+\phi^{\text{in}}_{-\mu}(z) = A e^{i\mu z} + e^{-i\mu z}$ for $z > 0$. For negative z, Eq. (3.2b) applies, since $\omega + V > 1$ (V being positive by convention). Let $C = 0$ and solve Eqs. (3.4) for A and D. On the other hand, for every $\xi > 0$ one can define functions ω and μ through Eqs. (3.3) with $\omega + V > 0$. Let ${}^+\phi^{\text{in}}_\xi(z) = e^{i\xi z} + D e^{-i\xi z}$ for $z < 0$. If $\omega > 1$, let $B = 0$ in Eq. (3.2a) and solve for A and D. If $|\omega| \le 1$, Eqs. (3.5) and (3.6) yield $D = (1 - i\eta)/(1 + i\eta)$ and $A' = 2/(1 + i\eta)$. Finally, if $\omega < -1$ (as can happen, since $V > 2$), one sets $A = 0$ and solves for B and D:

$$B = \frac{2\xi}{\xi - \mu}, \qquad D = \frac{\xi + \mu}{\xi - \mu}. \qquad (3.8)$$

The point of this construction, of course, is that a solution of the form

$$\int_{-\infty}^\infty dp\, {}^+\phi^{\text{in}}_p(z) e^{-i\omega t} g(p) \qquad (3.9)$$

has a simple form in the far past $(t \to -\infty)$. Here $\omega = \omega(p)$ is defined through Eqs. (3.3) with the sign conventions just described, and with $p = \xi$ or $p = -\mu$ as appropriate. The integral (3.9) can be written as a sum of various terms, typical of which is

$$\int_{-\infty}^{0} dp\, A e^{i|p|z} e^{-i\omega t}\, g(p)\theta(z), \qquad \omega = (p^2 + 1)^{\frac{1}{2}} > 0.$$

Comparing with expression (3.7) and the discussion following it, one sees that this term gives a vanishingly small contribution as $t \to -\infty$. In fact, one ean verify that in all the various solutions ${}^{+}\phi_p^{\text{in}}$ the only terms which do contribute to the integral in this limit are those whose coefficients have been chosen equal to 1; consequently, the expression (3.9) can be replaced by

$$\int_{-\infty}^{\infty} dp\, e^{ipz} e^{-i\omega t}\, g(p) \tag{3.10}$$

when t is sufficiently large and negative. It should be borne in mind that for positive p, ω is not $(p^2 + 1)^{\frac{1}{2}}$ but $(p^2 + 1)^{\frac{1}{2}} - V$, which is negative for small p.

The norm of the solution (3.9) can now be calculated by substituting the limiting form (3.10) into the formula for the indefinite scalar product [Eq. (2.15) of Part One, with $eA_0 = -V\theta(-z)$]. We may assume that the support of $g(p)$ is either entirely positive or entirely negative, since in the general case the packets corresponding to positive and negative p are located on opposite halves of the z axis and hence have zero scalar product. In either case the operator $\partial/\partial t + ieA_0$ simply contributes a factor $-i(p^2 + 1)^{\frac{1}{2}}$ to the integrand. Consequently, performing the x-space integration first, we obtain the norm

$$\epsilon = 2(2\pi)^s \int_{-\infty}^{\infty} dp\,(p^2 + 1)^{\frac{1}{2}} |g(p)|^2,$$

where $s = 1$ for the two-dimensional electrostatic problem, but $s = 3$ for the four-dimensional gravitational problem reduced via Eq. (2.11) etc. It follows that the normalization of the continuum solutions ${}^{+}\phi_p^{\text{in}}$ is

$$\langle {}^{+}\Phi_p^{\text{in}}, {}^{+}\Phi_{p'}^{\text{in}} \rangle = 2(2\pi)^s (p^2 + 1)^{\frac{1}{2}} \delta(p - p'). \tag{3.11a}$$

[In the four-dimensional case, factors $\delta(k - k')\delta(q - q')$ should be included.] In particular, these vectors are all of positive norm.

The analogous solutions of negative norm, $^-\phi_p^{in}$, are constructed in a similar way. In defining $\omega(p)$, one now takes in Eqs. (3.3) the negative root for ω or for $\omega + V$, as the case may be; $\omega + V$ will be positive for some modes. Again one chooses the coefficients in Eqs. (3.2) so that the only term which survives in the early-time limit is normalized to unity. If the motion of that incident packet is rightwards, p is negative according to the convention we shall adopt for negative-norm modes. Thus in all cases the incident wave is of the form e^{ipz}. The $^-\phi_p^{in}$ are normalized as in Eq. (3.11a), but with the negative sign, and also satisfy

$$\langle ^+\Phi_p^{in}, {}^-\Phi_{p'}^{in}\rangle = 0. \tag{3.11b}$$

We have evidently constructed a complete basis for all solutions of the field equation which asymptotically become free, since all possible values of incident momentum p have been taken into account, for both signs of the norm. Since it is easy to see that for the step-function potential there are no normalizable solutions (which would represent bound states), it is reasonable to assume that the $^\pm\Phi_p^{in}$ form a complete set, although, to the best of the author's knowledge, there is no spectral theorem covering this type of problem. Furthermore, all the complex-frequency solutions and inhomogeneous solutions [see Eq. (3.20) of Part One] are unbounded, and hence it is consistent with the physicist's usual rule of thumb not to accept them as "eigenfunctions", even in the generalized sense. The disappearance of complex modes in the limit $\theta \to \infty$ has been predicted already in Sec. VI of Part One. It is less obvious that there is not a Jordan "associated eigenfunction" Φ_0 attached to each real ω. However, the argument in Part One [below Eq. (3.26)] can be applied to wave packets to suggest that that situation would contradict Eq. (3.11a).

So far we have been labelling the basis solutions by the incident momentum, p, and we have found precisely what one would expect in a one-dimensional scattering problem without bound states. It is only when one thinks of the solutions as labelled by the frequency, ω, that the situation perhaps looks strange. For $\omega > 1$ (recapitulating the results of the construction above) we have two positive-norm solutions,

$$\psi_\omega^{in}(z) \equiv {}^+\phi_{-\mu}^{in}(z)$$
$$= [R^{in}(\omega)e^{i\mu z} + e^{-i\mu z}]\theta(z) + T^{in}(\omega)e^{-i\xi z}\theta(-z) \tag{3.12}$$

and

$$\psi_\omega^{\mathrm{up}}(z) \equiv {}^+\phi_\xi^{\mathrm{in}}(z)$$
$$= T^{\mathrm{up}}(\omega)e^{i\mu z}\theta(z) + [e^{i\xi z} + R^{\mathrm{up}}(\omega)e^{-i\xi z}]\theta(-z). \quad (3.13)$$

Here μ and ξ (always positive) are given by Eqs. (3.3); the reflection and transmission coefficients $R^{\mathrm{in}}(\omega)$ and $T^{\mathrm{in}}(\omega)$ were denoted in the previous discussion by A and D respectively, while R^{up} and T^{up} were called D and A. The "in"–"up" notation[18] is inspired by the black hole application, in which a left-moving incident wave packet ($p < 0$ for $\epsilon > 0$) is incoming from spatial infinity, but a right-moving one ($p > 0$) is coming up out of the black hole at $z = -\infty$. To continue the catalogue, for $-1 < \omega < 1$ there is one mode, with $\epsilon > 0$,

$$\psi_\omega^{\mathrm{up}}(z) \equiv {}^+\phi_\xi^{\mathrm{in}}(z)$$
$$= A'(\omega)e^{-\eta z}\theta(z) + [e^{i\xi z} + D(\omega)e^{-i\xi z}]\theta(-z). \quad (3.14)$$

where $|D| = 1$. These solutions are bound from the right; packets formed from them rise up out of the "black hole" and then are totally reflected back in. Their antiparticle counterparts appear when $-1 - V < \omega < 1 - V$:

$$\psi_\omega^{\mathrm{in}}(z) \equiv {}^-\phi_\mu^{\mathrm{in}}(z)$$
$$= [e^{i\mu z} + B(\omega)e^{-i\mu z}]\theta(z) + C'(\omega)e^{\lambda z}\theta(-z), \quad (3.15)$$

where $\lambda = [1 - (\omega + V)^2]^{\frac{1}{2}}$ and $|B| = 1$. (The incident packet here is left-moving, since the group velocity $\mu/\omega \equiv p/\omega$ is negative.) These functions correspond to packets which do not succeed in penetrating the black hole from outside. For $\omega < -1 - V$ there are two modes, with $\epsilon < 0$:

$$\psi_\omega^{\mathrm{in}}(z) \equiv {}^-\phi_\mu^{\mathrm{in}}(z)$$
$$= [e^{i\mu z} + R^{\mathrm{in}}(\omega)e^{-i\mu z}]\theta(z) + T^{\mathrm{in}}(\omega)e^{i\xi z}\theta(-z), \quad (3.16)$$

$$\psi_\omega^{\mathrm{up}}(z) \equiv {}^-\phi_{-\xi}^{\mathrm{in}}(z)$$
$$= T^{\mathrm{up}}(\omega)e^{-i\mu z}\theta(z) + [R^{\mathrm{up}}(\omega)e^{i\xi z} + e^{-i\xi z}]\theta(-z). \quad (3.17)$$

[18]   P. L. Chrzanowski and C. W. Misner, *Phys. Rev. D* **10**, 1701 (1974), especially Fig. 1.

Finally, the most interesting interval is $1 - V < \omega < -1$, which falls in the Klein region of the $V - \omega$ plane. Here we have one positive-norm mode,

$$\psi_\omega^{\mathrm{up}}(z) \equiv {}^+\phi_\xi^{\mathrm{in}}(z)$$
$$= T^{\mathrm{up}}(\omega)e^{-i\mu z}\theta(z) + [e^{i\xi z} + R^{\mathrm{up}}(\omega)e^{i\xi z}]\theta(-z), \quad (3.18)$$

in which the wave is incident from the left (from inside the black hole), and one of negative norm,

$$\psi_\omega^{\mathrm{in}}(z) \equiv {}^-\phi_\mu^{\mathrm{in}}(z)$$
$$= [e^{i\mu z} + R^{\mathrm{in}}(\omega)e^{-i\mu z}]\theta(z) + T^{\mathrm{in}}(\omega)e^{-i\xi z}\theta(-z), \quad (3.19)$$

where the wave is incident from the right. T^{up} and R^{up} in Eq. (3.18) are respectively B and D as given in Eqs. (3.8). Note that $|T^{\mathrm{up}}|$ and $|R^{\mathrm{up}}|$ are greater than 1; that is, this mode is superradiant, as is the mode (3.19). Perhaps the most important feature of these modes, though, is that the sign of the norm is tied to the direction of incidence of the wave on the potential step: There are no antiparticle "up" modes in the Klein interval, and no particle "in" modes.

The decision to choose basis functions characterized by simple behavior in the far past is entirely arbitrary. An equally good alternative basis consists of functions whose square-integrable linear combinations consist of single packets in the limit $t \to +\infty$. These functions, naturally, we denote by ${}^\pm\phi_p^{\mathrm{out}}$, or by $\psi_\omega^{\mathrm{out}}$ for $p/\omega > 0$ and $\psi_\omega^{\mathrm{down}}$ for $p/(\omega + V) < 0$ (the latter class forming packets which end up falling down the black hole entirely). They have the same normalization as the ϕ_p^{in}, Eqs. (3.11).

Since the ${}^\pm\phi_p^{\mathrm{in}}$ are a complete set, each element of the *out*-basis must be linearly expressible in terms of the *in*-functions of the same ω. In this context the normalization condition (3.11a) is inconvenient, since two different values of p^2 correspond to the same ω. However, since $\delta(p - p') = |d\omega/dp|\,\delta(\omega - \omega')$, it is easy to see from Eqs. (3.3) that Eqs. (3.11) may be rewritten as

$$\langle {}^\pm\Phi_p, {}^\pm\Phi_{p'}\rangle = \pm 2(2\pi)^s|p|\,\delta(\omega - \omega'),$$
$$\langle {}^\pm\Phi_p, {}^\mp\Phi_{p'}\rangle = 0, \quad (3.20)$$

regardless of whether $|p| = \mu$ or $|p| = \xi$. (In the massless case this complication does not arise, since $|d\omega/dp| = 1$.) Equations (3.20) are valid for both the *out*-basis and the *in*-basis. The considerations of Part One, Sec. VI, can now be applied to the functions $[2(2\pi)^s|p|]^{-\frac{1}{2}\pm}\Phi_p$, which satisfy the continuous analogue of the orthonormality equations (6.2) of that section.

We shall now examine the relationship between *in*- and *out*-modes for all the categories of solutions. In the case of a semi-bound mode like (3.14), one sees immediately that $\psi_\omega^{\text{down}} = \psi_\omega^{\text{up}*}$ and that it is the same as ψ_ω^{up} except for a phase factor, so there is nothing more to be said.

As representative of the ordinary "free" modes we consider

$$\psi_\omega^{\text{out}} \equiv {}^+\phi_\mu^{\text{out}} \qquad (\omega > 1)$$
$$= [e^{i\mu z} + R^{\text{out}}e^{-i\mu z}]\theta(z) + T^{\text{out}}e^{i\xi z}\theta(-z). \qquad (3.21)$$

Since the coefficients are uniquely determined, and the wave equation satisfied by the $\phi(z)$ is real, we see from Eq. (3.12) that $\psi_\omega^{\text{out}} = \psi_\omega^{\text{in}*}$,

$$R^{\text{out}} = R^{\text{in}*}, \qquad T^{\text{out}} = T^{\text{in}*}. \qquad (3.22)$$

We must have

$$\mu^{-\frac{1}{2}}\psi_\omega^{\text{out}} = \alpha\xi^{-\frac{1}{2}}\psi_\omega^{\text{up}} + \beta\mu^{-\frac{1}{2}}\psi_\omega^{\text{in}}, \qquad (3.23)$$

where, to preserve the normalization,

$$|\alpha|^2 + |\beta|^2 = 1. \qquad (3.24)$$

(If the potential vanishes, then β is 0 and α is 1.) Hence, from Eqs. (3.12), (3.13), and (3.21), we obtain

$$\mu^{-\frac{1}{2}} = \alpha\xi^{-\frac{1}{2}}T^{\text{up}} + \beta\mu^{-\frac{1}{2}}R^{\text{in}}, \qquad \mu^{-\frac{1}{2}}R^{\text{in}*} = \beta\mu^{-\frac{1}{2}},$$

$$\mu^{-\frac{1}{2}}T^{\text{in}*} = \alpha\xi^{-\frac{1}{2}}, \qquad 0 = \alpha\xi^{-\frac{1}{2}}R^{\text{up}} + \beta\mu^{-\frac{1}{2}}T^{\text{in}}.$$

These four equations yield

$$\alpha = (\xi/\mu)^{\frac{1}{2}}T^{\text{in}*}, \qquad \beta = R^{\text{in}*}, \qquad (3.25)$$

and, as consistency conditions,

$$|R^{\text{in}}|^2 + T^{\text{up}}T^{\text{in}*} = 1, \qquad (3.26)$$

$$R^{\mathrm{up}} T^{\mathrm{in}*} = -T^{\mathrm{in}} R^{\mathrm{in}*}. \tag{3.27}$$

Equation (3.24) leads to

$$|R^{\mathrm{in}}|^2 + \frac{\xi}{\mu}|T^{\mathrm{in}}|^2 = 1, \tag{3.28}$$

whence Eqs. (3.26) and (3.27) are seen to be equivalent to

$$T^{\mathrm{up}} = \frac{\xi}{\mu} T^{\mathrm{in}}, \tag{3.29}$$

$$R^{\mathrm{in}*} T^{\mathrm{up}} = -\frac{\xi}{\mu} T^{\mathrm{in}*} R^{\mathrm{up}}. \tag{3.30}$$

Equations (3.28)–(3.30) also follow directly from the constancy (in z) of the Wronskians $W(\psi_\omega^{\mathrm{out}}, \psi_\omega^{\mathrm{in}})$, $W(\psi_\omega^{\mathrm{in}}, \psi_\omega^{\mathrm{up}})$, and $W(\psi_\omega^{\mathrm{out}}, \psi_\omega^{\mathrm{up}})$.

The case of a superradiant mode is closely parallel to the above in algebra, but vastly different in implications. We have, for example,

$$\begin{aligned} \psi_\omega^{\mathrm{out}} &\equiv {}^-\phi_{-\mu}^{\mathrm{out}} \qquad (1 - V < \omega < -1) \\ &= [R^{\mathrm{out}} e^{i\mu z} + e^{-i\mu z}]\theta(z) + T^{\mathrm{out}} e^{i\xi z}\theta(-z) \end{aligned} \tag{3.31}$$

and

$$\begin{aligned} \mu^{-\frac{1}{2}} \psi_\omega^{\mathrm{out}} &= \alpha\mu^{-\frac{1}{2}} \psi_\omega^{\mathrm{in}} + \beta\xi^{-\frac{1}{2}} \xi_\omega^{\mathrm{up}} \\ &= \alpha\mu^{-\frac{1}{2}-}\phi_\mu^{\mathrm{in}} + \beta\xi^{-\frac{1}{2}+}\phi_\xi^{\mathrm{in}}, \end{aligned} \tag{3.32}$$

where [Part One, Eq. (6.5)]

$$|\alpha|^2 - |\beta|^2 = 1. \tag{3.33}$$

Since there is no negative-norm up-mode for this value of ω, the mixing in of the positive-norm function $\psi_\omega^{\mathrm{up}}$ is unavoidable. Similarly, $\psi_\omega^{\mathrm{down}} \equiv {}^+\phi_{-\xi}^{\mathrm{out}}$ is a mixture of $\psi_\omega^{\mathrm{up}}$ with a bit of the negative-norm in-mode, $\psi_\omega^{\mathrm{in}}$ [see Eqs. (6.4) of Part One]. Continuing the calculation as before, one finds that Eqs. (3.22), (3.26), and (3.27) still apply, while Eqs. (3.25) and (3.28) are replaced by

$$\alpha = R^{\mathrm{in}*}, \qquad \beta = (\xi/\mu)^{\frac{1}{2}} T^{\mathrm{in}*}, \tag{3.34}$$

$$|R^{\mathrm{in}}|^2 - \frac{\xi}{\mu}|T^{\mathrm{in}}|^2 = 1. \tag{3.35}$$

[The latter in turn induces sign changes in Eqs. (3.29) and (3.30).] Since ξ/μ is always positive, these modes are always superradiant ($|R| > 1$), and the modes considered previously are not. These Wronskian relations are quite general, applying to any potential which becomes constant asymptotically; the calculations for black holes (Refs. 5, 11, 1) are based on them.

In the context of second quantization, the significance of Eq. (3.32) is entirely different from that of Eq. (3.23). The latter, and the comparison formulas for $\psi_\omega^{\text{down}}$ and for the modes with $\omega < -1 - V$, represent just a relabelling, by unitary transformation of the basis vectors, of the states in the single-particle and single-antiparticle Hilbert spaces. The annihilation operator for particles in the mode ψ_ω^{out} is a linear combination of the annihilation operators for ψ_ω^{up} and ψ_ω^{in}. Thus the stipulation that the vacuum state be annihilated by these operators means the same thing, regardless of whether the in-basis or the out-basis is used.

For the modes of the Klein interval, on the other hand, the passage from in-functions to out-functions corresponds to a mixing of annihilation and creation operators, as explained in Sec. VI of Part One. If one attempts to define a vacuum, and particles, by the usual formal characterizations, the result depends on which basis of modes has been adopted. Indeed, there is a whole family of rival vacuum states, since any transformation of the form [cf. Eq. (6.4), Part One]

$$
\begin{aligned}
\mu^{-\frac{1}{2}}\psi_\omega^{\text{right}} &= \alpha(\omega)\mu^{-\frac{1}{2}}\psi_\omega^{\text{in}} + \beta(\omega)\xi^{-\frac{1}{2}}\psi_\omega^{\text{up}}, \\
\xi^{-\frac{1}{2}}\psi_\omega^{\text{left}} &= \beta(\omega)^*\mu^{-\frac{1}{2}}\psi_\omega^{\text{in}} + \alpha(\omega)^*\xi^{-\frac{1}{2}}\psi_\omega^{\text{up}},
\end{aligned}
\tag{3.36}
$$

$$
|\alpha|^2 - |\beta|^2 = 1, \qquad 1 - V < \omega < -1,
$$

leaves invariant the orthonormality relations (3.20) but mixes creation and annihilation operators. This situation is characteristic of a spontaneously broken symmetry. The symmetry that is being broken here is time reversal; this, of course, is the resolution of the time reversal "paradox" raised in Sec. I. Incidentally, the Fock representations based on two vacuum states related through Eqs. (3.36) are not unitarily equivalent, since a diagonal Bogolubov transformation acting on a continuous index cannot have a Hilbert–Schmidt kernel.[19]

---

[19]   Literature on Bogolubov transformations and their unitary implementability is listed as Ref. 21 of B. L. Hu, S. A. Fulling, and L. Parker, *Phys. Rev. D* **8**, 2377 (1973).

Two essentially equivalent methods have been used to give to this vacuum degeneracy a physical interpretation in terms of particle creation. Both rest on the assumption that the vacuum associated with the "in" and "up" modes is a state in which no physical particles are present initially (i.e., as $t \to -\infty$). In the case of a massless field, one can say that there are no particles on $\mathcal{I}^-$ and the past horizon.

The approach of Hawking and Gibbons (Refs. 10 and 11) is to study the coefficients α and β in the Bogolubov transformation (3.32) relating the *in*-basis to the *out*-basis. These give the content of the *in*-vacuum in terms of the particle concept (conjugate to the "out" and "down" modes) which presumably is physically relevant at $t = +\infty$. Since the initial and final representations are not unitarily equivalent, the calculation leads to infinities. Heuristic arguments are used to convert the expressions into finite rates of particle emission per unit time.

The other method, used by Unruh and DeWitt (Refs. 5 and 1), is to calculate the expectation value in the *in*-vacuum of the fluxes of energy and of angular momentum or charge. In the asymptotic flat region of space-time, these nonvanishing fluxes are attributed to a current of real particles. We shall outline here the flux calculation in the two-dimensional massive model. For large positive z, where the potential vanishes, the energy flux density operator is

$$T^{tz} = -\frac{1}{2}\left[\left\{\frac{\partial \phi^\dagger}{\partial t}, \frac{\partial \phi}{\partial z}\right\} + \left\{\frac{\partial \phi^\dagger}{\partial z}, \frac{\partial \phi}{\partial t}\right\}\right], \qquad (3.37)$$

and the electric current density operator is

$$J^z = -\frac{1}{2}ie\left[\left\{\phi^\dagger, \frac{\partial \phi}{\partial z}\right\} - \left\{\frac{\partial \phi^\dagger}{\partial z}, \phi\right\}\right], \qquad (3.38)$$

where $\{\dots\}$ denotes the anticommutator.[20] In the rectilinear gravitational model (2.1) one would study T^{xz} (a flux of transverse momentum) instead of J^z; in the Kerr model the interesting quantities are

---

[20]   The overall minus signs in these formulas come from the raising of the z index. Apart from that, the expressions are spatial components of the objects whose time components are the integrands of Eqs. (2.28) and (2.26) of Part One.

the radial flux of energy, T^{tr}, and of angular momentum, $T^{\phi r}$. The calculations in all these cases are nearly identical.[21]

The expansion of the field operator in normal modes, as follows from the foregoing discussion, is

$$\phi(t,z) = (4\pi)^{-\frac{1}{2}} \int_{-\infty}^{\infty} dp \, (p^2+1)^{-\frac{1}{4}}$$
$$\times [a_p^{(in)} + \phi_p^{in}(z)e^{-i\omega_+(p)t} + b_p^{(in)\dagger} - \phi_p^{in}(z)e^{-i\omega_-(p)t}], \quad (3.39a)$$

where $[a_p^{(in)}, a_{p'}^{(in)\dagger}] = \delta(p-p')$, etc., and $\omega_\pm(p)$ are two functions rather tedious to describe [see Eqs. (3.3) and later references to them]. Alternatively, one may write

$$(4\pi)^{\frac{1}{2}}\phi(t,z) = \qquad\qquad\qquad\qquad\qquad\qquad\qquad (3.39b)$$

$$\int_1^\infty d\omega \, \mu^{-\frac{1}{2}} a_\omega^{in} \psi_\omega^{in}(3.12)e^{-i\omega t} + \int_{1-V}^{-1} d\omega \, \xi^{-\frac{1}{2}} a_\omega^{up} \psi_\omega^{up}(3.18)e^{-i\omega t}$$

$$+ \int_{-1}^1 d\omega \, \xi^{-\frac{1}{2}} a_\omega^{up} \psi_\omega^{up}(3.14)e^{-i\omega t} + \int_1^\infty d\omega \, \xi^{-\frac{1}{2}} a_\omega^{up} \psi_\omega^{up}(3.13)e^{-i\omega t}$$

$$+ \int_{-\infty}^{-1-V} d\omega \, \xi^{-\frac{1}{2}} b_\omega^{up\dagger} \psi_\omega^{up}(3.17)e^{-i\omega t} + \int_{1-V}^{-1} d\omega \, \mu^{-\frac{1}{2}} b_\omega^{in\dagger} \psi_\omega^{in}(3.19)e^{-i\omega t}$$

$$+ \int_{-1-V}^{1-V} d\omega \, \mu^{-\frac{1}{2}} b_\omega^{in\dagger} \psi_\omega^{in}(3.15)e^{-i\omega t} + \int_{-\infty}^{-1-V} d\omega \, \mu^{-\frac{1}{2}} b_\omega^{in\dagger} \psi_\omega^{in}(3.16)e^{-i\omega t}$$

where the normalization is $[a_\omega^{in}, a_{\omega'}^{in\dagger}] = \delta(\omega - \omega')$, etc. In Eq. (3.39b) the numbers after the ψ functions indicate the equations giving their explicit forms.

Substituting Eq. (3.39a) into Eq. (3.37), one obtains, for the expectation value in the in-vacuum,

$$\langle T^{tz} \rangle = -i(8\pi)^{-1} \int_{-\infty}^{\infty} dp \, (p^2+1)^{-\frac{1}{2}}$$
$$\times [\omega_+(p)(^+\phi^* {}^+\phi' - {}^+\phi {}^+\phi^{*\prime}) + \omega_-(p)(^-\phi^* {}^-\phi' - {}^-\phi {}^-\phi^{*\prime})], \quad (3.40)$$

where $^+\phi'$ means $d(^+\phi_p^{in})/dz$, etc. The variable of integration can be changed to ω, whereupon the integration element $(p^2+1)^{-\frac{1}{2}} dp$ becomes $|p|^{-1} d\omega$, with $|p| = \mu(\omega)$ or $\xi(\omega)$, whichever is appropriate.

---

[21] No divergences arise in these calculations, but, as pointed out in Ref. 5, the integrals are only conditionally convergent. In other words, we are forced to assume that $\infty - \infty = 0$, but not that $\infty = 0$.

The expression for $\langle J^z \rangle$ is like Eq. (3.40) with $\omega_\pm$ replaced by e. The Wronskians appearing in the integrand are easily calculated from Eqs. (3.12)–(3.19):

(3.12) and (3.9):	$-2i\xi|T^{\text{in}}|^2$
(3.13):	$2i\mu|T^{\text{up}}|^2$
(3.14) and (3.15):	0
(3.16):	$2i\xi|T^{\text{in}}|^2$
(3.17) and (3.18):	$-2i\mu|T^{\text{up}}|^2.$

$$(3.41)$$

But from Eq. (3.29) and its counterpart for Klein modes one has

$$\frac{\mu}{\xi}|T^{\text{up}}|^2 = \frac{\xi}{\mu}|T^{\text{in}}|^2. \tag{3.42}$$

Thus all classes of modes except the semibound ones satisfy

$$|p|^{-1} \times (\text{Wronskian}) = \pm 2i(\xi/\mu)|T^{\text{in}}|^2, \tag{3.43}$$

where the sign is $+$ for (3.13) and (3.16) and $-$ for (3.12), (3.17), (3.18), and (3.19).

When this result is put into Eq. (3.40) [the range of the ω integration for each class of mode functions being read off from Eq. (3.39b)], the contribution of the in-modes precisely cancels that of the up-modes both for ordinary particles [(3.12) and (3.13)] and for ordinary antiparticles [(3.16) and (3.17)]. However, both varieties of Klein mode come in with the same sign, yielding

$$\begin{aligned}
\langle T^{tz} \rangle &= -(2\pi)^{-1} \int_{1-V}^{-1} d\omega \, (\xi/\mu)\,\omega|T^{\text{in}}|^2 \\
&= -(2\pi)^{-1} \int_{1-V}^{-1} d\omega \, \omega(|R^{\text{in}}|^2 - 1) > 0
\end{aligned} \tag{3.44}$$

and

$$\langle J^z \rangle = -e(2\pi)^{-1} \int_{1-V}^{-1} d\omega \, (|R^{\text{in}}|^2 - 1) < 0. \tag{3.45}$$

That is, there is a strictly positive flux of energy away from the potential, suggesting a beam of particles. The charge carried by these particles is negative (if e is positive); this was to be expected, since the negative potential well [sign convention of Part One, Eq. (2.31)] should eject the negatively charged member of each created pair.

The functions of the *out*-basis, being complex conjugates of the "in" functions, have Wronskians of the opposite sign. Therefore, in the *out*-vacuum the expectation values of T^{tz} and J^z are the negatives of the quantities (3.44) and (3.45), as one would expect from the proposed physical interpretation of that state. It is also possible to find a vacuum state, among the infinitude of them indicated by Eqs. (3.36), in which there is no net flux at all. Indeed, any basis consisting of real functions and satisfying Eqs. (3.20) would yield that result. If $\alpha e^{-2i\theta}$ is an imaginary number, where α is the coefficient so denoted in Eq. (3.32), then

$$\psi_\omega^{\text{right}} = \sqrt{2}\,\text{Re}(e^{i\theta}\psi_\omega^{\text{in}}),$$
$$\psi_\omega^{\text{left}} = \sqrt{2}\,\text{Re}(e^{-i\theta}\psi_\omega^{\text{up}}),$$

$$(3.46)$$

is a basis with those properties.

IV. Physical discussion

In the foregoing discussion we tentatively accepted the interpretation of the *in*-vacuum as a state in which the system in some sense is initially devoid of particles, so that at finite times there are particles being radiated from the black hole, but no particles impinging on the hole from infinity. Statistical arguments[22] then indicate that this quantum state is a physically reasonable initial condition[23] for a Kerr black hole (under the assumption that a complete black hole with past horizon exists at all). First, among the possible states of the ambient scalar field, either a beam of incoming particles radially converging on the black hole or compensating fluxes inward and outward would be much less plausible than a purely outgoing flux, causally traceable to the presence of the black hole itself. Second, states such as that defined

---

[22] Compare the modern understanding of the second law of thermodynamics and of retardation of electromagnetic radiation from accelerated charges. See P. C. W. Davies, *The Physics of Time Asymmetry* (Surrey Univ. Press, London, 1974).

[23] The inequivalence of the associated Fock representations makes the choice of the "correct" vacuum state less a matter of convention than it would otherwise be. One cannot depend on building up a given physically realizable state from an arbitrary vacuum state by applying creation operators.

through Eqs. (3.46), which can be interpreted as exhibiting no particle creation, are quite exceptional because of the precise cancellation of inward and outward fluxes; the *in*-vacuum is qualitatively more typical of the general situation.

This particle interpretation needs to be examined critically, however. The virtual particles of quantum field theory cannot always be reasoned about as if they were real classical particles. A standing wave is a superposition of two running waves, and hence a flux calculated from it will consist of compensating terms; does this mean that a state annihilated by annihilation operators conjugate to standing-wave modes contains two beams of particles in opposite directions? One does not normally think of the ground state of a free field in a box in those terms. The state associated with Eqs. (3.46) is of that type. It exhibits no flux, just a tremendous vacuum polarization that amounts to a spreading of the energy and the charge or spin of the black hole out over a wide area, once and for all. Its interpretation as a physical no-particle state appears defensible. It is the natural limit of the ground states of the models treated in Part One, where the absence of any stationary nonzero flux is forced by the condition of reflection at the wall or walls of the box. (Presumably, however, a state exhibiting nontrivial flux *for a finite time* can be formed in the theory of a sufficiently large box.) Furthermore, it may well be the physical ground state of a black hole in various equilibrium configurations; consider, for example, a universe uniformly filled with negatively charged black holes at absolute zero temperature.

Resolution of this uncertainty will require a firmer physical interpretation of the states of a quantum field theory in curved spacetime. Such an improved understanding may come from an analysis of the action of particle detectors (see Ref. 26), or from formulation of the observable consequences of the theory in terms other than particle language. For example, attention has centered on the energy-momentum tensor in much recent work.

An argument in favor of the emission picture is that virtual particles may become real in a time-dependent situation. That is, the stationary calculations of Unruh and others probably represent, at least approximately, what happens in a more realistic but less soluble model. References 10 and 11 reached that conclusion in regard to gravitational collapse of a spinning or charged body. Even if one starts with a fully formed black hole, the influence of created pairs, as sources, on the gravitational and electromagnetic potentials would be expected to degrade or neutralize those potentials to some extent. The emitted flux would be left over, in almost empty (or at least static) space, as real particles.

Another basic physical issue is whether external-potential quantum field theories manifesting vacuum instabilities are really physically relevant. There seems to be a general feeling among field theorists that they do not deserve serious study, their unusual features being due to neglect of part of the physics of the problem. For instance, it is argued[24] that the Klein paradox merely marks the breakdown of the external-potential approximation, showing that a strong potential is necessarily self-destructive because of pair creation. A more subtle observation is that a theory whose energy spectrum is not bounded below is *unstable* in a metatheoretical sense: Its qualitative predictions must be expected to change violently if one adds new interactions. For example, if one adds to the Hamiltonian the energy of the external field itself, or of an interaction among the particles of the quantized field (as in Refs. 6 and 7 of Part One), then the energy may become bounded below and a stable vacuum exist.

Clearly, these arguments are just a plea to study the reaction problem mentioned earlier — that is, to calculate the mutual influence of the quantized "matter" field on the gravitational or electromagnetic field (itself quantized, ideally). This, of course, is precisely what researchers in this area will do, as soon as they know how to do it. In the meantime it seems worthwhile to pursue the external-potential approximation all the way, for several reasons: (1) As explained above, there is reason to believe that the results have some relevance to more adequate, but more difficult, models. (2) Full knowledge of the consequences of a model is interesting and valuable, if only for the purpose of understanding in what ways the model is wrong. (3) The unsolved mathematical questions raised by these theories, which have been alluded to, but not attacked, on several occasions in this paper, are fascinating in their own right.

A final problem is that of boundary conditions on the horizons of the black hole. In this work those surfaces have been treated in

---

[24]   J. M. Jauch and F. Rohrlich, *The Theory of Photons and Electrons* (Addison–Wesley, Cambridge, Mass., 1955), pp. 311–312.

complete parallel with the null surfaces at infinity, although in truth
they are perfectly nonsingular boundaries between the exterior region
of the black-hole geometry and another region of space-time which has
been excluded from consideration. The prototype of this situation is
the horizon in Minkowski space which develops in the Fermi coordinate
system of an accelerated observer. A continuing body of work[25] has
shown that going through the motions of static field quantization in this
context yields a "no-particle state" different from the usual Poincaré-
invariant vacuum. Recently Unruh[26] has elucidated the significance of
this phenomenon for black hole physics. Invoking statistical arguments
very similar to those stated at the beginning of this section, he argues
that the most plausible initial state of a quantized field is not a static
vacuum, but a state which manifests the Hawking particle creation,
even though the black hole is of the "eternal", externally stationary
variety. When the Kerr metric is treated from this point of view, the
black-body radiation of Hawking will be combined with the radiation
associated with quantization of superradiant modes.

---

[25]   S. A. Fulling, *Phys. Rev. D* **7**, 2850 (1973); W. G. Unruh, *Proc. Roy.
Soc. A* **338**, 517 (1974); A. diSessa, *J. Math. Phys.* **15**, 1892 (1974); C. M.
Sommerfield, *Ann. Phys.* (N.Y.) **84**, 285 (1974); P. C. W. Davies, *J. Phys.
A* **8**, 609 (1975); D. G. Boulware, *Phys. Rev. D* **11**, 1404 (1975); Ref. 1; P.
Candelas and D. J. Raine, *J. Math. Phys.* **17**, 2101 (1976); S. A. Fulling and
P. C. W. Davies, *Proc. Roy. Soc. A* **348**, 393 (1976).
[26]   W.G. Unruh, *Phys. Rev. D* **14**, 870 (1976).

Acknowledgments

Arthur Wightman several years ago introduced me to the work of Schiff et al. and Schroer and Swieca and pointed out the need for a clarification of the status of the Klein paradox in field theory. The immediate stimulus of the present work was provided by Larry Ford, who communicated a preliminary account of his research on the Kerr model. I am greatly indebted to David Robinson for some discussions of the Kerr solution and other time-independent space-times (see Part Two, Sec. II). Other valuable conversations involved Chris Isham, Lars-Erik Lundberg, Abhay Ashtekar, Paul Davies, Gary Gibbons, Brandon Carter, Bill Unruh, and, in earlier days, Barry Simon.

The computer facilities used were those of the University of London Computer Centre. The drawings were made with the program DIMFILM; I am most grateful to Olivia Rogers of ULCC and to the King's College Computer Advisory Staff for assistance.

Bibliography

Literature citations in the appendix appear there as footnotes and are not repeated here.

Abramowitz, M., & Stegun, I. A., eds. (1968) *Handbook of Mathematical Functions with Formulas, Graphs, and Mathematical Tables* (National Bureau of Standards Applied Mathematics Series, No. 55), 7th printing, U.S. Government Printing Office, Washington.

Adler, S. L., & Lieberman, J. (1978) Trace anomaly of the stress-energy tensor for massless vector particles propagating in a general background metric, *Ann. Phys.* (N.Y.) **113**, 294–303.

Adler, S. L., Lieberman, J., & Ng, Y. J (1977) Regularization of the stress-energy tensor for vector and scalar particles propagating in a general background metric, *Ann. Phys.* (N.Y.) **106**, 279–321.

Allen, B., Folacci, A., & Ottewill, A. C. (1988) Renormalized graviton stress-energy tensor in curved vacuum space-times, *Phys. Rev. D* **38**, 1069–1082.

Allendoerfer, C. B. (1940) The Euler number of a Riemann manifold, *Amer. J. Math.* **62**, 243–248.

Ambjørn, J., & Wolfram, S. (1983) Properties of the vacuum. 2, *Ann. Phys.* (N.Y.) **147**, 33–56.

Anderson, P. R., & Parker, L. (1987) Adiabatic regularization in closed Robertson-Walker universes, *Phys. Rev. D* **36**, 2963–2969.

Ashtekar, A., & Magnon, A. (1975a) Quantum fields in curved space-times, *Proc. Roy. Soc. A* **346**, 375–394.

Ashtekar, A., & Magnon-Ashtekar, A. (1975b) Sublimation d'ergosphères, *Compt. Rend. Acad. Sci.* (Paris) *A* **281**, 875–878.

Atiyah, M., Bott, R., & Patodi, V. K. (1973) On the heat equation and the index theorem, *Invent. Math.* **19**, 279–330.

Avis, S. J., Isham, C. J., & Storey, D. (1978) Quantum field theory in anti-deSitter space-time, *Phys. Rev. D* **18**, 3565–3576.

Balian, R., & Bloch, C. (1970) Distribution of eigenfrequencies for the wave equation in a finite domain. I, *Ann. Phys.* (N.Y.) **60**, 401–447; erratum ibid. **84**, 559 (1974).

Balian, R., & Bloch, C. (1972) Distribution of eigenfrequencies for the wave equation in a finite domain. III, *Ann. Phys.* (N.Y.) **69**, 76–160.

Balian, R., & Bloch, C. (1974)   Solution of the Schrödinger equation in terms of classical paths, *Ann. Phys.* (N.Y.) **85**, 514–545.

Banks, T. (1985)   *TCP*, quantum gravity, the cosmological constant and all that, *Nucl. Phys. B* **249**, 332–360.

Barth, N. H., & Christensen, S. M. (1983)   Quantizing fourth order gravity theories: The functional integral, *Phys. Rev. D* **28**, 1876–1893.

Beals, R. (1967)   On eigenfunction expansions for elliptic operators, *Ill. J. Math.* **11**, 663–668.

Bekenstein, J. D., & Meisels, A. (1977)   Einstein A and B coefficients for a black hole, *Phys. Rev. D* **15**, 2775–2781.

Bell, J. S., & Leinaas, J. M. (1983)   Electrons as accelerated thermometers, *Nucl. Phys. B* **212**, 131–150.

Bender, C. M., & Hays, P. (1976)   Zero-point energy of fields in a finite volume, *Phys. Rev. D* **14**, 2622–2632.

Berezanskiĭ, Ju. M. (1968)   *Expansions in Eigenfunctions of Selfadjoint Operators* (Transl. Math. Monogr., Vol. 17), American Mathematical Society, Providence, R.I.

Bilodeau, A. (1977)   *Creation de Particules dans un Modele Rectiligne de la Metrique de Kerr*, thesis (Doctorat de troisième cycle), Université Louis Pasteur de Strasbourg.

Birrell, N. D. (1978)   The application of adiabatic regularization to calculations of cosmological interest, *Proc. Roy. Soc. A* **361**, 513–526.

Birrell, N. D., & Davies, P. C. W. (1982)   *Quantum Fields in Curved Space*, Cambridge University press.

Bisognano, J. J., & Wichmann, E. H. (1975)   On the duality condition for a Hermitian scalar field, *J. Math. Phys.* **16**, 985–1007.

Bjorken, J. D., & Drell, S. D. (1965)   *Relativistic Quantum Fields*, McGraw–Hill, New York.

Bogolubov, N. N., Logunov, A. A., & Todorov, I. T. (1975)   *Introduction to Axiomatic Quantum Field Theory*, Benjamin, Reading, Mass.

Bogoliubov, N. N., & Shirkov, D. V. (1959)   *Introduction to the Theory of Quantized Fields*, Wiley, New York.

Bokobza-Haggiag, J. (1969)   Operateurs pseudo-différentiels sur une variété différentiable, *Ann. Inst. Fourier* (Grenoble) **19**, 125–177.

Bonic, R. A. (1969)   *Linear Functional Analysis*, Gordon and Breach, New York.

Boulware, D. G. (1975a)   Quantum field theory in Schwarzschild and Rindler spaces, *Phys. Rev. D* **11**, 1404–1423.

Boulware, D. G. (1975b) Spin $\frac{1}{2}$ quantum field theory in Schwarzschild space, *Phys. Rev. D* **12**, 350–367.

Boulware, D. G. (1979) Quantum field theory renormalization in curved space-time, *Recent Developments in Gravitation: Cargèse 1978*, ed. by M. Lévy and S. Deser, Plenum, New York, pp. 175–217.

Boyer, T. H. (1970) Quantum zero-point energy and long-range forces, *Ann. Phys.* (N.Y.) **56**, 474–503.

Brown, M. R. (1978) Actions and anomalies, unpublished (preprint, Center for Relativity, University of Texas).

Brown, M. R. (1984) Symmetric Hadamard series, *J. Math. Phys.* **25**, 136–140.

Brown, M. R., & Ottewill, A. C. (1986) Photon propagators and the definition and approximation of renormalized stress tensors in curved space-time, *Phys. Rev. D* **34**, 1776–1786.

Brownell, F. H. (1955) An extension of Weyl's asymptotic law for eigenvalues, *Pacific J. Math.* **5**, 483–499.

Bunch, T. S., Christensen, S. M., & Fulling, S. A. (1978) Massive quantum field theory in two-dimensional Robertson–Walker space-time, *Phys. Rev. D* **18**, 4435–4459.

Bunch, T. S., & Davies, P. C. W. (1978) Quantum field theory in de Sitter space: Renormalization by point-splitting, *Proc. Roy. Soc. A* **360**, 117–134.

Bunch, T. S., Panangaden, P., & Parker, L. (1980) On renormalization of $\lambda\phi^4$ field theory in curved spacetime. I, *J. Phys. A* **13**, 901–918.

Callan, C. G., Coleman, S., & Jackiw, R. (1970) A new improved energy-momentum tensor, *Ann. Phys.* (N.Y.) **59**, 42–73.

Campbell, J. A. (1972) Computation of a class of functions useful in the phase-integral approximation. I, *J. Comput. Phys.* **10**, 308–315.

Candelas, P. (1980) Vacuum polarization in Schwarzschild spacetime, *Phys. Rev. D* **21**, 2185–2202.

Candelas, P. (1982) Vacuum polarization in the presence of dielectric and conducting surfaces, *Ann. Phys.* (N.Y.) **143**, 241–295.

Candelas, P., & Deutsch, D. (1977) On the vacuum stress induced by uniform acceleration, *Proc. Roy. Soc. A* **354**, 79–99.

Candelas, P., & Deutsch, D. (1978) Fermion fields in accelerated states, *Proc. Roy. Soc. A* **362**, 251–262.

Candelas, P., & Howard, K. W. (1984) Vacuum $\langle\phi^2\rangle$ in Schwarzschild space-time, *Phys. Rev. D* **29**, 1618–1625.

Casimir, H. B. G. (1948) On the attraction between two perfectly conducting plates, *Konink. Nederl. Akad. Weten., Proc. Sec. Sci.* **51**, 793–795.

Castagnino, M. A., & Harari, D. D. (1984) Hadamard renormalization in curved space-time, *Ann. Phys.* (N.Y.) **152**, 85–104.

Chakraborty, B. (1973) The mathematical problem of reflection solved by an extension of the WKB method, *J. Math. Phys.* **14**, 188–190.

Chern, S.-S. (1944) A simple intrinsic proof of the Gauss–Bonnet formula for closed Riemannian manifolds, *Ann. Math.* **45**, 747–752.

Chern, S.-S. (1945) On Riemannian manifolds of four dimensions, *Bull. Amer. Math. Soc.* **51**, 964–971.

Chern, S.-S. (1962) Geometry of a quadratic differential form, *J. Soc. Indust. Appl. Math.* **10**, 751–755.

Chernoff, P. R. (1973) Essential self-adjointness of powers of generators of hyperbolic equations, *J. Func. Anal.* **12**, 401–414.

Choquet-Bruhat, Y. (1968) Hyperbolic partial differential equations on a manifold, *Battelle Rencontres: 1967 Lectures in Mathematics and Physics*, ed. by C. M. DeWitt and J. A. Wheeler, Benjamin, New York, pp. 84–106.

Choquet-Bruhat, Y., DeWitt-Morette, C., & Dillard-Bleick, M. (1977) *Analysis, Manifolds and Physics*, North-Holland, Amsterdam.

Christensen, S. M. (1976) Vacuum expectation value of the stress tensor in an arbitrary curved background: The covariant point-separation method, *Phys. Rev. D* **14**, 2490–2501.

Christensen, S. M. (1978) Regularization, renormalization, and covariant geodesic point-separation, *Phys. Rev. D* **17**, 946–963.

Christensen, S. M., & Fulling, S. A. (1977) Trace anomalies and the Hawking effect, *Phys. Rev. D* **15**, 2088–2104.

Christensen, S. M., & Parker, L. (1989) *MathTensor, A System for Doing Tensor Analysis by Computer*, manual and software written in the *Mathematica* computer mathematics system, to appear.

Clark, C. (1967) The asymptotic distribution of eigenvalues and eigenfunctions for elliptic boundary value problems, *SIAM Rev.* **9**, 627–646.

Coleman, S., & Weinberg, E. (1973) Radiative corrections as the origin of spontaneous symmetry breaking, *Phys. Rev. D* **7**, 1888-1910.

Davies, B. (1972) Quantum electromagnetic zero-point energy of a conducting spherical shell, *J. Math. Phys.* **13**, 1324–1328.

Davies, P. C. W. (1977) Quantum vacuum stress without regularization in two-dimensional spacetime, *Proc. Roy. Soc. A* **354**, 529–532.

Davies, P. C. W., Fulling, S. A., Christensen, S. M., & Bunch, T. S. (1977) Energy-momentum tensor of a massless scalar quantum field in a Robertson–Walker universe, *Ann. Phys.* (N.Y.) **109**, 108–142.

Dean, C. E., & Fulling, S. A. (1982)   Continuum eigenfunction expansions and resonances: A simple model, *Amer. J. Phys.* **50**, 540–544.

Deutsch, D., & Candelas, P. (1979)   Boundary effects in quantum field theory, *Phys. Rev. D* **20**, 3063–3080.

Deutsch, D., & Najmi, A.-H. (1983)   Construction of states for quantum fields in nonstatic space-times, *Phys. Rev. D* **28**, 1907–1915.

DeWitt, B. S. (1965)   *The Dynamical Theory of Groups and Fields*, Gordon and Breach, New York.

DeWitt, B. S. (1975)   Quantum field theory in curved spacetime, *Phys. Reports* **19**, 295–357.

DeWitt, B. S. (1984)   The spacetime approach to quantum field theory, *Relativity, Groups and Topology II*, ed. by B. S. DeWitt and R. Stora, North-Holland, Amsterdam, pp. 381–738.

DeWitt, B. S., & Brehme, R. W. (1960)   Radiation damping in a gravitational field, *Ann. Phys.* (N.Y.) **9**, 220–259.

Dimock, J. (1979)   Scalar quantum field in an external gravitational field, *J. Math. Phys.* **20**, 2549–2555.

Dimock, J. (1980)   Algebras of local observables on a manifold, *Commun. Math. Phys.* **77**, 219–228.

Dimock, J. (1982)   Dirac quantum fields on a manifold, *Transac. Amer. Math. Soc.* **269**, 133–147.

Dimock, J., & Kay, B. S. (1986)   Classical and quantum scattering theory for linear scalar fields on the Schwarzschild metric. II, *J. Math. Phys.* **27**, 2520–2525.

Dimock, J., & Kay, B. S. (1987)   Classical and quantum scattering theory for linear scalar fields on the Schwarzschild metric. I, *Ann. Phys.* (N.Y.) **175**, 366–426.

Dowker, J. S. (1978)   Thermal properties of Green's functions in Rindler, de Sitter, and Schwarzschild spaces, *Phys. Rev. D* **18**, 1856–1860.

Drager, L. D. (1978)   *On the Intrinsic Symbol Calculus for Pseudo-Differential Operators on Manifolds*, Ph.D. dissertation, Brandeis University.

Duff, M. J. (1981)   Inconsistency of quantum field theory in curved spacetime, *Quantum Gravity 2: A Second Oxford Symposium*, ed. by C. J. Isham, R. Penrose, and D. W. Sciama, Oxford University Press, Oxford, pp. 81–105.

Duistermaat, J. J., & Guillemin, V. W. (1975)   The spectrum of positive elliptic operators and periodic bicharacteristics, *Invent. Math.* **29**, 39–79.

Dunford, N., & Schwartz, J. T. (1963)   *Linear Operators*, Part II, Wiley, New York.

Eguchi, T., Gilkey, P. B., & Hanson, A. J. (1980)   Gravitation, gauge theories and differential geometry, *Phys. Reports* **66**, 213–393.

Eidel'man, S. D. (1969)   *Parabolic Systems*, North-Holland, Amsterdam.

Epstein. D. B. A. (1975)   Natural tensors on Riemannian manifolds, *J. Diff. Geom.* **10**, 631–645.

Fierz, M. (1960)   Zur Anziehung leitender Ebenen im Vakuum, *Helv. Phys. Acta* **33**, 855–858.

Figari, R., Höegh-Krohn, R., & Nappi, C. R. (1975)   Interacting relativistic boson fields in the de Sitter universe with two space-time dimensions, *Commun. Math. Phys.* **44**, 265–278.

Folland, G. B. (1976)   *Introduction to Partial Differential Equations*, Princeton University Press, Princeton, N.J.

Fredenhagen, K., & Haag, R. (1987)   Generally covariant quantum field theory and scaling limits, *Commun. Math. Phys.* **108**, 91–115.

Friedlander, F. G. (1975)   *The Wave Equation on a Curved Space-Time*, Cambridge University Press.

Friedlander, F. G. (1982)   *Introduction to the Theory of Distributions*, Cambridge University Press.

Friedman, A. (1964)   *Partial Differential Equations of Parabolic Type*, Prentice–Hall, Englewood Cliffs, N.J.

Fröman, N. (1966)   Outline of a general theory for higher order approximations of the JWKB-type, *Arkiv Fysik* **32**, 541–548.

Fulling, S. A. (1972)   *Scalar Quantum Field Theory in a Closed Universe of Constant Curvature*, Ph. D. dissertation, Princeton University.

Fulling, S. A. (1973)   Nonuniqueness of canonical field quantization in Riemannian space-time, *Phys. Rev. D* **7**, 2850–2862.

Fulling, S. A. (1976)   Varieties of instability of a boson field in an external potential, *Phys. Rev. D* **14**, 1939–1943.

Fulling, S. A. (1977)   Alternative vacuum states in static space-times with horizons, *J. Phys. A* **10**, 917–951.

Fulling, S. A. (1979)   Remarks on positive frequency and Hamiltonians in expanding universes, *Gen. Rel. Grav.* **10**, 807–824.

Fulling, S. A. (1982)   The local geometric asymptotics of continuum eigenfunction expansions. I, *SIAM J. Math. Anal.* **13**, 891–912.

Fulling, S. A. (1983)   Two-point functions and renormalized observables, *Gauge Theory and Gravitation* (Lecture Notes in Physics, Vol. 176), ed. by K. Kikkawa, N. Nakanishi, and H. Nariai, Springer, Berlin, pp. 101–106.

Fulling, S. A. (1984)   What have we learned from quantum field theory in curved space-time?, *Quantum Theory of Gravitation*, ed. by S. M. Christensen, Hilger Ltd., Bristol, pp. 42–52.

Fulling, S. A. (1989)   The analytic approach to recursion relations, *J. Symbolic Comput.*, in press.

Fulling, S. A., & Kennedy, G. (1988)   The resolvent parametrix of the general elliptic linear differential operator: A closed form for the intrinsic symbol, *Transac. Amer. Math. Soc.* **310**, 583–617.

Fulling, S. A., Narcowich, F. J., & Wald, R. M. (1981)   Singularity structure of the two-point function in quantum field theory in curved spacetime. II, *Ann. Phys.* (N.Y.) **136**, 243–272.

Fulling, S. A., & Parker, L. (1974)   Renormalization in the theory of a quantized scalar field interacting with a Robertson–Walker space-time, *Ann. Phys.* (N.Y.) **87**, 176–204.

Fulling, S. A., Parker, L., & Hu, B. L. (1974)   Conformal energy-momentum tensor in curved spacetime: Adiabatic regularization and renormalization, *Phys. Rev. D* **10**, 3904–3924; erratum ibid. **11**, 1714.

Fulling, S. A., & Ruijsenaars, S. N. M. (1987)   Temperature, periodicity and horizons, sl Phys. Reports **152**, 135–176.

Fulling, S. A., Sweeny, M., & Wald, R. M. (1978)   Singularity structure of the two-point function in quantum field theory in curved spacetime, *Commun. Math. Phys.* **63**, 257–264.

Garabedian, P. R. (1964)   *Partial Differential Equations*, Wiley, New York.

Gårding, L. (1954)   Eigenfunction expansions connected with elliptic differential operators, *Medd. Lunds Univ. Mat. Sem.* **12**, 44–55.

Gel'fand, I. M., & Shilov, G. E. (1964)   *Generalized Functions*, Vol. 1, *Properties and Operations*, Academic Press, New York.

Gel'fand, I. M., & Shilov, G. E. (1967)   *Generalized Functions*, Vol. 3, *Theory of Differential Equations*, Academic Press, New York.

Geroch, R. (1970)   Domain of dependence, *J. Math. Phys.* **11**, 437–449.

Gibbons, G. W., & Hawking, S. W. (1977)   Cosmological event horizons, thermodynamics, and particle creation, *Phys. Rev. D* **15**, 2738–2751.

Gibbons, G. W., & Perry, M. J. (1978)   Black holes and thermal Green's functions, *Proc. Roy. Soc.* **A358**, 467–494.

Gilkey, P. B. (1974)   *The Index Theorem and the Heat Equation*, Publish or Perish, Boston.

Gilkey, P. B. (1975)   The spectral geometry of a Riemannian manifold, *J. Diff. Geom.* **10**, 601–618.

Gilkey, P. B. (1979)   Recursion relations and the asymptotic behavior of the eigenvalues of the Laplacian, *Compos. Math.* **38**, 201–240.

Gilkey, P. B. (1984)   *Invariance Theory, the Heat Equation, and the Atiyah-Singer Index Theorem*, Publish or Perish, Wilmington.

Glimm, J., & Jaffe, A. (1981)   *Quantum Physics: A Functional Integral Point of View*, Springer, New York.

Green, M. B., Schwarz, J. H., & Witten, E. (1987)   *Superstring Theory*, Vols. 1–2, Cambridge University Press.

Greiner, P. (1971)   An asymptotic expansion for the heat equation, *Arch. Rat. Mech. Anal.* **41**, 163–218.

Greiner, W., Müller, B., & Rafelski, J. (1985)   *Quantum Electrodynamics of Strong Fields*, Springer, Berlin.

Grossman, A., Loupias, G. & Stein, E. M. (1968)   An algebra of pseudodifferential operators and quantum mechanics in phase space, *Ann. Inst. Fourier (Grenoble)* **18**, 343–368.

Hadamard, J. (1952)   *Lectures on Cauchy's Problem in Linear Differential Equations*, Dover, New York.

Hájíček, P. (1977)   Theory of particle detection in curved spacetimes, *Phys. Rev. D* **15**, 2757–2774.

Halliwell, J. J. (1987)   Correlations in the wave function of the universe, *Phys. Rev. D* **36**, 3626–3640.

Halmos, P. R., & Sunder, V. S. (1978)   *Bounded Integral Operators on L^2 Spaces*, Springer, Berlin.

Harary, F. (1969)   *Graph Theory*, Addison–Wesley, Reading, Mass.

Hardy, G. H. (1949)   *Divergent Series*, Oxford University Press, Oxford.

Hartle, J. B., & Hawking, S. W. (1976)   Path-integral derivation of black-hole radiance, *Phys. Rev. D* **13**, 2188–2203.

Hawking, S. W. (1974)   Black hole explosions?, *Nature* **248**, 30–31.

Hawking, S. W. (1975)   Particle Creation by Black Holes, *Commun. Math. Phys.* **43**, 199–220.

Hawking, S. W., & Ellis, G. F. R. (1973)   *The Large Scale Structure of Space-Time*, Cambridge University Press.

Hilgevoord, J. (1960)   *Dispersion Relations and Causal Description*, North-Holland, Amsterdam.

Hill, E. L. (1951)   Hamilton's principle and the conservation theorems of mathematical physics, *Rev. Mod. Phys.* **23**, 253–260.

Hörmander, L. (1968)   The spectral function of an elliptic operator, *Acta Math.* **121**, 193–218.

Hörmander, L. (1979)   The Weyl calculus of pseudodifferential operators, *Comm. Pure Appl. Math.* **32**, 359–443.

Horowitz, G. T. (1981)   Is flat spacetime unstable?, *Quantum Gravity 2: A Second Oxford Symposium*, ed. by C. J. Isham, R. Penrose, and D. W. Sciama, Oxford University Press, Oxford, pp. 106–130.

Horowitz, G. T., & Wald, R. M. (1978)   Dynamics of Einstein's equation modified by a higher order derivative term, *Phys. Rev. D* **17**, 414–416.

Howard, K. W. (1984)   Vacuum $\langle T_\mu{}^\nu \rangle$ in Schwarzschild spacetime, *Phys. Rev. D* **30**, 2532–2547.

Howard, K. W., & Candelas, P. (1984)   Quantum stress tensor in Schwarzschild space-time, *Phys. Rev. Lett.* **53**, 403–406.

Hu, B. L. (1978)   Calculation of the trace anomaly of the conformal energy-momentum tensor in Kasner spacetime by adiabatic regularization, *Phys. Rev. D* **18**, 4460–4470.

Hu, B. L. (1979)   Trace anomaly of the energy-momentum tensor of quantized scalar fields in Robertson–Walker spacetime, *Phys. Lett. A* **71**, 169–173.

Hu, B. L., & O'Connor, D. J. (1987)   Symmetry behavior in curved space-time: Finite-size effect and dimensional reduction, *Phys. Rev. D* **36**, 1701–1715.

Ikebe, T. (1960)   Eigenfunction expansions associated with the Schroedinger operators and their applications to scattering theory, *Arch. Rat. Mech. Anal.* **5**, 1–34.

Isham, C. J. (1978)   Quantum field theory in curved space-times: A general mathematical framework, *Differential Geometrical Methods in Mathematical Physics II* (Lec. Notes in Math., Vol. 676), ed. by K. Bleuler, H. R. Petry, and A. Reetz, Springer, Berlin.

Israel, W. (1976)   Thermo-field dynamics of black holes, *Phys. Lett. A* **57**, 107–110.

Kac, M. (1966)   Can one hear the shape of a drum?, *Amer. Math. Monthly* **73**, 1–23.

Kay, B. S. (1978)   Linear spin-zero quantum fields in external gravitational and scalar fields. I, *Commun. Math. Phys.* **62**, 55–70.

Kay, B. S. (1985)   The double-wedge algebra for quantum fields on Schwarzschild and Minkowski spacetimes, *Commun. Math. Phys.* **100**, 57–81; erratum ibid., in press.

Kennedy, G., Critchley, R., & Dowker, J. S. (1980)   Finite temperature field theory with boundaries: Stress tensor and surface action renormalization, *Ann. Phys.* (N.Y.) **125**, 346–400.

Kodaira, K. (1949)   The eigenvalue problem for ordinary differential equations of the second order and Heisenberg's theory of S-matrices, *Amer. J. Math.* **71**, 921–945.

Krein, S. G. (1971) *Linear Differential Equations in Banach Space* (Transl. Math. Monogr., Vol. 29), American Mathematical Society, Providence, R.I.

Kristensen, P., Mejlbo, L., & Poulsen, E. T. (1967) Tempered distributions in infinitely many dimensions. III, *Commun. Math. Phys.* **6**, 29–48.

Lee, H. W., & Pac, P. Y. (1986) Higher-derivative operators and DeWitt's WKB ansatz, *Phys. Rev. D* **33**, 1012–1017.

Lee, H. W., Pac, P. Y., & Shin, H. K. (1987) New algorithm for asymptotic expansions of the heat kernel, *Phys. Rev. D* **35**, 2440–2447.

Leray, J. (1953) *Hyperbolic Differential Equations*, lecture notes, Institute for Advanced Study, Princeton, N.J.

Lichnerowicz, A. (1964) Propagateurs et quantification en relativité générale, *Conference on Relativistic Theories of Gravitation* (Warsaw and Jabłonna, 1962), ed. by L. Infeld, Gauthier–Villars, Paris, pp. 177–188.

Littlewood, J. E. (1963) Lorentz's pendulum problem, *Ann. Phys.* (N.Y.) **21**, 233–242.

Lukash, V. N., Novikov, I. D., Starobinsky, A. A., & Zel'dovich, Ya. B. (1976) Quantum effects and evolution of cosmological models, *Nuovo Cim. B* **35**, 293–307.

Mandl, F. (1959) *Introduction to Quantum Field Theory*, Wiley, New York.

Manogue, C. A. (1988) The Klein paradox and superradiance, *Ann. Phys.* (N.Y.) **181**, 261–283.

Maslov, V. P. & Fedoriuk, M. V. (1981) *Semi-Classical Approximation in Quantum Mechanics*, Reidel, Dordrecht.

Maurin, K. (1967) *Methods of Hilbert Spaces*, Polish Scientific Publishers, Warsaw.

Maurin, K. (1968) *General Eigenfunction Expansions and Unitary Representations of Topological Groups*, Polish Scientific Publishers, Warsaw.

Mautner, F. J. (1953) On eigenfunction expansions, *Proc. Natl. Acad. Sci. USA* **39**, 49–53.

McKean, H. P., & Singer, I. M. (1967) Curvature and the eigenvalues of the Laplacian, *J. Diff. Geom.* **1**, 43–69.

Messiah, A. (1961) *Quantum Mechanics*, Vols. 1–2, North-Holland, Amsterdam.

Minakshisundaram, S. (1953) Eigenfunctions on Riemannian manifolds, *J. Indian Math. Soc.* **17**, 159–165.

Minakshisundaram, S., & Pleijel, Å. (1949) Some properties of the eigenfunctions of the Laplace-operator on Riemannian manifolds, *Canad. J. Math.* **1**, 242–256.

Moyal, J. E. (1949)   Quantum mechanics as a statistical theory, *Proc. Camb. Phil. Soc.* **45**, 99–124.

Najmi, A.-H., & Ottewill, A. C. (1985)   Quantum states and the Hadamard form. III, *Phys. Rev. D* **32**, 1942–1948.

Newton, T. D., & Wigner, E. P. (1949)   Localized states for elementary systems, *Rev. Mod. Phys.* **21**, 400–406.

Olver, F. W. J. (1961)   Error bounds for the Liouville–Green (or WKB) approximation, *Proc. Camb. Phil. Soc.* **57**, 790–810.

O'Neill, B. (1966)   *Elementary Differential Geometry*, Academic Press, New York.

O'Neill, B. (1983)   *Semi-Riemannian Geometry with Applications to Relativity*, Academic Press, New York.

Osborn, T. A., & Fujiwara, Y. (1983)   Time evolution kernels: Uniform asymptotic expansions, *J. Math. Phys.* **24**, 1093–1103.

Osborn, T. A., Papiez, L., & Corns, R. (1987)   Constructive representations of propagators for quantum systems with electromagnetic fields, *J. Math. Phys.* **28**, 103–123.

Padmanabhan, T. (1989)   Semi-classical approximations for gravity and the issue of backreaction, to appear.

Pais, A. (1982)   *"Subtle is the Lord ... "*, Oxford University Press, Oxford.

Panangaden, P. (1979)   Positive and negative frequency decompositions in curved spacetime, *J. Math. Phys.* **20**, 2506–2510.

Papiez, L., Osborn, T. A., & Molzahn, F. H. (1988)   Quantum systems with external electromagnetic fields: The large mass asymptotics, *J. Math. Phys.* **29**, 642–659.

Parker, L. E. (1966)   *The Creation of Particles in an Expanding Universe*, Ph. D. thesis, Harvard University.

Parker, L. (1968)   Particle creation in expanding universes, *Phys. Rev. Lett.* **21**, 562–564.

Parker, L. (1969)   Quantized fields and particle creation in expanding universes. I, *Phys. Rev.* **183**, 1057–1068.

Parker, L. (1971)   Quantized fields and particle creation in expanding universes. II, *Phys. Rev. D* **3**, 346–356.

Parker, L. (1976)   Thermal radiation produced by the expansion of the universe, *Nature* **261**, 20–23.

Parker, L. (1979)   Aspects of quantum field theory in curved spacetime: Effective action and energy-momentum tensor, *Recent Developments in Gravitation: Cargèse 1978*, ed. by M. Lévy and S. Deser, Plenum, New York, pp. 219–273.

Parker, L., & Fulling, S. A. (1974)   Adiabatic regularization of the energy-momentum tensor of a quantized field in homogeneous spaces, *Phys. Rev. D* **9**, 341–354.

Peierls, R. E. (1952)   The commutation laws of relativistic field theory, *Proc. Roy. Soc. A* **214**, 143–157.

Petersen, B. E. (1983)   *Introduction to the Fourier Transform and Pseudo-Differential Operators*, Pitman, Boston.

Powers, R. T. (1967)   Absence of interaction as a consequence of good ultraviolet behavior in the case of a local Fermi field, *Commun. Math. Phys.* **4**, 145–156.

Putnam, C. R. (1967)   *Commutation Properties of Hilbert Space Operators and Related Topics*, Springer, New York.

Rafelski, J., Fulcher, L. P., and Klein, A. (1978)   Fermions and bosons interacting with arbitrarily strong external fields, *Phys. Reports* **38**, 227–361.

Reed, M., & Simon, B. (1972)   *Methods of Modern Mathematical Physics I: Functional Analysis*, Academic Press, New York.

Reed, M., & Simon, B. (1975)   *Methods of Modern Mathematical Physics II: Fourier Analysis, Self-Adjointness*, Academic Press, New York.

Reed, M., and Simon, B. (1978)   *Methods of Modern Mathematical Physics IV: Analysis of Operators*, Academic Press, New York.

Richtmyer, R. D. (1978)   *Principles of Advanced Mathematical Physics*, Vol. 1, Springer, New York.

Rodionov, A. Ya., & Taranov, A. Yu. (1987)   Computation of covariant derivatives of the geodetic interval with coincident arguments, *Class. Quantum Grav.* **4**, 1767–1775.

Ruijsenaars, S. N. M. (1981)   On Newton–Wigner localization and superluminal propagation speeds, *Ann. Phys.* (N.Y.) **137**, 33–43.

Rumpf, H. (1976)   Covariant description of particle creation in curved spaces, *Nuovo Cim. B* **35**, 321–332.

Rumpf, H. (1980)   Vacuum state and particle creation in external electromagnetic fields, *Helv. Phys. Acta* **53**, 85–111.

Rumpf, H. (1981)   Self-adjointness-based quantum field theory in de Sitter and anti-de Sitter space-time, *Phys. Rev. D* **24**, 275–289.

Rumpf, H. (1983)   Mass-analytic quantization, uniform acceleration, and black-hole space-times, *Phys. Rev. D* **28**, 2946–2959.

Rumpf, H., & Urbantke, H. K. (1978)   Covariant "in–out" formalism for creation by external fields, *Ann. Phys.* (N.Y.) **114**, 332–355.

Sakai, T. (1971)   On eigen-values of Laplacian and curvature of Riemannian manifold, *Tôhoku Math. J.* **23**, 589–603.

Schimming, R. (1981) Lineare Differentialoperatoren zweiter Ordnung mit metrischem Hauptteil und die Methode der Koinzidenzwerte in der Riemannschen Geometrie, *Beitr. Anal.* **15**, 77–91.

Schrader, R., & Taylor, M. E. (1984) Small $\hbar$ asymptotics for quantum partition functions associated to Yang–Mills potentials, *Commun. Math. Phys.* **92**, 555–594.

Schrödinger, E. (1939) The proper vibrations of the expanding universe, *Physica* **6**, 899–912.

Schutz, B. F. (1985) *A First Course in General Relativity*, Cambridge University Press.

Schwinger, J. (1951) On gauge invariance and vacuum polarization, *Phys. Rev.* **82**, 664–679.

Seeley, R. T. (1967) Complex powers of an elliptic operator, *Singular Integrals* (Proc. Symp. Pure Math., Vol. 10), American Mathematical Society, Providence, R.I., pp. 288–307.

Segal, I. E. (1967) Representations of the canonical commutation relations, *Cargèse Lectures in Theoretical Physics: Application of Mathematics to Problems in Theoretical Physics*, ed. by F. Lurçat, Gordon and Breach, New York, pp. 107–170.

Segal, I. E. (1974) Symplectic structures and the quantization problem for wave equations, *Symposia Mathematica*, Vol. 14 (Conference on Symplectic Geometry and Mathematical Physics, Rome, 1973), Academic Press, London, pp. 99–117.

Segal, I. E., & Goodman, R. W. (1965) Anti-locality of certain Lorentz-invariant operators, *J. Math. Mech.* **14**, 629–638.

Sewell, G. L. (1980) Relativity of temperature and the Hawking effect, *Phys. Lett. A* **79**, 23–24.

Sewell, G. L. (1982) Quantum fields on manifolds: PCT and gravitationally induced thermal states, *Ann. Phys.* (N.Y.) **141**, 201–224.

Soffel, M., Müller, B., & Greiner, W. (1982) Stability and decay of the Dirac vacuum in external gauge fields, *Phys. Reports* **85**, 51–122.

Streater, R. F., & Wightman, A. S. (1964) *PCT, Spin and Statistics, and All That*, Benjamin, New York.

Suen, W.-M. (1987) Back-reaction calculation of quantum fields in curved spacetime with arbitrary curvature coupling, *Phys. Rev. D* **35**, 1793–1797.

Synge, J. L. (1960) *Relativity: The General Theory*, North-Holland, Amsterdam.

Tabor, D., & Winterton, R. H. S. (1969) The direct measurement of normal and retarded van der Waals forces, *Proc. Roy. Soc. A* **312**, 435–450.

Taylor, M. E. (1981) *Pseudodifferential Operators*, Princeton University Press, Princeton. N.J.

Titchmarsh, E. C. (1937) *Introduction to the Theory of Fourier Integrals*, Oxford University Press, Oxford.

Titchmarsh, E. C. (1962) *Eigenfunction Expansions Associated with Second-order Differential Equations*, Part I, 2nd ed., Oxford University Press, Oxford.

Treves, F. (1975) *Basic Linear Partial Differential Equations*, Academic Press, New York.

Treves, F. (1980) *Introduction to Pseudodifferential and Fourier Integral Operators*, Vol. 1, Plenum, New York.

Unruh, W. G. (1976) Notes on black-hole evaporation, *Phys. Rev.* D **14**, 870–892.

Vilenkin, A. (1978) Pauli–Villars regularization and trace anomalies, *Nuovo Cimento A* **44**, 441–450.

Wald, R. M. (1975) On particle creation by black holes, *Commun. Math. Phys.* **45**, 9–34.

Wald, R. M. (1976) Stimulated-emission effects in particle creation near black holes, *Phys. Rev.* D **13**, 3176–3182.

Wald, R. M. (1977) The back reaction effect in particle creation in curved spacetime, *Comm. Math. Phys.* **54**, 1–19.

Wald, R. M. (1978a) On the trace anomaly of a conformally invariant quantum field in curved spacetime, *Phys. Rev.* D **17**, 1477–1484.

Wald, R. M. (1978b) Axiomatic renormalization of the stress tensor of a conformally invariant field in conformally flat spacetime, *Ann. Phys.* (N.Y.) **110**, 472–486.

Wald, R. M. (1979a) On the Euclidean approach to quantum field theory in curved spacetime, *Comm. Math. Phys.* **70**, 221–242.

Wald, R. M. (1979b) Existence of the S-matrix in quantum field theory in curved spacetime, *Ann. Phys.* (N.Y.) **118**, 490–510.

Weinberger, H. F. (1965) *A First Course in Partial Differential Equations with Complex Variables and Transform Methods*, Blaisdell, Waltham, Mass.

Weyl, H. (1927) Quantenmechanik und Gruppentheorie, *Zeit. Physik* **46**, 1–46.

Whitehead, J. H. C. (1932) Convex regions in the geometry of paths, *Quart. J. Math.* (Oxford) **3**, 33–42.

Widom, H. (1978) Families of pseudodifferential operators, *Topics in Functional Analysis*, ed. by I. Gohberg and M. Kac, Academic Press, New York, pp. 345–395.

Widom, H. (1980)   A complete symbolic calculus for pseudodifferential operators, *Bull. Sci. Math.* **104**, 19–63.

Wightman, A. S. (1962)   On the localizability of quantum mechanical systems, *Rev. Mod. Phys.* **34**, 845–872.

Wightman, A. S. (1967)   Introduction to some aspects of the relativistic dynamics of quantized fields, *Cargèse Lectures in Theoretical Physics: High Energy Electromagnetic Interactions and Field Theory*, ed. by M. Lévy, Gordon and Breach, New York, pp. 171–291.

Wightman, A. S., & Schweber, S. S. (1955)   Configuration space methods in relativistic quanntum field theory. I, *Phys. Rev.* **98**, 812–837.

York, J. W. (1985)   Black hole in thermal equilibrium with a scalar field: The back-reaction, *Phys. Rev. D* **31**, 775–784.

Zel'dovich, Ya. B., & Starobinskiĭ, A. A. (1971)   Particle creation and vacuum polarization in an anisotropic gravitational field, *Zh. Eksp. Teor. Fiz.* **61**, 2161–2175 [*Sov. Phys. – JETP* **34**, 1159–1166].

Index

Bibliographical items appear in italics.

Printed in the United States
By Bookmasters